The Role of the Environment in Poverty Alleviation

PEOPLE AND THE ENVIRONMENT LECTURE SERIES

The Sustainable Development Legal Initiative (SDLI) of the Leitner Center for International Law and Justice at Fordham University School of Law organizes an annual lecture series on environment and development, in cooperation with the United Nations Development Programme (UNDP), The Wildlife Conservation Society (WCS), and Fordham University's Graduate School of Arts & Sciences. Every spring, the partners select a broad issue within the sustainable development theme and organize a series of lectures with expert panelists.

Past series, held in cooperation with The Nature Conservancy, examined Conserving Biodiversity in the Developing World, The Role of Environment in Poverty Alleviation, and (Mis)Conceptions and Changing Perceptions: The Role of Markets, Education, and Media. The 2008 lecture series focuses on Reducing Emissions from Deforestation and Degradation (REDD), a rapidly evolving market for the payment of ecosystem services that has several critical implications for climate, poverty, and biodiversity.

Fordham University Press publishes a companion series based on the annual lecture series, beginning with The Role of the Environment in Poverty Alleviation. The Press will publish new volumes based on the corresponding panel discussions; the books draw on the lectures and also include contributions from other experts in the field.

The Role of the Environment in Poverty Alleviation

PAOLO GALIZZI, EDITOR

ALENA HERKLOTZ, ASSOCIATE EDITOR

FORDHAM UNIVERSITY PRESS NEW YORK 2008

Copyright © 2008 Fordham University Press

All rights reserved. No part of this publication may be reproduced, stored in a retrieval system, or transmitted in any form or by any means—electronic, mechanical, photocopy, recording, or any other—except for brief quotations in printed reviews, without the prior permission of the publisher.

Fordham University Press has no responsibility for the persistence or accuracy of URLs for external or third-party internet websites referred to in this publication and does not guarantee that any content on such websites is, or will remain, accurate or appropriate.

Library of Congress Cataloging-in-Publication Data

The role of the environment in poverty alleviation / Paolo Galizzi, editor.—1st ed.
    p. cm.
  Includes bibliographical references and index.
  ISBN 978-0-8232-2802-7 (cloth : alk. paper)—
  ISBN 978-0-8232-2803-4 (pbk. : alk. paper)
  1. Economic development—Environmental aspects.
2. Sustainable development. 3. Poverty—Environmental aspects. 4. Poverty—Prevention—Environmental aspects. I. Galizzi, Paolo.
  HD75.6.R644 2008
  362.5'6—dc22   2008022283

Printed in the United States of America
10 09 08 5 4 3 2 1
First edition

To Jim Leitner, for his friendship, vision, and humanity
—PG

To Adam and Brittany Levinson, with gratitude and recognition for their support
—AH

CONTENTS

LIST OF CONTRIBUTORS   xi

FOREWORD
Jeanmarie Fenrich, Olav Kjørven,
Steve McCormick, and Joseph M. McShane   xxiii

PREFACE   xxvii
Paolo Galizzi and Alena Herklotz

INTRODUCTION   xxxi
Paolo Galizzi and Alena Herklotz

PART I
Poverty Reduction and the Environment
Are Not Opposing Goals

1. Advancing a New Paradigm: Institutional and Policy Breakthroughs Toward Poverty Reduction and Sound Environmental Management
   Elspeth Halverson and Charles I. McNeill   3

2. Managing Environmental Wealth for Poverty Reduction: A Review
   David Pearce   30

3. Ecoagriculture: Agriculture, Environmental Conservation, and Poverty Reduction at a Landscape Scale
   Sara Scherr, Jeffrey A. McNeely, and Seth Shames   64

4. Conserving Biodiversity and Ensuring Sustainable Livelihoods:
   Sustainable Production and Consumption in Global Supply Chains
   Tensie Whelan   87

5. Restoring Ecosystems and Renewing Lives:
   The Story of Mhaswandi, a Once-Poor Village in India
   Marcella D'Souza and Crispino Lobo   121

PART II
Investing in the Environment, Protecting Communities from Natural Disasters

6. The International Disaster Risk Reduction Regime: Escaping the Cycle of Poverty and Tragedy
   Alena Herklotz   145

7. Unanticipated Consequences of the Tsunami Response
   Annie Maxwell   169

8. Lessons from Hurricane Katrina
   Sacha Thompson   187

PART III
Knowledge Necessary to Meet Poverty-Alleviation Goals

9. Knowledge Necessary to Meet Poverty-Alleviation Goals
   Gillian Martin Mehers, Elisabeth Crudgington, and Keith Wheeler   217

10. Conservation Information
    Jean-Louis Ecochard   230

11. Building Enterprises to Reach Low-Income Markets
    Yasmina Zaidman, Helen Ng, and Adrien Couton   253

12. Including Natural Capital in Environmental Decision Making
    Heather Tallis, Gretchen C. Daily, Joy Grant, Peter Kareiva, and Taylor Ricketts   272

PART IV
Legal Empowerment of the Poor

13. The Commission on Legal Empowerment of the Poor
    Naresh Singh   305

14. Opportunities in Environmental Stewardship:
    Climate Change and Legal Empowerment of the Rural Poor
    Olav Kjørven and Estelle Fach   329

CONCLUSION   356
Paolo Galizzi and Alena Herklotz

INDEX   359

CONTRIBUTORS

ADRIEN COUTON is the Water Portfolio Manager at Acumen Fund. He has experience studying and working in eight different countries. His professional background includes private equity, an Internet startup, service with the World Bank in water and sanitation consulting, and four years with McKinsey & Company, working mostly in the energy and transportation sectors. He holds master's degrees in management from HEC Paris, in political sciences from the University of Paris, and in public administration from the Kennedy School of Government at Harvard University.

ELISABETH CRUDGINGTON currently holds the position of Learning and Leadership Officer at the World Conservation Union (IUCN), working to promote and build capacities for effective learning and leadership within the IUCN community and beyond. A strong interest in the role of communication, learning, and leadership in driving positive change is at the core of her learning and professional practice. Her previous projects with IUCN include coediting the publications *Communicating Protected Areas* and *Achieving Environmental Objectives*, helping develop the World Conservation Learning Network, and establishing Web-based communication platforms to stimulate knowledge sharing and learning.

GRETCHEN DAILY is Senior Fellow at the Woods Institute for the Environment of Stanford University. An ecologist whose work ranges from conservation science to environmental policy analysis to public outreach, she is one of three founders of the Natural Capital Project and serves as its chief emissary to financial and government leaders. She is working to develop a scientific basis and political and institutional support for managing Earth's life-support systems. She has published more than 150 scientific and popular articles. Her most recent book is *The New Economy of Nature: The Quest to Make Conservation Profitable*, coauthored with journalist Katherine Ellison. She serves on the boards of The Nature Conservancy and the Beijer International Institute for Ecological Economics, and is Director of the Center for Conservation Biology at Stanford University.

MARCELLA D'SOUZA is Executive Director of the Watershed Organization Trust (WOTR). Prior to her work with WOTR, D'Souza organized a community-led indigenous-based healthcare system in Peru; coordinated the Women's Promotion Programme of the Indo-German Watershed Development Programme in Maharashtra; and founded and managed as its first executive director the Sampada Trust, a microfinance institution. D'Souza has published several articles, books, and training manuals and has consulted for the German Agency for Technical Cooperation, as well as the International Fund for Agricultural Development. She was recently awarded the ICONGO-Khemka Foundation Karmaveer Noble Laureate Award. Marcella is an alumna of the Government Medical College, Nagpur, and a Takemi Fellow of the Harvard School of Public Health.

JEAN-LOUIS ECOCHARD joined The Nature Conservancy as Chief Information Officer in 2002. He is an information-technology executive with more than twenty-five years' experience in developing successful information solutions in both the private and public sector. As CIO, Ecochard is responsible for developing both short- and long-range strategies for using information technology throughout the Conservancy. A native of France, Ecochard began his career by working as an independent consultant for major corporations and governments, then founded Network Research Group, a software consulting company that led national and international projects, including trading systems for Salomon Brothers and Chase Manhattan Bank. He has served as President of Physicians Online, the world's largest medical information and communications network, and of Electronic Direct Internet Transactions, a healthcare business-to-business company based in Atlanta.

ESTELLE FACH is an international law specialist and a consultant with the Energy and Environment Group of the United Nations Development Programme. She holds an MS in chemistry from Ohio State University and an MS in global affairs from New York University, where she specialized in international law. She has written on the recognition of local communities in environmental law and their role in development. At UNDP, she has worked on climate change, biodiversity, and HIV/AIDS as global issues calling for local solutions. She has been an environmental, human rights, and community activist since her childhood in France.

PAOLO GALIZZI is Associate Clinical Professor of Law and Director of the Sustainable Development Legal Initiative (SDLI) at the Leitner Center for International Law and Justice at Fordham Law School. He is also a Member of the Commission on Environmental Law of the World Conservation Union (IUCN). He joined Fordham from Imperial College, University of London, where he was Marie Curie Fellow in Law and Lecturer in Law. He previously held academic positions at the University of Nottingham, the University of Verona, and the University of Milan. He graduated from the Faculty of Law of the University of Milan in 1993 and continued his legal education at the School of Oriental and African Studies, University of London, where he received an LLM in public international law in 1995. He earned a PhD in international environmental law from the University of Milan in 1998. His research interests lie in international law, environmental law, and the law of sustainable development, and he has published extensively in these areas.

JOY GRANT is the Director of Global Partnerships for The Nature Conservancy, where she focuses on sustainable development and poverty issues. She has been compiling and synthesizing guidance for stakeholder identification and engagement. As cofounder and former Chief Executive Officer of the Programme for Belize, Grant secured title to more than 300,000 acres of tropical forest, about 4 percent of the country, and supervised one of the world's first carbon sequestration projects. Grant has also served as Deputy Head of Mission at the Embassy of Belize in Washington, D.C., and has more than ten years of experience in international and development banking, including in foreign trade, credit, and financial analysis, working at Barclays Bank International and at the Caribbean Development Bank in Barbados.

ELSPETH HALVERSON is a Program Officer for the United Nations Development Programme Equator Initiative, working in communications and knowledge management. The Equator Initiative is a partnership that brings together the United Nations, governments, civil society, business, and local initiatives to build the capacity and raise the profile of communities throughout the tropics that are working to improve livelihoods through the sustainable use and conservation of biodiversity. In addition to her role at the Equator Initiative, Halverson supports UNDP's Biodiversity Global Programme and the linkages between the UNDP's Environment and Energy Group and the Convention on Biological Diversity. Halverson has a MPh in environmental policy from the University of Cambridge and a BA in international relations from the University of British Columbia.

ALENA HERKLOTZ is the Adam and Brittany Levinson Fellow in International Law of Sustainable Development with the Sustainable Development Legal Initiative (SDLI) of the Leitner Center for International Law and Justice at Fordham Law School. She joined the Leitner Center as a Centennial Fellow in 2006–2007, during which she was also a Legal Fellow at the Commission on Legal Empowerment of the Poor. Herklotz graduated from Fordham Law in 2006, having obtained a BA from Barnard College in 2002. She has written in the fields of sustainable development and international environmental law and is currently working on a research project on the clean development mechanism (CDM), capacity building, and sustainability in West Africa.

PETER KAREIVA is Chief Scientist for The Nature Conservancy and cofounder of the Natural Capital Project, for which he serves as the project's liaison with TNC, while also offering strategic vision and leadership. Kareiva's interests encompass agriculture, conservation, ecology, and the interface of science and policy. In addition to a long academic career, including faculty positions at Brown University and the University of California at Santa Barbara, he has worked for NOAA Fisheries and was Director of the Northwest Fisheries Science Center Conservation Biology Division. Academically, Kareiva is best known for contributions to insect ecology, landscape ecology, risk analysis, mathematical biology, and conservation. His current projects emphasize the interplay of human land use and biodiversity, resilience in the face of global change, and evidence-based conservation.

OLAV KJØRVEN is Assistant Secretary General and Director of the Bureau for Development Policy at the United Nations Development Programme (UNDP). He is responsible for UNDP's key practice areas: democratic governance, HIV/AIDS, poverty reduction, and environment and energy. He also manages the integration of capacity development and gender equality throughout the work of the organization. From 2005 to 2007, Kjørven led UNDP's Environment and Energy Group, promoting sound environmental management and access to energy and tapping new financial mechanisms for development. Before joining UNDP, Kjørven served as the Norwegian State Secretary for International Development. Together with the Minister of International Development, Kjørven was responsible for policymaking and the overall management of Norwegian international development efforts. He also held the post of Political Advisor to the Minister of International Development and Human Rights (1997–2000). Kjørven has also worked as Director of International Development at ECON–Centre for Economic Analysis in Oslo (2000–2001) and as an environmental specialist at the World Bank (1992–1997).

CRISPINO LOBO is the Executive Director of the Sampada Trust, a microfinance and entrepreneurship development center in Maharashtra, India. He has long been involved in community-led natural resources management. In 1989, he co-organized the Indo-German

Watershed Development Program, and he is a cofounder of the Watershed Organization Trust, established in 1993. He played a significant role in the establishment of India's National Watershed Development Fund, and he later cofounded the Geneva-based International Rainwater Harvesting Alliance. Lobo has been a member of several state- and national-level planning and development bodies. He has received the Outstanding Social Entrepreneurship Award of the Schwab Foundation and the ICONGO-Khemka Foundation Karmaveer Noble Laureate Award. He has consulted extensively for the World Bank and the International Fund for Agriculture and Development, and published several books and articles. Lobo is an alumnus of the Kennedy School of Government, Harvard University.

GILLIAN MARTIN MEHERS is the Head of Learning and Leadership at IUCN, where she is responsible for initiating the development and integration of learning and leadership activities within IUCN and its partners. She is also the focal point for the IUCN Commission on Education and Communication, the Union's expert network of sustainability education and communication practitioners. She previously worked at LEAD International as Director of Capacity Development, building LEAD's global training curriculum and signature interactive methodology. Her expertise is in creating experiential learning environments, curriculum and training materials development, and interactive training design, primarily within the context of sustainable-development learning. She is an experienced process facilitator and has developed and implemented training-of-trainers activities internationally. She is coauthor of *Training Across Cultures: A Handbook for Trainers and Facilitators Working Around the World*.

ANNIE MAXWELL is Chief Operating Officer of Direct Relief International, overseeing day-to-day activities of the organization. From August 2005 to October 2006, Maxwell was seconded to the United Nation's Office of the Special Envoy for Tsunami Recovery, led by Special Envoy President Bill Clinton. She served as Partnerships and Outreach Officer, focusing on environmental issues and the role of NGOs in the recovery effort. Maxwell is a Phi Beta Kappa and magna cum laude graduate of the University of Michigan, Ann Arbor. She also earned her master's degree in public policy from Michigan's Gerald R. Ford School of Public Policy. Maxwell serves as Vice Chair on the Alumni Board of Governors at the Ford School and is a member of the founding board of directors for the nonprofit Wizzy Digital. In December 2006, Maxwell was selected for the 2007 Marshall Memorial Fellowship.

JEFFREY A. MCNEELY, Chief Scientist for the World Conservation Union (IUCN), has been at IUCN since 1980, when he was appointed Executive Officer of the Commission on National Parks and Protected Areas. He served as Director of the Biodiversity Programme Division from 1983 to 1987, when he became Deputy Director General. He was named Chief Conservation Officer in 1988 and appointed Chief Scientist in 1996, responsible for overseeing all of IUCN's scientific work. As Secretary General of the 1992 IV World Congress on National Parks and Protected Areas, McNeely helped develop new concepts relating people to protected areas and edited *Parks for Life*, the official report of the Congress. He has published forty books and more than five hundred articles on conservation issues, linking natural resource conservation to the maintenance of cultural diversity and economic sustainability. He serves on the editorial advisory board of fourteen international journals, has advised more than fifty governments, and was a founder of the Global Biodiversity Forum.

CHARLES MCNEILL is Manager of the Environment Programme Team within the United Nations Development Programme (UNDP), with oversight of UNDP's Environment Global Program. In addition, he manages UNDP's Biodiversity Conservation and Poverty Reduction program, as well as UNDP's participation in the Convention on Biological Diversity. He also oversees UNDP's work on the Equator Initiative, a multipartner effort to promote and build capacity for community initiatives in the equatorial belt to reduce poverty through the protection and wise use of biodiversity. Before undertaking these responsibilities in 2001, McNeill helped establish UNDP's first major environment program as Coordinator of the Sustainable Energy and

Environment Division's Center of Experimentation. From 1992 to 1996, he worked with UNDP's Global Environment Facility (GEF) program, building its Africa program on biodiversity and climate change and subsequently leading UNDP's policy and strategic planning for the GEF. Before joining UNDP, McNeill worked in the nongovernmental sector, focusing on hunger eradication and sustainable development in Africa and South Asia.

HELEN NG is the Portfolio Manager for Housing at Acumen Fund, managing a portfolio of innovative housing investments in South Asia and East Africa. Before joining Acumen Fund, Ng was the principal and founder of Domus A.D., a consulting firm providing economic and financial advisory services for global urban planning and infrastructure projects that improve sustainable habitation. Previously, she engaged in project financing in telecommunications, transportation, and water at Dresdner Kleinwort Benson and Barclays Capital. Ng's board affiliations currently include Domus A.D. (Chairperson), City College Architecture Center and New Housing New York (steering committee). She has a BS in civil engineering and an AB in international relations from Stanford University, as well as an MBA from the MIT Sloan School of Management.

DAVID PEARCE, OBE, was Emeritus Professor in the Department of Economics at University College London (UCL), where he established the MS program in environmental and resource economics. A pioneer and specialist in environmental economics, Pearce published more than fifty books and more than three hundred academic articles on the subject, including the Blueprint for a Green Economy series. He also served as an adviser to the United Kingdom Secretary of State for the Environment between 1989 and 1992, as well as Codirector of the Centre for Social and Economic Research on the Global Environment (CSERGE) from 1991 to 2001. UNEP named him in the 1989 Global 500 Roll of Honour for Services to the World Environment. In 2000, he was awarded the Order of the British Empire. The European Association of Environmental and Resource Economists honored him with a Lifetime Achievement Award in 2004 in recognition of his many contributions to the development of environmental economics.

TAYLOR RICKETTS is the Director of Conservation Science for the World Wildlife Fund and cofounder of the Natural Capital Project, serving as the project's liaison with the WWF and providing strategic guidance. His interests span a broad range of topics in ecology and conservation biology, from global analyses of biodiversity patterns to field studies on the ecological and economic effects of land-use change. Ricketts led WWF's conservation assessment of North American ecoregions, the first in a continuing series published by Island Press. His current research focuses on the agricultural value of wild pollinators and their habitats, and on mapping the economic costs and benefits of conservation. He received his PhD from Stanford University and has been recognized with awards from the Society for Conservation Biology, the National Science Foundation, the Summit Foundation, and others.

SARA J. SCHERR is President of Ecoagriculture Partners, a nongovernmental organization that supports agricultural communities that manage landscapes both to increase production and incomes and to enhance wild biodiversity and ecosystem services. She is a member of the United Nations Millennium Project Task Force on Hunger and of the board of directors of the World Agroforestry Centre, the Katoomba Group, and REBRAF-USA. Until recently, she was also the Director of Ecosystem Services for Forest Trends, which promotes forest conservation through improved markets for forest products and ecosystem services. She has published more than thirty-five articles and eleven books.

SETH SHAMES is Policy Project Manager for Ecoagriculture Partners. He has studied agroforestry systems in Peru and conservation tillage for small farmers in Ethiopia, and organized community-supported agriculture groups in New York City. He holds an MESc from the Yale School of Forestry and Environmental Studies, focusing on policy issues at the intersection of environment and agriculture, and a BA from

Columbia in anthropology and environmental science. Shames also has experience as an organic farmer and pickler.

NARESH SINGH is Executive Director of the Commission on Legal Empowerment of the Poor. He has had a distinguished career in international development that spans forty countries. The author of more than one hundred publications on governance, poverty, and sustainable livelihoods, he is currently an adjunct professor at the Boston University School of Public Health and a visiting scholar at Fordham Law School. He has also been a visiting scholar at Harvard's Global Equity Initiative. He has served as Director General of Governance and Social Development at the Canadian International Development Agency, Principal Advisor to UNDP, and Director of the Poverty and Empowerment Program at the International Institute for Sustainable Development in Winnipeg.

HEATHER TALLIS is Lead Scientist for the Natural Capital Project, a collaboration between the Nature Conservancy, the World Wildlife Fund, Stanford University, and other universities to develop a set of tools to model and map ecosystem service values in given landscapes. Her research interests include marine ecology, ecosystem science, biogeochemistry, and traditional knowledge systems. She has explored resource-management options with Penan communities in Borneo, villagers in Cambodia, timber and aquaculture industries in Washington, and a hydropower company in New Zealand. Her academic research has focused on the biological consequences of nutrient and carbon flows in the open ocean, coastal zone, rivers, and forests. She received her PhD from the University of Washington.

SACHA THOMPSON is a Centennial Fellow with the Sustainable Development Legal Initiative (SDLI) of the Leitner Center for International Law and Justice at Fordham Law School. Thompson graduated from Fordham Law in 2006, having obtained a BA in international relations from Tufts University in 2000. She is a member of the New York State Bar. Her primary area of interest is international economic development, specifically emerging markets and microfinance.

KEITH WHEELER is Chair of the IUCN Commission on Education and Communication. In addition, Wheeler is the President of the Foundation for Our Future, whose mission is to support capacity development for individuals and institutions in the area of sustainable development, organizational development, knowledge management, education and training, and strategic communication planning. Wheeler is also Chief Executive Officer for ZedX, an international agroenvironmental knowledge management and IT company, and is a senior partner of WorldViews LLC, an international sustainable development and training consultancy. In addition, he is President of the board of trustees of the Paul F. Brandwein Institute, an international conservation science education nonprofit, and a Director of the Keystone Center, an international NGO focused on environmental conflict resolution.

TENSIE WHELAN is the Executive Director of the Rainforest Alliance (RA). She joined RA as a board member in 1990 and later served as a consultant, becoming Executive Director in 2000. Whelan served as Executive Director of the New York League of Conservation Voters from 1992 to 1997, before which she was Vice President of Conservation Information at the National Audubon Society. She has worked as a journalist and magazine editor, as well as a management and environmental communications consultant to various environmental and political groups. Whelan's published work includes one of the first books on eco-friendly tourism, *Nature Tourism: Managing for the Environment*. She holds an MA in international communication from American University's School of International Service and a BA in political science from New York University.

YASMINA ZAIDMAN is the Director of Portfolio Strategies at Acumen Fund. She works with diverse audiences and strategic partners in the international development, philanthropic, academic, and corporate sectors to distill and share insights and replicable models from Acumen Fund's investment portfolio. Before joining Acumen Fund, Zaidman worked in the arena of international environmental protection and social entrepreneurship for seven years. She led the Environmental Innovations Initiative at Ashoka, where, as Acting Director, she headed an effort to capture and disseminate the best practices of leading environmental innovators throughout the developing world. She holds a BA from Vassar College and an MBA from Stanford University.

FOREWORD

The current state of human affairs is disparate. Some people enjoy a high standard of living and a long and prosperous life, while others live in dismal poverty, suffer from treatable diseases, and have little access to healthcare services, educational facilities, nutritious food, or clean water. At the same time, the natural environment, common to all the people of the earth, is in peril. Global climate change, pollution, loss of plant and animal species, overuse of natural resources, and natural disasters threaten to reverse hard-won progress toward the creation of an equitable and sustainable world system. Improving the human condition for people all around the world is the challenge of our new millennium, so much so that, in the year 2000, world leaders agreed to the achievement of the Millennium Development Goals (MDGs) by 2015. The MDGs are the world's timebound and quantified targets for addressing extreme poverty in its many dimensions—income poverty, hunger, disease, lack of adequate shelter, and exclusion—while promoting gender equality, education, and environmental sustainability. With the adoption of the MDGs, ensuring a high quality of human life on this planet has come to be one of the top priorities of multilateral institutions, governments, nongovernmental organizations, community-based organizations, citizen's groups, universities, scientists, researchers, and humanitarians alike.

We face two major challenges. We must find a way to improve the state of human development so that all people are lifted from poverty.

We must also find a way to protect our natural environment to ensure that future generations will continue to benefit from the health of our planet's ecosystem. The world's poor, especially those living in rural areas, rely disproportionately on the environment for the ecosystem services it provides, such as food, fuel, shelter, medicine, and sources of drinking water. Inasmuch as rural poverty accounts for nearly 63 percent of poverty worldwide, the protection of the environment is clearly central to sustainable human development and the eradication of poverty. Once thought to be separate and irreconcilable agendas, protection of the environment and the eradication of poverty through human development are now seen as interwoven branches of one larger goal, securing a future worth having for all of humanity. If we fail, the mere survival of the human species—along with myriad other life forms—will be at stake.

The level of human development can be raised and the environment can be protected if a concerted effort is made to align these objectives into clear and consistent set of practices, policies, and behaviors. This agenda, called Environment and Development for the 21st Century, has begun to be established by policy makers, government officials, environmentalists, development specialists, indigenous groups, civil society, NGOs, academics, researchers, and students across the globe. A range of new ideas and insights about simultaneously protecting the environment and promoting human development has been developed through cooperation, coordinated activity, research, planning, and discussion among a wide range of stakeholders. These new ideas are changing opinions about the use of resources and the nature of wealth. By taking what have been regarded as conflicting goals and combining them into a single cohesive focus, a new paradigm for the future of humanity is being created, one in which the riches of nature are the greatest wealth, and one that is based on the value of human cooperation and quality of life. This new way forward is reflected in the Millennium Development Goals, the Kyoto Protocol, the Convention on Biological Diversity, and many other international, national, and local policies from around the globe.

As a contribution to these efforts, Fordham University, The Nature Conservancy, and the United Nations Development Programme have forged a partnership and are conducting an ongoing lecture series, "People and the Environment." The lecture series has raised many questions and examined the relationships, approaches, and syncrgics that are currently being promoted to ensure worldwide environmental protection, development, and eradication of poverty. By drawing on the insights of policymakers, scientists, scholars, local and indigenous community leaders, and practitioners from a wide variety of disciplines and backgrounds, the lecture series has produced a wealth of knowledge and is changing the "environment versus development debate" into the "environment and development dialogue" as a means to eradicate poverty. The topics covered in the lecture series, as selected by the three partner institutions, include conserving biodiversity in the developing world (2005), the role of the environment in poverty alleviation (2006), and perceptions and misperceptions of protecting the environment and promoting development (2007).

This series of books is the result of these discussion sessions, and it is hoped that it will become a useful tool for all those interested in ways to protect the environment, promote development, and eradicate poverty.

Jeanmarie Fenrich
Executive Director
Leitner Center for International Law and Justice,
Fordham University School of Law

Olav Kjørven
Assistant Secretary-General and Director
of the Bureau for Development Policy
United Nations Development Programme

Steve McCormick
President and Chief Executive Officer
The Nature Conservancy

Joseph M. McShane
President
Fordham University

PREFACE

PAOLO GALIZZI AND ALENA HERKLOTZ

Growing environmental threats and degradation emphasize the urgent need to conserve natural resources to safeguard our planet. Over the years, the international community has strived to address the many environmental challenges we face. The first major international gathering to address environmental degradation was the 1972 United Nations Conference on the Human Environment. The Stockholm Conference, as it is commonly known, was followed by several international gatherings devoted to finding solutions to environmental degradation, notably the United Nations Conference on Environment and Development and the World Summit on Sustainable Development.

Concern for the state of the environment, however, has not been the only issue at the forefront of global affairs. With more than half of the global population still living on USD 2 a day or less, there is also a glaring need for development initiatives to eradicate poverty. Indeed, the ever-expanding gap between rich and poor peoples and nations has come to dominate international debates. Numerous conferences have been convened the world over with the goal of finding successful strategies to promote development and poverty eradication, including the two United Nations Conferences on Human Settlements, the World Conference on Education for All, the International Conference on Nutrition, the International Conference on Population and Development, the Copenhagen World Summit for Social Development, the World Conference on Women, and the World Food Summit.

These and the many other international conferences devoted to development have set out the key objectives of the international development agenda.

Development and environmental protection are undoubtedly two fundamental areas for international cooperation. Their prominence on the international plane was confirmed at the Millennium Summit, a global gathering at the dawn of the twenty-first century that tried to reconcile both objectives and set out an ambitious program for the new millennium designed to promote, among other things, a more sustainable development. The international community's efforts to address jointly the needs of the poor and of the planet are the focus of the People and the Environment Lecture Series, which presents arguments and evidence to suggest that poverty eradication and environmental protection, in fact, represent two sides of the same coin. The lecture series owes its origins to the initiative of Nancy Gillis, then a graduate student in international political economy and development at Fordham University. With invaluable support and guidance from Dean Nancy Busch of the Graduate School of Arts and Sciences, Gillis established a partnership between Fordham University, The Nature Conservancy, and the United Nations Development Programme to create a venue for identifying and exploring the challenges in addressing environmental protection and developmental concerns.

Every year, the partner institutions select a set of issues within the broader theme of environment and development and organize a series of lectures delivered by leading experts, from a variety of disciplines and perspectives. The lectures provide the platform for an exciting dialogue and allow for information exchange among policymakers, scientists, scholars, students, practitioners, local and indigenous community leaders, and the greater public. The topics posed and explored at the annual lectures provide the basis for this volume, the first to be published in Fordham University Press's "People and the Environment" series.

While every volume draws extensively from the substance of the companion lectures, each volume and the chapters therein stand on their own, with complete scholarly apparatus, including notes and references, to provide a comprehensive analysis of greater depth and substance than can be achieved through panel discussion. The People and the Environment Series captures the current debate in a unique collection of perspectives and personal experience and, it is hoped, will contribute to further discussion and innovation in international policy and practice.

Acknowledgments

This lecture and publication series would not have been possible without the support, cooperation, and vision of a great many individuals and organizations. We wish to recognize Nancy Gillis for her exceptional intuition and efforts in conceiving the lecture series and gathering a group of extraordinary sponsors, speakers and authors. We would like to acknowledge the partner organizations behind this project, Fordham University, The Nature Conservancy, and the United Nations Development Programme; and, for their leadership and encouragement, Dean Nancy Busch of the Fordham University Graduate School of Arts and Sciences; Dr. Charles McNeill, Manager of the UNDP Environment Programme Team; and Steven Dennin, former Director of The Nature Conservancy's New York office. Our colleagues in the partner organizations, particularly Eileen de Ravin, Elspeth Halverson, and Nina Kantcheva of the UNDP Equator Initiative, and Molly Northrup, our liaison at The Nature Conservancy, contributed invaluable insight and assistance. We also extend our appreciation to those who have presented and contributed their papers and thank in particular Sue Pearce for granting permission for the inclusion of the exceptional work of her late husband, Professor David Pearce.

We are also enormously grateful for the generous support of the Leitner family and Adam and Brittany Levinson. We would like to recognize and thank Fordham Law School and Dean William Treanor and Associate Dean Matthew Diller. We owe a huge debt of gratitude to our invaluable colleagues and friends at the Leitner Center for International Law and Justice: Martin Flaherty, Tracy Higgins, Jeanmarie Fenrich, and Chi Mgbako.

We have been fortunate to work with Fordham University Press, and in particular with Robert Oppedisano, Nicholas Frankovich, and Eric Newman. We thank them all for their support, assistance, and patience throughout the editing process.

Last but not least, we wish to thank our families and friends for their support, most especially Joseph A. Brown, Elaine Brown, Donald and Vivian Herklotz, the Brown-Christenson Family, LACE, and Zachary Patten.

INTRODUCTION

PAOLO GALIZZI AND ALENA HERKLOTZ

"People and the Environment: The Role of the Environment in Poverty Alleviation," the 2006 lecture series on which this first volume in the Fordham University Press "People and the Environment" series is based, explored the environment as both a "cause of" and a "way out" for people living in poverty. Environmental degradation disproportionately affects poor communities. The natural resources that need to be protected are, in fact, often the same resources essential for the poor to improve their economic wellbeing. Efforts to promote environmental conservation are more effective when designed to also advance development. Measures aimed at protecting the environment can and should have both objectives in mind.

This inaugural volume in the "People and the Environment" series examines the many roles that the environment may play in promoting development and poverty eradication. Divided into four parts, the chapters explore the options for reconciling environmental and developmental objectives, the impact that natural disasters and the measures intended to prevent and limit their consequences have on poor communities, the role of information and knowledge sharing in promoting environmental awareness among development decision makers, and the role of legal empowerment in achieving sustainable development. Contributors to this first volume include leading experts in their fields, presenting theoretical and practical perspectives across diverse disciplines. While they all deserve praise, special mention must be made of David Pearce, late Emeritus Professor of Economics at University College London and the world's preeminent environmental economist. Originally a summary of a Poverty and Environment Partnership report for which Professor Pearce served as principal author, his chapter, which his former student Joshua Bishop of the World Conservation Union finalized for publication, is testament to his unique and unsurpassed scholarship and insight. It is with immeasurable honor and gratitude that we present to our readers his final work.

Part I, Poverty Reduction and the Environment Are Not Opposing Goals, addresses the crucial relationship between environment and development and argues that environmental policies can and do provide tangible economic benefits for the poor. The authors make the case for linked poverty and environment policy and planning, from the international level all the way down to national and local practice. The economic and governance dimensions of sound environmental management have a significant impact on poverty reduction. Mainstreaming conservation practices into agricultural production is shown to generate win-win solutions to poverty-environment challenges, in particular, thanks to the promising new paradigm of ecoagriculture. The use of consumer choice to catalyze sustainable production patterns is also proven highly successful, with the exciting emergence of new market-based instruments for sustainability, such as payments for ecosystem services. The creative use of markets is further explored in the context of environmental certification of consumer goods, specifically the impact that purchasing decisions as a lifestyle choice can have on the quality of the environment in the developing world. Part I also presents powerful illustrations of the symbiotic relationship between the poor and the ecosystems in which they live and the capacity of the environment to regenerate and thereby restore people's livelihoods and very lives.

Part II, Investing in the Environment, Protecting Communities from Natural Disasters, explores the relationship between natural hazards, poverty, and development, as well as the international regime for natural-disaster reduction and response, before moving on to examine the role of the environment in prevention and mitigation. Contributors review the aftermaths of the 2004 tsunami in South Asia and Hurricane Katrina in the United States, in which poor resource management made natural catastrophes worse for the peoples and ecosystems affected—in particular, the most disadvantaged. The best of environmental intentions are proven vulnerable to competing demands for swift reconstruction, lack of coordination among responders, and crippling deficits in local institutional and human capacity, while policies that fail to consider human and environmental tolls prove capable of inflicting massive devastation in even the most developed of nations. In closing, these chapters also offer recommendations for environmental preparedness, investment, and response for dealing with environmental disasters.

Part III, Knowledge Necessary to Meet Poverty Alleviation Goals, discusses not only the substance of information needed to promote poverty eradication but also the processes that are capable of increasing its impact and flow. Information networks are explored as outstanding options for making development and environmental learning and policy more process-oriented, collaborative, and targeted—and, therefore, effective. The poverty of information is also addressed as a factor limiting sustainable development, along with successful examples and promising proposals for providing conservation information to poor people seeking guidance on how best to manage the natural resources on which they depend. Contributors also explore models for empowering social entrepreneurs with the knowledge needed to create financially and environmentally sustainable enterprises and communities. Finally, Part III presents the case for including ecological and social information in development decision making, specifically the economic valuation of ecosystem services, to improve environmental, social, and financial outcomes alike.

Part IV, Legal Empowerment of the Poor, explores the ways in which legal systems can be used to fight poverty while promoting environmental protection as well. Without functioning legal structures, the poor lack the means to create sustainable environmental benefits. Contributors address the challenges faced in the promotion of effective legal empowerment to achieve sustainable environmental practices that alleviate poverty as well. This part examines the Commission on Legal Empowerment of the Poor, a global initiative focused on identifying the link between informality, poverty, and the law as well as the most promising measures for facilitating poor people's participation in the economic growth of their nations. Part IV also presents an example of legal empowerment through innovative international and national legal regimes in carbon finance that simultaneously endow people with environmental stewardship and provide them a way out of poverty.

PART I POVERTY REDUCTION AND THE ENVIRONMENT ARE NOT OPPOSING GOALS

1 | Advancing a New Paradigm

Institutional and Policy Breakthroughs Toward Poverty Reduction and Sound Environmental Management

ELSPETH HALVERSON AND CHARLES I. MCNEILL

The debate over whether efforts toward poverty reduction and sound environmental management are mutually supportive or exclusive is not new; it has been a fixture of the development and conservation communities for the last fifty years. Today, however, we find ourselves at a special time in history, having observed a truly significant shift in approach that aligns these two endeavors and demands that they be understood as sides of the same coin. This shift in understanding identifies the environmental agenda for what it really is, an effort to sustain life on earth and improve the quality of life for people everywhere. As such, the environmental agenda is not an impediment to human development. It is, in fact, integral to the development process. The overall goal of this section is to explore the evidence for the premise that poverty reduction and environmental conservation are not opposing goals and that they are, in fact, totally dependent on each other. In other words, investing in sound environmental management is essential for successful, long-term poverty reduction and conversely, environmental goals cannot be achieved without parallel development gains. A number of very recent developments have occurred to help facilitate this shift, and the United Nations system and

---

Elspeth Halverson is a Programme Officer with the United Nations Development Programme (UNDP) Equator Initiative. Charles McNeill is the Manager of the Environment Program Team within UNDP and of UNDP's Biodiversity Conservation and Poverty Reduction programs.

United Nations Development Programme are involved in many of them. Yet, many key actors in both developed and developing countries continue to overlook the vital policy links between poverty reduction and the environment. This chapter will review some of the encouraging integration that is happening throughout the poverty reduction and environmental management communities and document why it is becoming increasingly difficult to treat these issues as separate, unrelated challenges.

To make the case for linked poverty and environment policy and planning, at the international level and all the way down to the national and local level, it is important to look at the economic and governance dimensions of sound environmental management as it relates to poverty reduction, examining the specific cases of mainstreaming conservation practices into the agricultural sector and using consumer choice to catalyze sustainable production patterns, and exploring the emergence of new market-based instruments for sustainability such as payments for ecosystem services. This chapter will provide an overview of the issues related to environmental sustainability and poverty reduction in international development policy and planning by looking at recent shifts at the international level and at the conclusions of key reports released in the last two years by major environment and development institutions. This overview will conclude with a case study of the Equator Initiative partnership, illustrating concrete examples of the impact that can be made at the local level to create sustainable livelihoods given the right policy, governance, and market conditions.

The United Nations Millennium Summits and Beyond: Tracking the Institutional Paradigm Shift

In September 2005, world leaders gathered at UN headquarters in New York City at the 2005 World Summit to reconfirm their commitment to achieve the Millennium Development Goals (MDGs) by 2015. In so doing, all developing countries pledged to establish MDG-based national development strategies, and the United Nations Development

Programme (UNDP), the UN's primary development program, and the United Nations Environment Programme (UNEP), the UN's environment program, took on the task of supporting global and national efforts to create and implement these strategies. The MDGs are drawn from the actions and targets contained in the Millennium Declaration that was adopted by 189 nations and signed by 147 heads of state and government during the UN Millennium Summit in September 2000 (UNDP, n.d.a). The MDGs synthesize, in a single package of eight development goals (see Table 1.1), many of the most important commitments made separately at the international conferences and summits of the 1990s[1] and recognize explicitly the interdependence between economic growth, poverty reduction, and sustainable development.

In order to develop a strategy to achieve the MDGs, Secretary General Kofi Annan commissioned the Millennium Project, overseen by the economist Jeffrey Sachs, to develop a concrete action plan for the world to reverse the poverty, hunger, and disease that affects billions of people. The Millennium Project, in its four-year tenure, identified

Table 1.1. The Millennium Development Goals

MDG1	Eradicate extreme hunger and poverty
MDG2	Achieve universal primary education
MDG3	Promote gender equality and empower women
MDG4	Reduce child mortality
MDG5	Improve maternal health
MDG6	Combat HIV/AIDS, malaria, and other diseases
MDG7	Ensure environmental sustainability
MDG8	Develop a global partnership for development

1. For example: the World Summit for Children in Cairo, September 1990; the Earth Summit in Rio de Janeiro, June 1992; the World Conference on Human Rights in Vienna, June 1993; the International Conference on Population and Development in Cairo, September 1994; the World Summit for Social Development in Copenhagen, March 1995; and the Fourth World Conference on Women in Beijing, September 1995.

the main elements required for strategies to achieve the MDGs. With the expert advice of ten thematic task forces composed of more than 250 world-renowned experts in international development, environmental sustainability, and public health, the Millennium Project measured existing baseline conditions for each of the eight MDGs, estimated the incremental cost of reaching their related targets, and in January of 2005, published the findings in *Investing in development: A practical plan to achieve the Millennium Development Goals*. The report outlines a straightforward account of the kind and quantity of investment and the strategic steps that must be taken, by both developed and developing countries, to achieve the MDGs. The report is based on the premise that the technology and knowledge to achieve them exist already, and that only the scale and focus of application need to be amplified considerably (UN Millennium Project, 2005). High-income countries, the report recommends, need to increase official development assistance (ODA) from an average of 0.25 percent in 2003 to the 1 percent agreed to at the Millennium Summit by 2015.

The fact that "achieving environmental sustainability," MDG 7, is one of the eight MDGs demonstrates the centrality of environmental concerns to the development community. MDG 7 calls for the integration of the principles of sustainable development into national-level development policies and programs so as to reverse the loss of environmental resources; the halving, by 2015, of the proportion of people without sustainable access to safe drinking water and basic sanitation; and significant improvement, by 2020, in the lives of at least 100 million slum dwellers (UN Millennium Development Goals, n.d.). Most recently, the MDG framework has formally incorporated the Convention on Biological Diversity's 2010 Biodiversity Target, to achieve by 2010 a significant reduction of the current rate of biodiversity loss at the global, regional, and national levels as a contribution to poverty alleviation and to the benefit of all life on earth, as an official target for MDG 7 (Strategic Plan for the Convention on Biological Diversity, 2002). This is just one more example of how the environment and development communities are coming together, even within the primary "development" framework, the MDGs. It also makes biodiversity the concern of everyone, not just the environment and conservation community. Hence, the commitment to achieving the MDGs is, in effect, also a commitment to national and global environmental sustainability.

United Nations Reform

In addition to the refocusing of national development strategies around the MDGs, the years between the Millennium Summits, 2000 to 2005, saw significant institutional changes within the UN system. These changes are increasingly linking programming for environmental sustainability and poverty reduction. For example, the recently released Report of the Secretary General's High Level Panel on UN System-wide Coherence, entitled *Delivering as one* (2006), endorses the new close cooperation of UNEP and UNDP to mainstream environmental issues into these MDG-based national development plans. We see in the UN system itself, therefore, recognition of the close interrelationships of development and environmental issues since the two UN programs with responsibilities in each of these areas have a new mandate to collaborate closely and in a new way to assist developing countries integrate environmental concerns into development planning. In February 2007, Kemal Dervis, Administrator of UNDP, and Achim Steiner, Director General of UNEP, announced the creation of a joint Poverty-Environment Facility that is mandated to provide technical products and services to UNDP Regional Centres and UNEP Regional Offices to support programs designed to mainstream environment into development (UNDP Newsroom, 2007). According to the announcement, the aim will be to strengthen support delivered through UN Country Teams for environment mainstreaming activities.

Poverty-Environment Partnership: Two Worlds Collide

In 2002, ten years after the Rio Earth Summit, and between the two UN Millennium Summits of 2000 and 2005, the world gathered in Johannesburg, South Africa, for the World Summit on Sustainable Development (WSSD) to further articulate concrete commitments for implementing Agenda 21. Agenda 21 is a comprehensive plan of action to be taken at all levels (global, national, local, and corporate) to address human impact on the environment that was agreed in Rio (1992). Among many agreements captured in the Johannesburg Plan of Implementation (JPOI) there is the recognition by world leaders that sound and equitable management of natural resources and ecosystem services is critical to sustained poverty reduction and achievement of the Millennium Development Goals (Johannesburg Plan of Implementation, 2002). It is also worth noting that the Equator Initiative (described later in this overview) was created before the Johannesburg World Summit as an effort to assess and learn from the record of achievement toward sustainable development of local level actors and communities.

In advance of the WSSD, a group of like-minded agencies and institutions came together to form the Poverty Environment Partnership. Involving UNDP, UNEP, the World Bank, with national governments, regional development banks, and the major environmental and development NGOs, this partnership is an informal but influential network that aims to improve the coordination of work on poverty reduction and the environment within the framework of internationally agreed principles and processes for sustainable development. The partnership's goal is to build a consensus on the critical links between poverty and the environment, particularly that better environmental management is essential to lasting poverty reduction by sharing insights and best practice around mainstreaming environmental dimensions into development (UNDP/UNEP Poverty and Environment Initiative [PEI], n.d.). In particular, the Poverty Environment Partnership has the goals of (1) sharing knowledge and operational experience; (2) identifying ways and means to improve coordination and collaboration at country and policy levels; and (3) developing and implementing joint activities (PEI, n.d.). The PEP alliance launched a major policy document at the WSSD, entitled "Linking Poverty Reduction and Environmental Management: Policy Challenges and Opportunities," which emphasized the need for policy and institutional changes to improve governance, increase the assets of the poor, improve the quality of economic growth, and reform international and industrial country practices (Department for International Development, 2002). With this report, the PEP alliance put forward pro-poor integrated poverty-environment policy approaches as a centerpiece for the MDG agenda.

Between 2003 and 2005, the PEP alliance prepared itself to contribute substantially to the international discourse and agreements reached at the September 2005 UN World Summit through analytical work (described later), advocacy and by expanding the global coalition to promote the key role of environment in achieving the MDGs. During the Summit, PEP organized high visibility policy events to display the results of the analytical work undertaken on the economic case for sound environmental management and worked with governments to ensure the central role of environment in the MDGs was recognized in the formal agreements reached. One of the PEP events on environment for the MDGs, a dinner for heads of state, provided an opportunity for public announcements of major commitments by international organizations of significant investments on the order of several billion dollars over the next ten years to ensure that environmental management and conservation are central to efforts to achieve the MDGs (International Institute for Sustainable Development, 2005). In 2005, UNDP and UNEP brought their long-standing individual initiatives on poverty-environment linkages together into one joint effort now called the UNDP/UNEP Poverty-Environment Initiative (PEI). Growing out of the Poverty-Environment Partnership, this alliance recognizes, from an institutional standpoint, the interdependency of poverty alleviation and environmental conservation goals in the developing world.

MDG Support Initiative

Internally, UNDP recently launched the MDG Support Initiative after the UN Millennium Project, which concluded in 2005, to support and build capacity in developing countries to prepare and implement their MDG-based national strategies. The MDG Support Initiative will be rolled out in approximately eighty countries by 2007. Initial countries will include eleven in Africa (Ethiopia, Ghana, Kenya, Liberia, Malawi, Mali, Nigeria, Rwanda, Senegal, Tanzania, and Uganda), six in Asia (Afghanistan, Bhutan, Cambodia, Lao PDR, Mongolia, and Pakistan); and five in Europe/CIS (Moldova, Tajikistan, Armenia, Georgia, and Kyrgyzstan). The UN Millennium Summit commitment made by all developing nations to draft MDG-based poverty reduction strategy plans is a significant and challenging task, requiring governments to understand their baseline conditions with respect to each of the MDGs and to assess the types of interventions required and to estimate the cost of reaching the MDG targets. As with the Millennium Project, the premise of the MDG Support process is that countries should not be limited to basing their planning for the MDGs on the resources currently available to them, but rather should assess the amount of resources actually needed to reach the MDGs. (UNDP, n.d.b).

The initiative's mandate, therefore, is to provide MDG-based diagnostics, needs assessment, and planning to achieve the MDGs, as well as to support the widening of policy choices and options for inclusion in MDG-based development strategies. In order to deliver this mandate, the MDG Support Initiative also has the express goal of strengthening national capacity to deliver such development strategies by building public administration skills, fostering effective governance, and creating an integrated progress monitoring system (UNDP, n.d.c). The MDG Support Initiative is helping countries identify what interventions are needed to achieve each of the MDGs, and it is developing costing tools to assist countries to calculate how much the interventions will cost so they can make a compelling case internally and externally to raise these funds. MDG 7, ensuring environmental sustainability is, by definition, an integrated part of this initiative's programming and UNDP and UNEP are working together to tackle the significant analytical challenges associated with "costing" environment interventions. The overall goal is that with the guidance of the MDG Support team, each participating country will be able to gauge the quantity and quality of investment/intervention required to achieve each of the MDGs, and thereby be able to present the international donor community with concrete requests supported by an analytical needs assessment.

Conceptual Advances on the Role of the Environment in Poverty Reduction

The 2005 Millennium Review Summit also marked a critical turning point in the analytical and substantive debate on environment and development through a series of reports from prominent researchers and development institutions. Leading environmental organizations, corporations, development and academic institutions, including UNDP, UNEP, the World Bank, World Conservation Union (IUCN), International Institute for Environment and Development (IIED), the World Resources Institute, World Wildlife Fund (WWF), Swiss Re, and Harvard University, recently published findings that firmly place the health of the environment as central to lasting economic and social development.

The Millennium Ecosystem Assessment

Perhaps most fundamentally, the Millennium Ecosystem Assessment (MA) has irrevocably altered our understanding of the threats to the health of the world's ecosystems and the stark implications for human society. The MA was commissioned by UN Secretary General Kofi Annan in 2000 and was authorized through four international environmental conventions: the Convention on Biological Diversity, the Ramsar Convention on Wetlands, the Convention to Combat Desertification and the Convention on Migratory Species (Millennium Ecosystem Assessment [MA], 2005). The MA brought together the work of 1,360 scientific, technical and policy experts to assess the consequences of ecosystem change for human well-being (MA, 2005). It took place over five years, documenting, analyzing and reporting findings on four thematic clusters: condition and trends, scenarios, responses, and subglobal assessments.

The MA findings were published in 2005. *Ecosystems and human well-being: Synthesis* summarizes the findings in the technical assessment reports of each thematic cluster. In a world that is becoming increasingly advanced technologically and, in the case of the developed world, detached from the natural world, it is easy to think that we are no longer dependent on natural systems. These natural systems, known as ecosystems, are defined as dynamic complexes of plant, animal, and microorganism communities and the nonliving environment interacting as a functional unit (Convention on Biological Diversity, 1992). The MA documents that this is not the case and that, in fact, society (both in the developed and developing world) is very much dependent on ecosystems for the services they deliver—for the food and fresh water that keep us alive, the wood that gives us shelter and furniture, even the climate and air we breathe are products of the planet's living systems. According to the MA, the benefits that humans obtain from ecosystems fall into four main categories that comprise the whole of ecosystem services:

- *Provisioning services*, such as food, fresh water, wood, fiber, and fuels. These services comprise the production of basic goods such as food crops and livestock, drinking and irrigation water, fuels, fodder, timber, and fiber such as cotton and wool.

- *Regulating services*, such as regulation of climate, floods, disease, and water purification. These services include flood protection and coastal protection supplied by mangroves and reefs, pollination, regulation of water and air quality, the modulation of disease, the absorption of wastes, and the regulation of climate.

- *Cultural services*, such as aesthetic, spiritual, and recreational values. These are the nonmaterial benefits people obtain from ecosystems through spiritual enrichment, recreation, and aesthetic experiences.

- *Supporting services*, such as photosynthesis, soil formation, and production of atmospheric oxygen. These are the services that are necessary for the production of all other ecosystem services.

The ecosystem services delivered by the environment are interdependent, increased access to one ecosystem service often means reduced access to another ecosystem service, and trade-offs need to be considered. For example, land conversion to expand agricultural land to increase food production (a provisioning service) can lead to decreased flood protection and regulation of water quality (both regulatory services) or reduced productivity of downstream fisheries (another provisioning service) due to siltation of rivers from soil erosion stimulated by the land conversion. Decreased recreation or aesthetic values (cultural services) are other ecosystem services that could be impacted. As such, increasing the use of one kind of ecosystem service can cause a reduction in the availability or quality of other kinds of ecosystem services. By looking at natural resources and the products and benefits derived from them by society, the concept of ecosystem services offers important insights into how ecosystem services can be managed to maximize benefits to society in a way that have enormous implications for our future. The four main findings of the Millennium Ecosystem Assessment reflect this fact. These findings are as follows:

- Over the past fifty years, humans have changed ecosystems more rapidly and extensively than in any comparable period of time in human history, largely to meet rapidly growing demands for food, fresh water, timber, fiber, and fuel. This has resulted in substantial and largely irreversible loss in the diversity of life on Earth.

- The changes that have been made to ecosystems have contributed to substantial net gains in human well-being and economic development, but these gains have been achieved at growing costs in the form of degradation of many ecosystem services, increased risks of nonlinear changes, and the exacerbation of poverty for some groups of people. These problems, unless addressed, will substantially diminish the benefits that future generations obtain from ecosystems.

- The degradation of ecosystem services could grow significantly worse during the first half of this century and is a barrier to achieving the MDGs.

- The challenge of reversing the degradation of ecosystems while meeting increasing demands for their services can be partially met under some scenarios that the MA considered, but these involve significant changes in policies, institutions, and practices that are not currently under way. (MA, 2005).

To address this sobering reality, the MA puts forward a series of potential responses, ranging from changes in institutional and environmental governance frameworks to the design of effective decision making processes. One important response relates to the economic and financial interventions to regulate the use of ecosystem goods and services and the need to remove harmful market mechanisms such as subsidies that promote the excessive use of ecosystem services. Artificial market support promotes the overuse of water, fertilizers and pesticides, and reduces the profitability of agriculture in developing countries (MA, 2005). A number of MA responses focus on assigning tangible economic value to ecosystem services, which we examine further here.

Generating New Resources for Change: Environmental Finance

The integration of environment and development issues is also happening in the financial and economic domains. For example, the MA describes a number of policy options for economic instruments and market mechanisms with the potential to manage ecosystem services, including schemes to establish payments for ecosystem services and markets for carbon emissions. Until recently, there has been no monetary value placed on ecosystem services such as carbon sequestration, regulating floods, soil protection, air and water purification, and so on. New financial mechanisms demonstrate the potential to generate revenue flows by placing a monetary value on these services and thereby slowing the rate of environmental degradation and creating financial incentives for sustainable development. Payment for ecosystem services presents promising potential for developing countries to profit from maintaining intact ecosystems. In the case of Costa Rica, the government established a system to direct payments to landowners to encourage the maintenance and provision of ecosystem services. The government brokers contracts between international and domestic buyers and sellers of sequestered carbon, biodiversity, watershed services, and scenic beauty. Further, landowners are paid for sustainable logging, forest conservation, and new plantations, funded in part by a tax on energy use.

The global climate change agenda has brought about a series of "cap-and-trade" systems that place value on the environmental cost of emissions from industrial processes and the consumption of fossil fuels. The markets created to facilitate the trade of carbon credits are already generating hundreds of millions of dollars per year and are expected to generate billions of dollars annually by 2010 (MA, 2005). In response, UNDP has launched the MDG Carbon Facility to expand and democratize access by developing countries to carbon finance, since currently only a very few countries are benefiting from Clean Development Mechanism (CDM) financial flows endorsed by the Kyoto Protocol. The Facility will provide a one-stop-shop for both Kyoto and voluntary markets to access carbon credits from projects in the developing world that are a balance of cost-effective carbon sequestration projects (such as landfill methane recovery projects) and high-development impact projects (small community renewable energy projects). This mechanism will create finances and incentives to drive sustainable development and progress toward achieving the MDGs.

Both payment for ecosystem services and carbon finance present promising solutions to reversing the trends identified by the MA team; however, both depend on a sound environmental management and economic development model. The MA makes a compelling case for linked poverty and environment policy and planning.

The Economic Case: Investment in the Environment as a Driver of Development

The Poverty Environment Partnership Report *Investing in environmental wealth for poverty reduction* makes a clear, concise, economic case for the centrality of sound environmental management for poverty reduction in the developing world (Pearce, 2005). The report was commissioned by the Poverty-Environment Partnership and prepared by the economist David Pearce as a lead-up to the 2005 Millennium Review Summit to assess the case for investment in the environment to reduce poverty. The report asks three framing questions: How important are environmental assets to poor people? Are pro-poor environmental investments economically attractive? How much must the world spend to stabilize environmental assets? (Pearce, 2005).

The report found empirical evidence that the environment matters disproportionately to the poor, in terms of both livelihoods and vulnerability. In fact, environmental assets, such as forest products, account for an average of twenty percent of household income in poor countries (Pearce, 2005). Making up one-fifth of household income, environmental assets, and their protection, are certainly consistent with economic development. Based on these findings the report concludes that environmental conservation is not a luxury good, a privilege only afforded to wealthy countries, as once thought by economists and policy makers. If environmental assets and ecosystem services are the base for a significant portion of economic activity in developing countries, it only makes sense that environmental investments will be economically attractive. However, in the current economic system, environmental assets and ecosystem services are often regarded as free, as they lack monetized value in the production process. David Pearce and his colleagues have found that economic rates of return on environmental investments can be high, and net economic benefits from investing in environmental assets are almost always positive. For example, investments in clean drinking water and sanitation systems yield benefit-cost ratios of 4:1 and 14:1 (Pearce, 2005). The MA provides another example of the comparative benefits and values associated with ecosystems converted for shrimp farming and those left intact as mangroves in southern Thailand. The assessment found that the ecosystems left intact as mangroves delivered three times more benefits and economic value than those converted to shrimp farms which quickly degraded, damaged surrounding ecosystems and eventually became unproductive (Pearce, 2005).

As additional evidence of the value of intact ecosystems, a study of coral reefs in the Caribbean indicates that sustainable harvesting of coral fish for food and industries such as the pet and aquaria trade may be worth USD 300 million a year, while coral-based tourism is worth over USD 2 billion annually and shoreline protection from reefs up to USD 2.2 billion a year (UNEP News Centre, 2005). This pattern holds for other ecosystems also. For example, similar findings have been recorded for intact wetlands vs. intensively farmed wetlands, sustainable forestry vs. small-scale farming, and traditional forest use vs. unsustainable timber harvest (Pearce, 2005). By demonstrating that environmental assets are integral to livelihoods in developing countries and that pro-poor environmental investments are economically attractive, the report opens the door to the question of how much must be invested in order to stabilize environmental assets and ecosystem services. The amount, not insurmountable, is estimated to be USD 60–90 billion per annum over the next ten to fifteen years (Pearce, 2005). This amount is much less than one-tenth of the annual global military expenditures and is roughly equivalent to the entire official aid budget proposed at the end of the 2005 G8 meeting (Pearce, 2005).

Similarly, the World Bank report *Where is the wealth of nations: Measuring capital for the 21st century* provides an analysis of the range and amount of assets held by developing countries and has found that natural capital makes up a large share of their wealth, and therefore needs to be managed properly to help address problems of poverty, hunger, and child mortality. The report finds that natural capital (referred to as environmental assets or ecosystem services) makes up over a quarter of overall wealth in low-income countries, and of that wealth, cropland, pastureland, timber resources, and subsoil assets make up almost 90 percent (World Bank, 2006). Exhaustible natural capital (nonrenewable natural resources), on one hand, can be transformed into other assets only through the reinvestments of profits gleaned from their extraction, while renewable natural capital can provide sustainable income streams, if properly managed. The report endorses the Hartwick rule for sustainability, which requires countries to invest their rents from natural resources in order to achieve sustainable consumption (World Bank, 2006). This recommendation implicitly requires that natural capital be directly incorporated into national accounting practices and, by default, all sectors of the economy and development planning. The report also endorses the adoption of a system of integrated environmental and economic accounting (SEEA) to provide a basis to monitor sustainability and its policy applications, necessary in a world that demands integrated poverty and environment planning (World Bank, 2006). At its core, the World Bank report is a rejection of the Environmental Kuznets Curve hypothesis that postulates that at low incomes, environmental impact increases significantly with income, but at higher incomes, it levels off or decreases. By documenting how heavily the poor in developing countries, especially the rural poor, rely on natural capital and environmental assets for their livelihoods, the *Where is the wealth of nations* negates the premise that environmental quality is by definition a luxury good only of value to the comparatively wealthy.

The Corporate Perspective: The Bottom Line for Sound Environmental Management

No issue more strikingly illustrates the links between environmental management and social and economic conditions than climate change. The compelling economic case for action to mitigate climate change made in October 2006 by Sir Nicholas Stern shows how shortsighted and damaging inaction will be (Cabinet Office, 2006). The International Panel on Climate Change (IPCC) Report of February 2007 underscores the serious risks we face by ignoring impacts of greenhouse gas pollution.

Work recently undertaken by Harvard's Medical School, Swiss Re, and UNDP entitled *Climate change futures: Health, ecological, and economic dimensions* has also contributed to this understanding, especially in the global business community (Epstein and Mills, 2005). The report lays out in clear terms the impacts of climate change on ecosystems, the economy, and human health, with reference to the fact that developing countries will be most vulnerable to these impacts. By modeling scenarios that predict gradual warming with growing variability and increased catastrophic weather events, *Climate change futures* estimates the economic, environmental and human health costs of global warming. This study, funded in part by Swiss Re, the largest reinsurance company in the world, differs from the others mentioned in this text in that it is driven by the corporate need to prepare for future business conditions in order to maintain profit margins. The report's findings paint a dire picture for the unabated continuation of the current rate of carbon dioxide emissions. For instance, rising temperatures favor the spread of infectious and respiratory disease (such as malaria, West Nile virus, Lyme disease, and asthma), and extreme weather events or sustained changes in temperature or moisture regimes often catalyze outbreaks of other communicable diseases, the treatment of which is costly and reduces economic productivity (Epstein and Mills, 2005). In fact, thirty new infectious diseases associated with changed or degraded ecosystems have been documented since 1970, more than in any other period in recorded history. Climate change will impact directly on the environment, causing pest infestations, drought, flooding, forest fires, and bleaching of coral reefs, which will upset the delicate balance of ecosystems, endangering many species and driving others to extinction and putting at risk ecosystem services such as the provision of water and food (Epstein and Mills, 2005). In terms of direct economic impact in the short term, the greatest risks of climate change are the further augmentation of vulnerabilities in the energy sector. According to the scenarios presented, continued reliance on fossil fuels will be affected by increased storms disrupting the operation of offshore oilrigs, pipelines, refineries, and distribution systems, and northern pipelines will be disabled by melting tundra (Epstein and Mills, 2005). This scenario paints a picture of our already tenuous relationship with oil being rendered increasingly more so, bringing with it implications of shortages and international conflict.

Not surprisingly, the poor, especially in developing countries, are most vulnerable to these threats. Accordingly, every effort to catalyze economic and social development must also address environmental concerns and create durable livelihoods that will be responsive to the impacts brought about by climate change. Climate change threatens to erase or even set back any development gains of the past several decades unless serious mitigation efforts are taken and unless significant investments in helping developing countries anticipate and respond to impending impacts are made. This critical body of work is called climate change adaptation, and it is further evidence of the global recognition that environmental issues can have enormous implications on the quality of life of people everywhere, and they need to be taken into account.

Swiss Re is not the only corporation joining civil society to prepare actively to deal with climate change. The environment and economic/finance communities are also coming together in another away. CERES is a movement to engage institutional investors in global environmental issues such as climate change. As a result, managers of trillions of dollars of investments are now demanding of the Fortune 500 companies that they invest in and take climate change risks and other environmental issues into account in their business practices (United Nations Foundation, n.d.). In 2005, CERES brought together investors, Wall Street, and corporate leaders at the United Nations to address the growing financial risks and opportunities posed by climate change. This ground-breaking meeting included twenty-eight U.S. and European investors, who approved a ten-point action plan seeking stronger analysis, disclosure, and action from companies, Wall Street, and regulators on climate change (UNF et al., 2005). These efforts are having a significant impact by creating yet another force for change toward ensuring the business sector recognizes and advances environmental concerns in their operations.

The concept of the ecological footprint is another major innovation and advance by creating a means to quantify the environmental impact of our production and consumption patterns. The ecological footprint is a resource-management tool that measures how much land and water a human population requires to produce the resources it consumes and to absorb its wastes under prevailing technology (Ecological Footprint, 2006). Estimates are that humanity's ecological footprint exceeds the planet's ability to regenerate by 23 percent. This tool demonstrates that our economic and social choices are intimately linked to the state of the natural environment. Since humanity is limited to one finite planet, we are obliged to consider environment and economic development issues together. The ecological footprint measure also clearly distinguishes between the relative environmental impacts of the lifestyles of the industrialized and developing worlds and helps assign responsibility accordingly. An ecological footprint can be measured at any scale, from global to national to individual. At the individual level, it helps people to understand how their personal lifestyle choices and consumption patterns directly impact the environment.

Grassroots Initiative: Communities and the Success of Holistic Environment and Development Approaches

The impacts of the approaches outlined here are most clearly visible when the focus is shifted to the local level, where the effects of investing in environmental assets, the devolution of environmental governance, the integration of environment into key productive sectors of the economy and of northern consumer choices, market access, and markets for environmental services, have been documented most carefully. Produced by the World Resources Institute (WRI), UNDP, UNEP, and the World Bank, the 2005 World Resources Report, *The wealth of the poor: Managing ecosystems to fight poverty*, offers a comprehensive analysis of development policy based on the thesis that income, goods and services derived from ecosystems can act as a fundamental stepping stone to the economic empowerment of the rural poor, but only if accompanied by the right kind of governance. The report documents how ecosystem management, empowered by pro-poor governance, can raise the household income of many poor people (WRI, 2005). *The wealth of the poor* draws on five case studies to document how local-level actors in developing countries are turning their natural assets into wealth. Three of these case studies look at the winners of the Equator Prize, awarded on a biennial basis to outstanding community-based initiatives working to reduce poverty through the conservation and sustainable use of biodiversity within the equatorial belt. These communities present a compelling case for community-based initiatives to simultaneously address the livelihood needs of the rural poor while conserving biological diversity. These communities provide evidence that the ability to benefit from ecosystem services is one of the most significant determinants of wealth and well-being for the rural poor in developing countries. These examples are important because they illustrate, in concrete terms, how the rural poor depend on the environment for their livelihoods and understand fundamentally the importance of healthy ecosystems for their well-being.

The Equator Prize is awarded by the Equator Initiative partnership, which brings together UNDP with the government of Canada, Conservation International, Fordham University, the German Federal Ministry of Economic Development and Cooperation (BMZ), the Convention on Biological Diversity, the International Development Research Centre (IDRC), IUCN, The Nature Conservancy (TNC), GROOTS International, the Television Trust for the Environment (TVE), and the United Nations Foundation. Launched in 2002, the Equator Initiative focuses on the region between 23.5 degrees north and 23.5 degrees south of the equator, a zone that holds the world's greatest concentrations of both human poverty and biological wealth. The Initiative is dedicated to celebrating successful local initiatives; creating opportunities for sharing community experiences and best practice; informing policy and forging an enabling environment for local action; and building the capacity of grassroots organizations to deliver results and scale-up impact.

This crosscutting partnership is based on four pillars: The Equator Prize, as described above, is an international award that recognizes outstanding local efforts to reduce poverty through the conservation and sustainable use of biodiversity. Equator Dialogues is a program of community and local-global dialogues, learning exchanges, and meetings that celebrate local successes, share experiences, and inform policy. Equator Knowledge is a comprehensive research and learning initiative dedicated to synthesizing lessons from local conservation and poverty reduction practice. Equator Ventures is an investment program focused on blended finance and capacity development for biodiversity enterprises in the most biodiversity-rich locations of the world. The Equator Initiative demonstrates the power of working in partnerships to influence policy at a local, regional, and global level. The Equator Initiative programs and the grassroots organizations that they work with are strengthened by the support of the multisector partners from civil-society organizations, governments, international agencies, and industry. The program is a direct response to the reality outlined in this overview: that poverty and the state of the environment are inextricably linked, progress toward the reduction of poverty cannot be achieved without attention to the health of the environment, and environmental conservation cannot be achieved without the creation of sustainable livelihoods for the world's poorest inhabitants.

The case of the Namibian Torra Conservancy, winner of the 2004 Equator Prize, documents how changes in national policy through the Nature Conservation Act of 1996 enabled the creation of conservancies on state land under the Community Based Natural Resource Management Programme (WRI, 2005). Thanks to this policy shift, the conflict between wildlife and people (threats to livestock, poaching, and so on) has been resolved in a way that protects species and enriches communities simultaneously. Conservancies hold the rights to a sustainable wildlife harvest quota set by the national government, and can enter into contracts with tourism operators (WRI, 2005). The Torra Conservancy, which covers 352,000 hectares of land in the Kunene region of northwest Namibia and is a particularly successful result of this devolution of environmental governance, was registered in 1998 and benefits a mixed-ethnicity community on an equitable distribution basis. Those who live on the conservancy receive cash payouts, jobs, game meat, livestock protection measures (fences and watering points) and improved access to social services such as education and health care (WRI, 2005). In 2003, the conservancy committee paid out 630 Namibian dollars to every member of the conservancy over the age of eighteen, an amount roughly equal to half the average annual household income for the area (Barker, 2003). Together with the private sector, they have also founded Damaraland Camp, a luxury tented lodge that has received recognition as an outstanding ecotourism destination. Damaraland Camp is fully managed and staffed by conservancy residents and has injected 1.6 million Namibian dollars into the community economy. As members of the Management Committee, community members monitor wildlife and human activity and ensure that policies for land and wildlife management are locally informed and ultimately successful (WRI, 2005).

Similarly, in Tanzania's Shinyaga region, the HASHI project has had broad success in reversing land degradation through a rebirth of traditional forms of conservation. Through the project's work, people have been able to reestablish their traditional Ngitili system of land management with huge dividends for both the natural environment and the livelihoods of communities, so much so that HASHI was awarded an Equator Prize in 2002. Through the Ngitili system of enclosures, farmers prolong the availability of fodder during dry periods to better ensure the survival of their cattle. Restored areas now support production of more food products, including fruits, meat, and milk. In addition, more tree varieties have been planted, soil conditions have improved, wells have been restored, and households now spend less time away from their farms searching for food and water. Biodiversity benefits have arisen from the restoration of ecosystems, regrowth of tree species and medicinal plants, and the return of species to the arid region, including bird and butterfly species (WRI, 2005). The success of this project rests, in large part, in the rich ecological knowledge and strong traditional institutions of the local agro-pastoralist Sukuma people and the fact that the rules governing the ownership, development and utilization of the Ngitili method are set by customary institutions and by village by-laws that are designed to meet, and be responsive to, the needs of local people (UNDP Practice, n.d.). The success of the HASHI project was facilitated by a national initiative, established in response to increasing desertification, to promote the adoption of the traditional Ngitili system. This case is another example of the importance of decentralized environmental governance and investment in environmental assets for the realization of long-term socioeconomic benefits in developing countries.

Another winner of the 2002 Equator Prize, the Fiji Locally-Managed Marine Area (LMMA) Network, founded in 1999, has grown to include communities in six districts and covers 10 percent of the inshore marine area of Fiji (WRI 2005). The network relies primarily on the use of closures, "tabu" areas, to allow marine resources to regenerate and seed adjacent waters. Typically, villages put aside 10–15 percent of their traditional fishing territory as tabu (WRI, 2005). These areas are marked and protected by a group of local people elected by the village. This initiative is driven by coastal communities working in partnership with local fisheries specialists, and has resulted in increased number and size of clams, crabs, and other species harvested just outside the tabu areas. As a result, household incomes have increased 35 percent over three years, and catches have tripled (WRI, 2005). Much of the success of the network can be attributed to its participatory and collaborative focus, which has ensured that local people are at the center of the network's operations. As a testament to the success of the network in protecting marine biodiversity and alleviating poverty in fishing communities, the government of Fiji has recently incorporated many of its approaches into national policies designed to protect the coastal resources of Fiji for future generations (WRI, 2005). For communities in Fiji, like so many elsewhere in the world, lasting socioeconomic development depends on their ability to benefit on a long-term basis from the goods and services provided by the natural environment. The national government's decision to adopt the traditional tabu system being used by the Fiji LMMA Network demonstrates how creating an open and inclusive process can create policies that work for both ecosystems and the people who depend on them.

Conclusion

The institutional and conceptual shifts documented in this chapter are encouraging examples of progress in a world that is coming to terms with the fact that the quality of future human life depends on the state of the environment. However, the adoption of a new lens that views environmental conservation and economic development as integrated and interdependent ends, while significant, is not, in and of itself, enough to change the course of environmental change that is affecting the lives of billions of people throughout the developing world. With the global adoption of the MDG framework to chart the path and measure progress towards better living conditions in the developing world, the global dialogue on development is at a turning point. The United Nations and the UNDP, the world's largest development institution, have implemented institutional changes that mandate an increasingly integrated approach to environment and development policy and planning. The Millennium Ecosystem Assessment team, through the concept of ecosystem services, has helped us identify, and therefore value, the goods, services, and benefits provided to humans by biological diversity, ecosystems and the environment. Markets for carbon emissions and mechanisms to channel associated profits to developing countries, like MDG Carbon, are presenting new and exciting financial incentives for joint poverty reduction and environmental conservation initiatives. Economists and the corporate world alike are recognizing and documenting the fact that investing in the health of the environment makes sounds financial sense. Communities and grassroots initiatives on the ground in developing countries are continuing to demonstrate that their future livelihoods are dependent upon the ecosystems in which they live and international and national policy makers are beginning to recognize the wisdom inherent in their approaches. Though these shifts are remarkable and positive steps toward truly sustainable development, further action is needed. The developed world needs to make good on its commitment to official development assistance to achieve the MDGs, making financial and technical resources available where they are needed most. Further, national policymakers should continue to take progressive steps to integrate environmental concerns into every aspect of the economy so that future decisions are based on more complete information. This way, the real environmental costs of production and consumption will be taken into account in the articulation of development strategies that are economically, socially, and environmentally robust.

References

Agenda 21 (1992). Retrieved June 14, 2007, from www.un.org/esa/sustdev/documents/agenda21/index.htm.

Barker, L. (2003, January 9). Torra conservancy pays dividends to members. *The Namibian*. Retrieved June 14, 2007, from www.namibian.com.na/2003/January/national/03A95B2B25.html.

Cabinet Office, Her Majesty's Treasury (2006). *The Stern review: The economics of climate change*. Cambridge: Cambridge University Press.

CERES (n.d.). About us. Retrieved June 14, 2007, from www.ceres.org/ceres.

Convention on Biological Diversity, Article 2: Use of Terms (1992). Retrieved June 14, 2007, from www.biodiv.org/convention/articles.shtml?a=cbd-02.

Department for International Development, European Commission, United Nations Development Programme, and World Bank (2002). *Linking poverty reduction and environmental management: Policy challenges and opportunities.* Washington, D.C.: World Bank.

Ecological Footprint (2006). Ecological footprint: Overview. Retrieved June 14, 2007, from www.footprintnetwork.org/gfn_sub.php?content=footprint_overview.

Epstein, P. R., and E. Mills (eds.) (2005). *Climate change futures: Health, ecological and economic dimensions.* Cambridge, Mass.: Center for Health and Global Environment, Harvard Medical School.

Intergovernmental Panel on Climate Change (2007). *Climate change 2007: Impacts, adaptation and vulnerability* (Contribution of Working Group II to the Fourth Assessment Report of the Intergovernmental Panel on Climate Change). Retrieved July 11, 2007, from www.ipcc-wg2.org.

International Institute for Sustainable Development and United Nations Development Programme (2005). Summary of the "Environment for the MDGs" high-level events: 14 September 2005. *Environment for the MDGs Bulletin*, 114 (1), 1.

Johannesburg Plan of Implementation (2002). Retrieved June 14, 2007, from www.un.org/esa/sustdev/documents/WSSD_POI_PD/English/POIToc.htm.

Millennium Ecosystem Assessment (2005). *Ecosystems and human well-being: Synthesis.* Washington, D.C.: Island Press.

Pearce, D. (2005). *Investing in environmental wealth for poverty reduction.* New York: United Nations Development Programme.

Strategic Plan for the Convention on Biological Diversity (COP 6, Decision VI/26, paragraph 11) (2002). Retrieved June 14, 2007, from www.biodiv.org/decisions/default.aspx?m=COP-06&id=7200.

United Nations Development Programme (n.d.a). About the MDGs: The basics. Retrieved June 14, 2007, from www.undp.org/mdg/basics.shtml.

United Nations Development Programme (n.d.b). MDG support: Overview. Retrieved June 14, 2007, from www.undp.org/poverty/mdgsupport.htm.

United Nations Development Programme (n.d.c). MDG tools & research. Retrieved on June 14, 2007, from www.undp.org/poverty/tools.htm.

United Nations Development Programme Newsroom (2007). *UNDP and UNEP cement their partnership with new poverty and environment facility.* Retrieved June 14, 2007, from http://content.undp.org/go/newsroom/february-2007/undp-and-unep-cement-their-partnership-with-new-poverty-and-environment-facility.en.

United Nations Development Programme Practice (n.d.). Tanzania. Retrieved June 14, 2007, from www.undp.org/biodiversity/biodiversitycd/practiceTanzania.htm.

United Nations Development Programme/United Nations Environment Programme Poverty and Environment Initiative (PEI) (n.d.). About PEP. Retrieved June 14, 2007, from www.undp.org/pei/aboutpep.html.

United Nations Environment Programme News Centre (2005, September 14). Investing in the environment gives "big bang for your buck," *UNEP Press Release.* Retrieved June 14, 2007, from www.unep.org/Documents.Multilingual/Default.asp?DocumentID=452&ArticleID=4 933&l=en.

United Nations Foundation, United Nations Fund for International Partnerships, State of Connecticut, and CERES (2005). *2005 Institutional investor summit on climate risk, final report.* Retrieved June 14, 2007, from www.ceres.org/pub/docs/Ceres_2005IISummit_finalreport_1005.pdf.

United Nations Millennium Development Goals (n.d.). Retrieved June 14, 2007, from www.un.org/millenniumgoals.

United Nations Millennium Project (2005). *Investing in development: A practical plan to achieve the millennium development goals. Overview.* New York: United Nations Development Programme.

United Nations Secretary-General's High-level Panel on UN System-wide Coherence in the Areas of Development, Humanitarian Assistance, and the Environment (2006). *Delivering as one: Report of the Secretary General's high level panel.* New York: United Nations.

World Bank (2006). *Where is the wealth of nations: Measuring capital for the 21st century.* Washington, D.C.: World Bank.

World Resources Institute, United Nations Development Programme, United Nations Environment Programme, and World Bank (2005). *World resources 2005—The wealth of the poor: Managing ecosystems to fight poverty.* Retrieved July 11, 2007, from www.wri.org/biodiv/pubs_description.cfm?pid=4073.

2 | Managing Environmental Wealth for Poverty Reduction

A Review

DAVID PEARCE

The Millennium Development Goals, endorsed by the United Nations (UN) General Assembly, are rooted in the concept of sustainable development. Environmental sustainability is not only acknowledged as a specific goal in its own right but is also integral to the achievement of most of the Millennium Development Goals (MDGs).[1] Sustainable development requires a rising average standard of living, *and* faster than average growth in the well-being of the poorest sections of society. Another way to say this is that sustainable development requires the overall stock of assets, or *wealth*, to increase in per capita terms (Atkins et al., 1997; Pearce and Barbier, 2000). Wealth comprises human-made capital, human capital, social capital, and environmental capital. The common feature of all forms of wealth is that each asset generates a flow of benefits through time.

The late David Pearce was Emeritus Professor of Economics at University College London. His former student, Joshua Bishop, who is currently Senior Adviser on Economics and Environment at the World Conservation Union (IUCN), adapted this chapter from the author's original summary of the report, *Investing in Environmental Wealth for Poverty Reduction* (2005). This report was prepared on behalf of the Poverty Environment Partnership, as a contribution to the 2005 United Nations Global Summit. The full report can be downloaded at www.unpei.org/PDF/InvestingEnvironmentalWealthPovertyReduction.pdf.

1. The linkages among the goals are explored in World Bank (2002); Department for International Development (2002); and UNDP (2002). They are emphasized again in the report of the United Nations Millennium Project's Environmental Sustainability Task Force (United Nations Millennium Project, 2005).

Investment is the act that increases wealth. For the asset base to increase, investment needs not only to cover depreciation but must add new assets as well. If decisions to increase assets are sound, then those benefits will exceed the costs of investment. Investment has a context, a set of factors that condition the use of the assets and without which investments will often fail or fail to occur. These include resource rights and the entire structure of incentives that face resource users. Changes in these contextual factors can significantly alter the value of natural assets, through, for instance, price fluctuation, modification of use rights, or outright expropriation.

A popular view is that environmental assets can be sacrificed during the early stages of economic development because the environment is a "luxury good," something that can be sacrificed now and restored later when people are richer. This conclusion is derived from the Environmental Kuznets Curve (EKC)[2] hypothesis, which purports to show that nations first surrender environmental quality in the name of economic growth, and then improve it later when average incomes are higher. Such a conclusion is unsound and potentially dangerous in policy terms. It ignores the damage done to human health by poor environmental quality, damage that shows up in long-run losses of economic growth.[3] It ignores the irreversibility of some forms of environmental damage. And it ignores the potential for adopting sound policies to secure rising well-being without environmental sacrifice.[4]

Poverty, Wealth, and the Environment

The links between poverty and the environment can, thus, best be understood by focusing on the diverse asset base of the poor, which

---

2. The EKC has nothing to do with Simon Kuznets, Nobel Prize winner in economics, but takes its inspiration from Kuznets's observation that income inequality tends to worsen in the early stages of economic development and then improves later. The EKC mimics the same worsening-improving sequence as development occurs. How far Kuznets's original postulate is true is open to question; see Deininger and Olinto (2000).

3. For example, see Atkinson et al. (1997); Gangadharan and Valenzuela (2001).

4. For example, see Panayotou (1997).

ranges from housing to access to infrastructure, from livestock to tools and cash, from community relationships to soil quality, forests, and clean air and water (Table 2.1).

Children are assets too, both as labor and as social security in old age.[5] But large family sizes dissipate other capital assets and can threaten sustainability. Hence the emphasis on per capita wealth is important, since average levels of well-being are unlikely to rise if extra wealth has to be shared among many more people.

Table 2.1. The diversity of wealth

	HOUSEHOLD LEVEL	COMMUNITY LEVEL	NATIONAL LEVEL +
Physical assets	Housing Tools Animals Machines	Schools Hospitals Infrastructure	Major infrastructure
Financial assets	Cash	Access to credit/insurance	Access to credit/insurance
Human assets	Labor Education Skills Health	Pooled labor	Labor markets
Environmental assets	Land Soil fertility Woodlots	Common land Fisheries Forests Water Sanitation Air quality Watersheds	Rivers, seas, lakes Large watersheds Minerals Fuels Global climate
Social assets	Family Trust	Community trust Security Governance Participation Cultural assets Rights Justice systems Markets	Inter-community links Trust Political freedoms Rights Justice systems Markets

5. For example, see Dasgupta (1992); Mäler (1997).

The poor are poor because they begin with a low asset base. They have limited surplus savings to invest in replacing assets as they depreciate, let alone building up their asset base. Ideally, incomes grow as wealth grows. But it is perfectly possible for incomes to grow through wealth reduction—the mining of wealth. In such cases, the growth of incomes is illusory in the sense that it is not sustainable (Table 2.2) (Hamilton, 2000).

Because wealth is the capacity to generate future well-being, the focus of poverty reduction efforts should be on building the asset base of the household or enterprise. Investments in technology are important, because they usually raise the productivity of wealth, that is, how much well-being can be derived from a unit of wealth.

An important part of the asset base of the poor comprises wealth from those natural resources to which they have access. The most studied resource is the forest. Around 1 billion of the world's poor

Table 2.2. Difference between income and wealth for selected nations

	INCOME PER CAPITA 1997 (US$)	POPULATION BELOW $1 DAY (%; 1990–2002)	WEALTH PER CAPITA 1997 (US$)	CHANGE IN WEALTH PER CAPITA (US$)	INCOME GROWTH RATE (%)
Mozambique	166	38	2,100	−12	+4.9
Sierra Leone	173	57	2,800	−74	−4.4
Niger	189	61	2,700	−76	+1.5
Nepal	221	38	2,900	−57	+5.1
Madagascar	251	49	3,600	−83	+0.9
Vietnam	324	18	4,000	−27	+8.6
India	396	35	4,800	−27	+6.0
Azerbaijan	579	16	8,600	−234	−15.1
China	735	17	6,800	+212	+11.6
Sri Lanka	814	7	11,200	−1	+5.3
Morocco	1,227	2	16,400	−74	+1.9
Guatemala	1,690	16	27,100	−572	+4.1
El Salvador	1,900	31	31,100	−433	+5.6
Costa Rica	2,748	2	32,700	+347	+3.8
Brazil	5,012	8	70,500	−36	+3.4

Shaded rows show where income growth is positive but wealth per capita is declining. Subject to uncertainties in the data, these figures suggest that some countries may not be growing sustainably. Adapted from Hamilton, 2000; United Nations Development Programme, 2004.

rely on forests for income or income supplements (United Nations Millennium Project, 2005). A recent review of more than fifty case studies found that forest products accounted for about 22 percent of total household income, or about USD 678 per annum, on average (Vedeld et al., 2004). In some cases the contribution was far higher.

The case-study evidence for coral reefs, mangroves, and wetlands tells a similar story. A study of a Ramsar wetland site in Cambodia, for example, revealed the substantial benefits to local communities from fisheries, water supply, transportation, poles and fuelwood, wild animals and plants, floodplain rice, and recreation (World Conservation Union, 2005). These benefits averaged USD 3,225 per household per year. Significantly, the poorer segments of the communities relied more heavily on income from fisheries.

National wealth accounting exercises, likewise, show that environmental wealth is more important as a fraction of total wealth in poor countries: 26 percent for non-OECD countries, *excluding* major oil producers, compared to 2 percent for OECD (Hamilton et al., 2005; Kunte et al., 1998). This is due to the high dependence of many developing countries on primary products (e.g., minerals and agriculture). And, yet, the absolute value of natural wealth is almost five times higher in rich countries, showing clearly that the environment is not a luxury that must be sacrificed to achieve development but, rather, an integral part of the wealth of nations (Table 2.3).

Household-level wealth accounts are not widely available. What evidence we have suggests that the poor not only have limited quantities of assets, but their quality is likely to be low, as reflected, for example, in relatively high exposure to pollution. Moreover, the income the poor derive from environmental assets is often insecure, because environmental assets themselves are at risk from overuse by the poor themselves due to open-access exploitation, expropriation by rent-seekers and the non-poor, and natural disasters and climate change.

Analysis of different ecosystems reveals the vulnerability of the environmental assets used by the poor. The poor tend to occupy more fragile soils, face more risks than the rich, suffer more from morbidity

Table 2.3. Composition of per-capita wealth (USD 2,000)

INCOME GROUP (EXCLUDING OIL STATES)	MAN-MADE, OR "PRODUCED" WEALTH	ENVIRONMENTAL OR "NATURAL" WEALTH	RESIDUAL OR "INTANGIBLE" WEALTH	OVERALL WEALTH PER CAPITA	ENVIRONMENTAL WEALTH (AS % OF TOTAL WEALTH)
Low income	1,174	1,925	4,433	7,532	26
Middle income	5,347	3,496	18,773	27,616	13
High income OECD	76,193	9,531	353,339	439,063	2
World	16,850	4,011	74,998	95,860	4

and premature mortality, and rely more heavily on biomass fuels, contaminated water, and increasingly scarce bushmeat and fish stocks. Poor people also appear to be most likely to suffer, and disproportionately so, from the effects of global climate change: one study predicts a loss of GDP in Africa in 2050 ranging from 11 percent (with no adaptation) to 2 percent (with maximum adaptation), and for Asia from 6 percent to 0.4 percent (Winters et al., 1998).

Gender differences matter substantially. Women are generally more vulnerable to environmental hazards than men, due to closer exposure to risks, such as indoor air pollution, contaminated water, and long distances traveled to collect water and fuel. An estimated 2.4 billion people rely on biomass for cooking and heating, for example, and the resulting indoor air pollution is thought to account for at least 1.6 million additional deaths each year, mainly among children and women, half of them in China and India (International Energy Agency, 2004; Warwick and Doig, 2004). Yet women's voices are not heard equally, or may be heard and ignored.

The observation of a resource curse suggests a further link between natural resources and poverty, albeit a perverse one. Experience suggests that the more resource-abundant a country is, the lower its rate of income growth (Sachs and Warner, 2001). There are many explanations for the resource curse, but a significant part of the problem is clearly the mismanagement of the revenues from resource exploitation.[6] Sensible management rules could have made some resource-rich countries significantly better off today (Hamilton et al., 2005; Hamilton and Hartwick, 2005).

Other explanations may be needed to explain the persistence of poverty among the rural poor. Political economy approaches suggest that a combination of resource dependency and a rich/poor divide results in worsening environmental degradation and increasing marginalization of the rural poor (Barbier, 2006). A study of mangrove benefits in Cambodia provides an illustration, revealing that fuelwood, charcoal, fishing, and construction materials added between 20 and 58 percent to other incomes, yet the mangroves are under threat from unsustainable shrimp production, with shrimp farms typically being abandoned after only a few years operation (Bann, 2002a).

Vicious and Virtuous Cycles

The reasons why the poor remain poor are to be found in the interactions between poverty, population, the environment, and the institutions and incentives that drive behavior. The non-poor contribute both to factors that directly perpetuate poverty and to environmental degradation that disproportionately affects the poor. While there is no *necessary* set of linkages—experience varies location by location—some of the linkages are likely to generate vicious cycles from which the poor cannot escape.

Poverty itself may make the poor agents of environmental destruction, which, in turn, makes poverty worse through ill health and low labor productivity. Environmental factors such as poor water quality, poor sanitation, and indoor air pollution, for example, are thought to account for almost 20 percent of adverse health outcomes in developing countries, compared to about 4 percent in developed countries (Lvovsky, 2001). Being an agent of environmental

---

6. A summary of the alternative explanations linking resource endowments to poor growth performance can be found in Gylfason (1999).

change implies nothing about responsibility: the poor all too often have few opportunities to behave otherwise.

Poverty perpetuates a short-term view of life, since the poor understandably focus on current necessities rather than what might be done to generate long-term benefits. Lack of property rights and the risks of expropriation by others further reinforce this view. Short-term thinking shows up in empirical evidence of high average discount rates among the poor (Anderson et al., 2004; Cuesta, Carlson, and Lutz, 1997; Holden, Shiferaw, and Wik, 1998; Poulos and Whittington, 2000). But the problem can be overcome with carefully designed credit and resource-rights policies. Some communities do invest in long-term conservation (Moseley, 2001). They may have better options to achieve it and stronger social capital, and be securing higher rates of return on their investments.

The interactions between environment and population change are also important. Poverty encourages larger families, which provide labor supply and social security; however, large families also contribute to asset dissipation, making poverty worse. Large populations tend to promote resource degradation, making the feedback to poverty worse still. An alternative view stresses the role of population growth in stimulating technological change, which raises the productivity of wealth (the "Boserup" hypothesis) (Boserup, 1980).

Once the various interactions are placed in a dynamic context, in which factors change over time and in response to each other, the picture becomes even more complex. Rapid population change tends to make resource degradation worse, which, in turn, requires institutional adaptation to cope with the changes; however, adaptation itself depends on many factors and is not automatic. A major form of adaptation lies in changing resource rights and local governance (Lopez, 1998). Some societies fail to respond while others succeed. Finding out why is essential for policy purposes. Even then, recipes for success may not be transferable everywhere. The important transition is from open-access situations to management regimes where the community has secure rights to land and resources. So long as social capital is strong, communal property will tend to avoid many of the problems associated with private or state ownership.

The Investment Response

If the asset base of the poor is to increase, there must be a combination of increased investment as well as a policy environment that reduces risk and supports effective investment. Estimating a global environmental investment budget is obviously subject to considerable uncertainty. Nonetheless, we need some idea of the financial challenge. The costs of meeting MDG7, and related environmental goals consistent with the MDGs as a whole, are additional to the cost of addressing global climate change, which itself threatens the poorest in a disproportionate way (Kahn, 2003; Mendelsohn et al., 2004; Parry et al., 2001; Winters et al., 1998). Surveying and supplementing several efforts to estimate these costs, it is estimated that around USD 60–90 billion per annum will be needed to address environment-poverty goals consistent with the MDGs (Table 2.4). In addition, a minimum of USD 80 billion per annum will be needed to tackle climate change

Table 2.4. Environmental investment needs for poverty reduction (USD billion per annum)

SECTOR	COST TO ACHIEVE MDG7 (BILLIONS)
Water and sanitation	11–26
Clean energy	28
Land degradation	10–23
Protected areas	8
Urban slums	4
Total	60–90

Author's estimate based on data in Hutton and Haller (2004); Martin-Hurtado (2002); International Energy Agency (2004); World Energy Council (1999); Sagar (2005); Bruner, Gullison, and Balmford (2004), based on previous estimates in Balmford et al. (2002), James, Green, and Paine (1999), and other sources; United Nations Secretary-General (2000).

over a fifty-year period, and more if efforts are to be concentrated in a twenty-year period.[7]

The resulting budgets are for environmental goals only, yet they are roughly equal to the *total* official aid budget proposed at the G8 Edinburgh Summit in June 2005. This suggests two conclusions: first, the scale of development assistance must be increased further; and, second, other non-aid sources of investment finance must be mobilized. It is important to note that the investment budgets suggested here amount to about 0.5 percent of current developed country income, or USD 0.50 on every USD 100 of income.

Within the hypothesized environmental budget, there is an uneasy tension between protected areas and poverty reduction. Several studies reveal that some protected areas have imposed unacceptable costs on surrounding communities (Emerton, 1999a; 1999b; 2001; Emerton and Mfunda, 1999: Ghate, 2003; Gong, 2004; Madhusudhan, 2003; Sander and Zeller, 2004; Weladji and Tchamba, 2003). Efforts to link conservation with development, through integrated conservation and development projects, for example, have not been very successful (Simpson, 2004; Wells and Brandon, 1992). While there is widespread agreement regarding the need to halt biodiversity loss and strengthen the world's system of protected areas, there is considerable debate about how best to achieve these goals.

Rates of Return to Environmental Investments

Investment of scarce funds in environmental assets must compete with investment in other assets. While some see the comparison of costs and benefits as irrelevant for such urgent issues, no economic case can be made for prioritizing environmental investments if their rate of return is below that of other opportunities for asset formation. While there is an extensive literature measuring the money value of the

---

7. Author's estimate, based on data in Intergovernmental Panel on Climate Change (2001).

nonmarket benefits associated with ecosystem services, comparatively little research compares benefits with costs.[8] What material does exist, however, is both informative and reassuring.

The social worth of investment can be measured by benefit-cost ratios or by internal rates of return. There are reasons to believe that, in some cases at least, these measures will understate the true social returns to environmental investments.[9] Rates of return for investment in water and sanitation are very high, with estimated benefit-cost ratios ranging from 4 to 14:1, or a return of 400 percent to 1400 percent (Hutton and Haller, 2004). Meeting the MDG targets for water and sanitation could save the lives of up to 1 billion children under five years of age over the period 2015–20 (Martin-Hurtado, 2002). About three-quarters of the benefit will accrue in the form of time savings from avoiding the need to make long journeys to collect water or for sanitation. Cost savings from avoided illness are also substantial. There is considerable regional variation in the rates of return, with ratios of around 10:1 typical for Africa and Asia.

Investments in access to electricity and in replacing traditional biomass fuels almost certainly have high rates of return. The costs of a basic program to improve access to modern energy services would average less than 10 percent of current household energy expenditures in the relevant countries.[10] Benefit-cost ratios for investments in improved household energy provision appear to be very large. A study in Kenya and Guatemala, for example, analyzed the net benefits of improved household stoves, in terms of the health of women and children (Larson and Rosen, 2004; von Schirnding et al., 2002). Using available epidemiological data, the authors report a benefit-cost ratio for Kenya ranging between 47:1 and 118:1, while for Guatemala it is estimated at 7:1. Studies of the net benefits of outdoor air pollution

---

8. The literature on ecosystem benefit estimates is surveyed in EFTEC (2005). However, the report rarely addresses the issue of the opportunity costs of conservation, that is, what is sacrificed.

9. See, for example, Bucknall, Kraus, and Pillai (2000).

10. See Table 2.4 and data on average energy expenditure by households in Energy Sector Management Assistance Programme (2003).

control, likewise, suggest high rates of return, in some cases (Li et al., 2004; Joh, 2000; World Bank, 1994).

Climate change investment is more debatable, in cost and benefit terms. The relevant investments are those designed to stabilize the long-run concentration of atmospheric carbon at 550 ppm, while the relevant cost-benefit comparison is between the social cost of carbon, that is, the damage done to the world by releasing an additional ton of carbon, and the cost of reaching the target (Intergovernmental Panel on Climate Change, 2001). The way the estimates are constructed makes this comparison difficult. Including potential catastrophic costs is likely to tip the balance in favor of achieving the 550-ppm goal. Climate change investments are likely to be especially beneficial for the poor.[11]

A hypothetical antidesertification policy yields benefit-cost ratios of 1.5 to 3.3, a very favorable rate of return (Martin-Hurtado, 2002). This estimate is based on a program to reduce on-site productivity losses only; the additional social, global, and off-site benefits of reversing soil degradation would imply somewhat higher rates of return (Gisladottir and Stocking, 2005; Pagiola, 1999). Available studies reveal wide variation in the costs and benefits of soil conservation, depending on labor costs, land scarcity, alternative income opportunities, and other factors.[12]

The economic values of forests have probably been more studied than for any other ecosystem.[13] Despite a popular literature that ascribes vast economic values to natural forests, the economic evidence is less clear-cut. Some forest conversion appears to pay, although comparisons are often difficult because of the baseline against which conservation is measured (Chomitz and Kumari, 1998; Secretariat of

11. For example, Pearce (2005); and Tol (2005).
12. For example, Bishop (2002); Knowler (2004); Lutz, Pagiola, and Reiche (1994); United Nations Environment Programme (1991).
13. For example, Bann (2002b); Batagoda et al. (2000); Bishop (2000); Carret and Loyer (2003); Grimes et al. (1994); Howard and Valerio (1996); Kishor and Constantino (1993); Kumari (1996); Ricker et al. (1999); van Beukering, Cesar, and Janssen (2003); World Bank (1996); Yaron (2002).

the Convention on Biological Diversity, 2001). For example, if the assumed alternative is sustainable forest management for timber production, then wholesale forest conversion often appears to be more profitable. If the alternative use includes other nontimber benefits, then the picture is more likely to favor conservation.

Of critical importance are payments for carbon storage and sequestration, since the evidence suggests that carbon dominates other forest ecosystem values. Another way of looking at the role of forest carbon is to ask what payment would tip the balance from forest conversion to conservation. Available studies suggest that this value is between USD 10 and 30 per ton of carbon (tC), which may not be far removed from market values under carbon trading in the near future (Smith and Scherr, 2002).

The value of forest genetic resources for developing pharmaceutical and other biotechnology products is also debated. Very different estimates emerge from various studies of the willingness to pay of researchers, companies, and other bio-prospectors for such information.[14] Unfortunately, the range of estimates appears to result largely from variation in analysts' choice of values for model parameters, rather than the logical construction of the valuation models themselves (Costello and Ward, 2006). This makes it difficult to resolve the debate until better information is forthcoming.

Available studies of the costs and benefits of coral-reef conservation suggest a strong result. Benefits exceed costs by a significant factor and reliable benefit-cost ratios of between 1.3 and 5 have been reported (Cesar, 1996). A study in Indonesia, for example, estimated that poison and blast fishing might generate incomes of USD 15,000 to 33,000 per square kilometer of reef area, but at a cost of USD 40,000 to 750,000 per square kilometer from forgone activities, including sustainable fisheries, tourism, and environmental protection (Cesar, 2000; IMM, 2002).

Wetlands, including mangroves, are the focus of increasing economic studies. To date, the studies are fairly consistent in showing

---

14. For example, Rausser and Small (2000); Simpson, Sedjo, and Reid (1996).

that wetland conservation pays.¹⁵ Notably, the conversion of mangroves to shrimp aquaculture is seen to be economically very unattractive.

Since the context of a great many fisheries is that they are either or both economically and ecologically overfished, it is hardly surprising that the benefits of reducing catch effort are high. A quarter of the world's fisheries are seriously overfished, and another half are on the verge of being overfished (Food and Agriculture Organization of the United Nations, n.d.). Since fisheries have effectively been open-access resources, however, the cost of reducing effort can be high in terms of unemployment.¹⁶ Around 20 percent of the world's fisheries are used mainly by small-scale household and communal enterprises. This illustrates the problems of letting resource depreciation go too far without taking corrective action. Once the resource is degraded, there may be heavy social costs in reversing the situation. Nonetheless, careful restriction of access to overfished fisheries together with the possible use of tradable quota schemes can minimize the loss of employment.¹⁷

Wildlife conservation illustrates some of the problems of environmental investments. Not only must the investment pay in aggregate terms, but local communities must also secure net benefits. Otherwise they will have little incentive to cooperate with the conservationists. Some wildlife schemes have not met this condition and the poor have suffered net losses, despite efforts to share the benefits. A study of the Lake Mburo National Park in Uganda, for example, estimated that the local community received benefits worth some USD 230,000 in a single year, while losses from wildlife damage to crops and livestock and restrictions on access to park resources amounted to USD 700,000 (Emerton, 1999b).

15. For example, Ayoo (1998); Bae (2002); Bann (2002a); Barbier and Thompson (1998); de Lopez et al. (2001); Emerton and Bos (2004); Gammage (1994); Hammitt, Liu, and Liu (2001); World Conservation Union (2001), as cited in World Conservation Union (2003b); Janssen and Padilla (1999); Loth (2004); Ruitenbeek (1994); Sathirathai (1998); Turpie et al. (1999), as cited in World Conservation Union (2003a).

16. See, for example, Israel and Banzon (1998).

17. See, for example, Rojat, Rajaosafara, and Chaboud (2004).

In contrast, evidence from the Namibian Conservancies suggests that such schemes can pay, and can be beneficial for poor, illustrating the need for careful design of benefit-sharing measures (Bandyopadhyay et al., 2004). First, the conservancies appear to have improved local well-being through cash and noncash income (especially meat distribution) and through community benefits. Second, in some cases, the poor gained proportionately more than the less poor. In other cases, the benefits have been neutral with respect to income groups. Third, those who participated in the schemes have gained, but not significantly more than those who did not participate. Fourth, the early conservancies attracted the most participation and local benefit, suggesting that as awareness increases and experience is gained, local benefits will increase. The analysis shows that such schemes can be designed to be at least "poor neutral," but with the possibility for them to be pro-poor as well. Finally, most of the conservancy schemes in Namibia have generated very high rates of return, with net benefits to local communities especially noteworthy at between 23 percent and 230 percent (Barnes, MacGregor, and Weaver, 2002).

While there is an extensive ecological literature on ecosystem diversity and resilience, the counterpart, economic valuation literature is missing. Yet ecosystem resilience is likely to be a local benefit rather than a global one, with the consequence that it is the poor who need insurance against the vagaries of external shocks and stresses. Holding a diversified portfolio of assets is the surest way to gain this insurance, but it is also the one way least open to the poor. Further substantial work on the economics of ecological resilience is needed.

The Policy Context for Successful Investment

The institutions and policies that need to be in place in order to make pro-poor environmental policy and pro-poor environmental investments "work" are generally well known. The focus must always be on creating assets for the poor. The role of social capital in development is still debated. Nevertheless, cohesive and cooperative communities are clearly essential prerequisites of common property regimes that can protect local environments and raise living standards. What is less clear is the role of public policy in creating or sustaining social capital. Much of the literature implies that policy interventions designed to increase social capital are often counterproductive (Paldam and Svendsen, 2000). In contrast, policies designed to remove constraints on the evolution of social capital may be important, such as resource rights and the rule of law. In Nepal and India, for example, the devolution of access rights and concessions for forest use stimulated the formation of more than twenty thousand local user groups (Pretty and Ward, 2001). Reallocating resource rights produced the social capital necessary to protect and manage those rights. Examples such as this suggest that the focus of analysis should be on identifying those factors that underpin (or undermine) social capital. Where these factors are under government control, action may be taken to remove significant obstacles.

A consensus is emerging that good governance matters for economic growth.[18] While the poor generally benefit from economic growth, they also benefit directly from good governance. Better governance implies that the concerns of the poor are heard and their interests are taken into account. Good governance is also about reducing corruption, which often feeds on the rents available from natural resources (Barbier, Strand, and Sathirathai, 2002).[19]

The value of a resource depends significantly on who has the right to use it. A study in Thailand, for example, found that the lack of effective management of a wetland, specifically open access conditions, led to overfishing that reduced the value of the wetland by more than one-third (Barbier, Strand, and Sathirathai, 2002). Pro-poor environmental investments are, thus, unlikely to succeed without clearer definition and enforcement of the resource rights of the poor. Incentives to conserve resources are seriously weakened if others, such as more powerful groups, private interests, or the state, can easily expropriate resources.

---

18. See, for example, Perotti (1996); Mauro (1995); Islam, Kaufman, and Pritchett (1995); Rodrik (2000).

19. For an extensive discussion, see Rose-Ackerman (1999).

Common property regimes have much to be commended for. The introduction of private property rights can result in resource grabs by the rich, to the detriment of the poor. Private rights also fail to cope with environmental externalities, unless a competent regulatory regime is in place. On the other hand, communal property requires strong social capital, which can be undermined by significant inward or outward migration, and may be weakened by inappropriate state intervention. Common property regimes are more likely to succeed when: (a) the community concerned is small and cohesive, (b) appropriate resource management rules and effective mechanisms for enforcing them are in place, (c) the returns on cooperation are high, (d) the state is more absent than present, and (e) the benefits of resource use are shared equitably (Heltberg, 2002).

Prices play a key role in investment decisions. Hence, the removal of environmentally harmful subsidies has to be one of the top policy priorities for stimulating pro-poor environmental investments.[20] While subsidies in developing countries frequently distort investment decisions, a higher priority is the reform of rich country subsidies, especially in agriculture, which generally, but not exclusively, discriminate against the developing world (Anderson et al., 2001; Tokarick, 2005). Total global subsidies now seem likely to amount to at least USD 1 trillion per annum, or about ten times the pro-poor environmental investment budget suggested earlier (Table 2.5).

The introduction of market-based environmental policies may be a long-run goal for the poorest countries, due to institutional capacity constraints (Greenspan Bell and Russell, 2002; 2003; Russell and Powell, 1996). However, this should not obscure the fact that such instruments are already used in a number of developing economies, in the form of environmental taxes and charges, and, to a small extent, tradable quota regimes, such as water rights in Chile and Mexico, and air pollution in Chile. In Madagascar, for example, concerns about the

---

20. See, for example, Organisation for Economic Co-operation and Development (1996; 1997; 1998); de Moor and Calamai (1997); Myers and Kent (1998); Porter (2003); van Beers and de Moor (2001); van Beers and van den Bergh (2001).

Table 2.5. Global subsidies 1994–1998 (USD billion per annum)

	OECD	NON-OECD	WORLD	OECD AS % OF WORLD
Natural resource sectors:				
Agriculture	335	65	400	84
Water	15	45	60	25
Forestry	5	30	35	4
Fisheries	10	10	20	50
Mining	25	5	30	83
Energy and industry sectors:				
Energy				
Road transport	80	160	240	33
Manufacturing	200	25	225	89
	55	negligible	55	100
Total	725	340	1065	68
Total as % GDP	3.4	6.3	4.0	

Adapted from van Beers and de Moor, 2001: 32.

state of the shrimp fishery led to the introduction of long-term, tradable licenses in 2000 (Rojat, Rajaosafara, and Chaboud, 2004). Early indications show that more sustainable management is beginning to emerge. A preliminary evaluation of the licensing scheme suggests a benefit-cost ratio of 1.5.

A great deal of attention is being paid to programs of payments for environmental services (PES).[21] Since the primary purpose of these schemes is environmental conservation, it is not surprising that their potential to reduce poverty has not figured prominently in their design. While there appears to be potential for multiple objectives of this kind, at the moment the record does not support the view that PES schemes are especially pro-poor (Wunder, 2005). Costa Rica, for example, has more experience of PES than most developing countries,

---

21. For example, Pagiola, Arcenas, and Platais (2005); Pagiola, Bishop, and Landell-Mills (2002); Pagiola et al. (2004); Pearce (2004); Scherr, White, and Kaimowitz (2004); Wunder (2005).

with more than four thousand beneficiaries receiving payments between 1997 and 2001. A recent study reveals that the majority of participants were relatively wealthy private landowners with other sources of income, such as lawyers and engineers (Miranda, Porras, and Moreno, 2003). Many landowners did not switch to PES because the payments did not compensate for the forgone uses of the land. Another study found that small farmers in one region were not eligible for participation in the PES scheme (Zbinden and Lee, 2005). Nonetheless, it is still very early and more work is needed to see how PES programs can be made more pro-poor. Access to credit is crucial for poverty reduction. By their nature, investments in asset creation involve a short-term sacrifice for longer-term benefits. But without access to credit on reasonable terms, the poor simply cannot afford the necessary short-run sacrifices, and asset-formation is severely reduced. While it is tempting to conclude that subsidized formal credit is needed, informal credit markets have many advantages since lenders and borrowers are closer to and often know each other better.[22] Experiments with subsidized formal credit have often failed to reach the poor. But whether formal or informal, it remains the case that most poor people do not enjoy access to credit, making it difficult for them to smooth their consumption during good and bad times, and exacerbating poverty.

There are similar issues with insurance. Because the poor have few tangible assets, their ability to self-insure is limited. A certain measure of self-insurance is achieved by diversifying sources of income, thereby reducing the risk of catastrophe if a single source of income fails. Mutual insurance, such as cooperative agreements to insure each member of the community, is widespread but is easily overwhelmed by adverse events such as natural disasters, which often affect entire communities. Evidence is emerging that natural ecosystems act as a kind of natural insurance, providing a reliable source of sustenance at times when agricultural incomes are low.[23]

---

22. A very detailed survey of incentive structures for effective credit can be found in Ray (1998).

23. See, for example, Pattanayak and Sills (2001).

Whatever source of finance is found to invest in the environment, there remains a need to improve the sustainability of financing itself. In particular, it is important that funds are available to cover costs beyond the initial period of capital formation, especially where the benefits are nonmonetary. Many environmental investments require a long-term commitment that few investors—public or private—are prepared to make (Gutman, 2003). Strengthening incentives for long-term investment thus remains a particular challenge, and not only for the environment.

Conclusion

While a great deal is known about poverty-environment relationships, there remain large gaps in the information base required to manage the environment effectively for poverty reduction. Priorities for new information include:

- geographic data on the spatial overlay of poverty, environmental quality, and resource rights;

- improved wealth accounting including a wider range of resources and ecosystem services, as well as household-level wealth accounts;

- more consistent evaluation of costs as well as benefits when assessing environmental investments;

- empirical evidence of the links between biological diversity per se, ecological resilience, and poverty;

- better understanding of how common property regimes evolve in the face of increasing resource scarcity;

- more analysis of the environmental impacts of subsidies for water, fishing, and land conversion;

- better understanding of the social (poverty) impacts of environmental policy and investments at a subnational level; and

- better understanding of the interactions and interdependencies among different forms of capital (social, natural, human, etc).

However, information gaps must never be allowed to divert attention from taking action now, even where there is uncertainty about the nature of the required policies and their effectiveness. There is no reason to postpone policy while better information is gathered, particularly because the policy interventions are themselves one of the main ways in which the required information is generated. Two major challenges are how to design effective environmental policies in the context of weak institutional capacity, and how to ensure that the interests of developing countries are adequately reflected in multilateral policy on global public goods, notably climate change.

Tackling poverty in practical ways that have a reasonable chance of success is immensely complex. This chapter has focused on one dimension of this complex issue—addressing the management of environmental assets to improve the lot of the poor. Drawing boundaries around this dimension is difficult, simply because there are so many interdependencies in poverty policy. Singling out environmental policy and investments as a major means of helping the poor will not work unless many other conditions are satisfied—notably governance and institutional change within developing economies, as well as changes in the way the rich world treats the poor world.

References

Anderson, C., Dietz, M., Gordon, A., and Klawitter, M. (2004). Discount rates in Vietnam. *Economic Development and Cultural Change* 52 (4): 873–888.

Anderson, K., Dimaranan, B., Francois, J., Hertel, T., Hoekman, B., and Martin, W. (2001). The cost of rich (and poor) country protection to developing countries. *Journal of African Economies* 10 (3): 227–257.

Atkinson, G., Dubourg, R., Hamilton, K., Munasinghe, M., Pearce, D. W., and Young, C. (1997). *Measuring sustainable development: Macroeconomics and the environment*. Cheltenham, England: Edward Elgar.

Ayoo, C. (1998). *A cost-benefit analysis of alternative wetland uses in Kenya: The case of Yala swamp* (Unit for Environmental Economics, Paper 1998:1). Gothenburg, Sweden: Gothenburg University.

Bae, J.-H. (2002). Wetland conversion in South Korea: The economics and political economy of Saemangeum tidal flats. MSc Thesis, University College London.

Balmford, A., Bruner, A., Cooper, P., Costanza, R., Farber, S., Green R., et al. (2002). Economic reasons for conserving wild nature. *Science* 297: 950–953.

Bandyopadhaya, S., Humavindu, M., Shyamsundar, P., and Wang, L. (2004). *Do households gain from community-based natural resources management? An evaluation of community conservancies in Namibia* (World Bank Policy Research Working Paper, No. 3337). Washington, D.C.: World Bank.

Bann, C. (2002a). Economic analysis of alternative mangrove management strategies in Cambodia. In D. W. Pearce, C. Pearce, and C. Palmer (Eds.), *Valuing the environment in developing countries: Case studies*, 501–535. Cheltenham, England: Edward Elgar.

Bann, C. (2002b). Economic analysis of tropical forest land use options in Cambodia. In D. W. Pearce, C. Pearce, and C. Palmer (Eds.), *Valuing the environment in developing countries: Case studies*, 536–569. Cheltenham, England: Edward Elgar.

Barbier, E. B. (2006). Natural capital, resource dependency and poverty in developing countries: The problem of "dualism within dualism." In R. Lopez and M. A. Toman (Eds.), *Economic development and environmental sustainability: New policy options*, 23–59. Oxford: Oxford University Press.

Barbier, E. B., Strand, I., and Sathirathai, S. (2002). Do open access conditions affect the valuation of an externality? Estimating the welfare effects of mangrove-fishery linkages in Thailand. *Environmental and Resource Economics* 21: 343–367.

Barbier, E. B., and Thompson, J. (1998). The value of water: Floodplain versus large-scale irrigation benefits in northern Nigeria. *Ambio* 27 (6): 434–440.

Barnes, J., MacGregor, J., and Weaver, C. (2002). Economic analysis of community wildlife use initiatives in Namibia. *World Development* 30 (4): 667–681.

Batagoda, B. M. S., Turner, K., Tinch, R., and Brown, K. (2000). Towards policy relevant ecosystem services and natural capital values: Rainforest non-timber products. *CSERGE Global Environmental Change Working Papers, 2000–06*. London: Centre for Social and Economic Research on the Global Environment.

Bishop, J. (2002). The economics of soil fertility management: Theory and evidence from West Africa. PhD thesis, University of London.

Bishop, J. (Ed.) (2000). *Valuing forests: A review of methods and applications in developing countries*. London: International Institute for Environment and Development.

Boserup, E. (1980). *Population and technological change*. Chicago: University of Chicago Press.

Bruner, A., Gullison, R., and Balmford, A. (2004). Financial costs and shortfalls of managing and expanding protected area systems in developing countries. *Bioscience* 54 (12): 1119–1126.

Bucknall, J., Kraus, C. M., and Pillai, R. (2000). Poverty and environment (Environment Strategy Background Paper). Washington, D.C.: World Bank.

Carret, J. C., and Loyer, D. (2003). *Madagascar protected area network sustainable financing: Economic analysis perspective*. Washington, D.C.: World Bank.

Cesar, H. (1996). *Economic analysis of Indonesian coral reefs* (Environment Department Work In Progress). Washington, D.C.: World Bank.

Cesar, H. (2000). Coral reefs: Their functions, threats and economic value. In H. Cesar (Ed.), *Collected essays on the economics of coral reefs*, 14–39. Kalmar, Sweden: Kalmar University.

Chomitz, K., and Kumari, K. (1998). The domestic benefits of tropical forests: A critical review. *The World Bank Research Observer* 13 (1): 13–35.

Costello, C., and Ward, M. (2006). Search, bioprospecting and biodiversity conservation. *Journal of Environmental Economics and Management* 52 (3): 615–626.

Cuesta, M., Carlson, G., and Lutz, E. (1997). *An empirical assessment of farmers' discount rates in Costa Rica and its implications for soil conservation* (Environment Department Work in Progress). Washington, D.C.: World Bank.

Dasgupta, P. (1992). Population, resources and poverty. *Ambio* 21: 95–101.

Deininger, K., and Olinto, P. (2000). *Asset distribution, inequality, and growth* (Policy Research Working Paper, No. 2375). Washington, D.C.: World Bank.

De Lopez, T., Vihol, K., Proeung, S., Dareth, P., Thea, S., Sarina, C., et al. (2001). *Policy options for Cambodia's Ream National Park: A stakeholder and economic analysis*. Singapore: Economy and Environment Program for Southeast Asia.

de Moor, A., and Calamai, P. (1997). *Subsidising unsustainable development, undermining the earth with public funds*. The Hague: Institute for Research on Public Expenditure.

Department for International Development, Directorate General for Development of the European Commission, United Nations Development Programme and World Bank (2002). *Linking poverty reduction and environmental management*. Washington, D.C.: World Bank.

EFTEC (2005). *The economic, social and ecological value of ecosystem services: A literature review*. London: Department for Environment, Food and Rural Affairs.

Emerton, L. (1999a.) *Mount Kenya: the economics of community conservation* (Evaluating Eden Discussion Paper, No. 4). London: International Institute for Environment and Development.

Emerton, L. (1999b). *Balancing the opportunity costs of wildlife conservation for communities around Lake Mburo National Park, Uganda* (Evaluating Eden Discussion Paper, No. 5). London: International Institute for Environment and Development.

Emerton, L. (2001). The nature of benefits and benefits of nature: Why wildlife conservation has not economically benefited communities in Africa. In D. Hulme and M. Muphree (Eds.), *African wildlife and* livelihoods: The promise and performance of community conservation, 208–226. Harare, Zimbabwe: Weaver Press.

Emerton, L., and Bos, E. (2004). *Value: Counting ecosystems as water infrastructure.* Gland, Switzerland: World Conservation Union.

Emerton, L., and Mfunda, I. (1999). *Making wildlife economically viable for the communities living around the Western Serengeti, Tanzania* (Evaluating Eden Discussion Paper, No. 1). London: International Institute for Environment and Development.

Energy Sector Management Assistance Program (2003). *Household energy use in developing countries: A multi-country study.* Washington, D.C.: World Bank.

Food and Agriculture Organization of the United Nations (n.d.). *State of world fisheries and aquaculture.* Retrieved August 1, 2007, from www.fao.org/sof/sofia/index_en.htm.

Gammage, S. (1994). *Estimating the total economic value of a mangrove ecosystem in El Salvador.* London: Overseas Development Administration.

Gangadharan, L., and Valenzuela, R. (2001). Interrelationships between income, health and the environment: Extending the environmental Kuznets curve hypothesis. *Ecological Economics* 36: 513–531.

Ghate, R. (2003). Global gains at local costs: Imposing protected areas: Evidence from Central India. *International Journal of Sustainable Development and World Ecology* 10 (4): 377–389.

Gisladottir, G., and Stocking, M. (2005). Land degradation control and its global environmental benefits. *Land Degradation and Development* 16: 99–112.

Gong, Y. (2004). *Distribution of benefits and costs among stakeholders of a protected area: An empirical study from China* (Economy and Environment Program for Southeast Asia Research Report, No. 2004-RR3). Singapore: Economy and Environment Program for Southeast Asia.

Greenspan Bell, R., and Russell, C. (2002). Environmental policy for developing countries. *Issues in Science and Technology,* Spring 2002, 63–70.

Greenspan Bell, R., and C. Russell (2003). Ill-considered experiments: the environmental consensus and the developing world. *Harvard International Review* 24 (4): 20–25.

Grimes, A., Loomis, S., Jahnige, P., Burnham, M., Onthank, K., Alarcón, R., et al. (1994). Valuing the rainforest: The economic value of non-timber forest products in Ecuador. *Ambio* 23 (7): 405–410.

Gutman, P. (Ed.). (2003). *From goodwill to payments for environmental services. A survey of financing options for sustainable natural resource management in developing countries.* Gland, Switzerland: World Wildlife Fund.

Gylfason, T. (1999). *Principles of economic growth.* Oxford: Oxford University Press.

Hamilton, K. (2000). *Sustaining economic welfare: Estimating changes in wealth per capita* (Environment Department Policy Research Working Paper, No. 2498). Washington, D.C.: World Bank.

Hamilton, K., and Hartwick, J. (2005). Investing exhaustible resource rents and the path of consumption. *Canadian Journal of Economics* 38 (2): 615–621.

Hamilton, K., Ruta, G., Markandya, A., Pedroso, S., Silva, P., Ordoubadi, M., et al. (2005). *Where is the wealth of nations? Measuring capital for the 21st century* (Draft). Washington, D.C.: World Bank.

Hammitt, J., Liu, J.-T., and Liu, J.-L. (2001). Contingent valuation of a Taiwanese wetland. *Environment and Resource Economics* 6: 259–268.

Heltberg, R. (2002). Property rights and natural resource management in developing countries. *Journal of Economic Surveys* 16 (2): 189–226.

Holden, S., Shiferaw, B., and Wik, M. (1998). Poverty, market imperfections and time preferences: Of relevance for environmental policy? *Environment and Development Economics* 3: 105–130.

Howard, A., and Valerio, J. (1996). Financial returns from sustainable forest management and selected agricultural land-use options in Costa Rica. *Forest Ecology and Management* 81: 35–49.

Hutton, G., and Haller, L. (2004). *Evaluation of the costs and benefits of water and sanitation improvements at the global level* (WHO/SDE/WSH/04.04). Geneva: World Health Organization. Retrieved April 14, 2007, from www.who.int/water_sanitation_health/en/wsh0404.pdf.

IMM Ltd. (2002). *Reef livelihoods assessment project: Global overview of reef dependent livelihoods and the poor.* London: Department for International Development.

Intergovernmental Panel on Climate Change. (2001).*Climate change 2001: Synthesis report.* Cambridge: Cambridge University Press.

International Energy Agency (2004). *World energy outlook 2004.* Paris: International Energy Agency.

Islam, J., Kaufmann, D., and Pritchett, L. (1995). *Governance and returns on investment* (World Bank Policy Research Working Paper, No. 1550). Washington, D.C.: World Bank.

Israel, D., and Banzon, C. (1998). *Over-fishing in the Philippine marine fisheries sector* (Research Report). Singapore: Economy and Environment Program for Southeast Asia.

James, A. N., Green, M. J. B., and Paine, J. R. (1999). *Global review of protected area budgets and staff.* Cambridge: UNEP World Conservation Monitoring Centre.

Janssen, R., and Padilla, J. (1999). Preservation or conversion? Valuation and evaluation of mangrove forest in the Philippines. *Environmental and Resource Economics* 14: 297–331.

Joh, S. (2000). *Studies of health benefit estimation of air pollution in Korea.* Seoul: Korea Environment Institute.

Kahn, M. (2003). The death toll from natural disasters: The role of income, geography and institutions. Unpublished document.

Kishor, N., and Constantino, L. (1993). *Forest management and competing land uses: An economic analysis for Costa Rica* (LATEN Dissemination Note No. 7). Washington, D.C.: World Bank.

Knowler, D. (2004). The economics of soil productivity: Local, national and global perspectives. *Land Degradation and Development* 15: 543–561.

Kumari, K. (1996). Sustainable forest management: Myth or reality? Exploring the prospects for Malaysia. *Ambio* 25 (7): 459–467.

Kunte, A., Hamilton, K., Dixon, J., and Clemens, M. (1998). *Estimating national wealth: Methodology and results* (Environmental Economics Series Paper, No. 57). Washington, D.C.: World Bank.

Larson, B., and Rosen, S. (2000). Household benefits of indoor air pollution control in developing countries. Paper presented at USAID/WHO Global Technical Consultation on the Health Impacts of Indoor Air Pollution and Household Energy in Developing Countries, Washington, D.C.

Li, J., Guttikunda, S., Carmichael, G., Streets, D., Chang, Y.-S., and Fung, V. (2004). Quantifying the human health benefits of curbing air pollution in Shanghai. *Journal of Environmental Management* 70: 49–62.

López, R. (1998). Where development can or cannot go: The role of poverty-environment linkages. In B. Pleskovic and J. Stiglitz (Eds.), *Annual World Bank conference on development economics 1997*, 285–306. Washington, D.C.: World Bank.

Loth. P. (2004). *The return of the water: Restoring the Waza Logone floodplain in Cameroon*. Gland, Switzerland: World Conservation Union.

Lutz, E., Pagiola, S., and Reiche, C. (1994). The costs and benefits of soil conservation: The farmer's viewpoint. *World Bank Research Observer* 9 (2): 273–295.

Lvovsky, K. (2001). *Health and environment* (Environment Department, Environment Strategy Papers, Strategy Series No. 1). Washington, D.C.: World Bank.

Madhusudhan, M. (2003). Living amidst large wildlife: Livestock and crop depredation by large mammals in the interior villages of Bhadra tiger reserve, South India. *Environmental Management* 31 (4): 466–475.

Mäler, K.-G. (1998). Environment, poverty and economic growth. In B. Pleskovic and J. Stiglitz (Eds.), *Annual world bank conference on development economics 1997*. Washington, D.C.: World Bank.

Martin-Hurtado, R. (2002). *Costing the 7th Millennium Development Goal: Ensuring environmental sustainability*. Unpublished document, Environment Department, World Bank.

Mauro, P. (1995). Corruption and growth. *Quarterly Journal of Economics* 110: 681–712.

Mendelsohn, R., Basist, A., Kurukulasuriya, P., and Dinar, A. (2004). Climate and rural income. In R. Mendelsohn, A. Dinar, A. Basist, P. Kurukulasuriya, M. Ajwad, F. Kogan, and C. Williams (2004), *Cross sectional analyses of climate change impacts* (Policy Research Working Paper 3350), 79–99. Washington, D.C.: World Bank.

Miranda, M., Porras, I., and Moreno, M. (2003). *The social impacts of payments for environmental services in Costa Rica*. London: International Institute for Environment and Development.

Moseley, W. (2001). African evidence on the relation of poverty, time preference and the environment. *Ecological Economics* 38: 317–326.

Myers, N., and Kent, J. (1998). *Perverse subsidies: Tax dollars undercutting our economies and environments alike.* Winnipeg: International Institute for Sustainable Development.

Organisation for Economic Co-operation and Development (1996). *Subsidies and the environment: Exploring the linkages.* Paris: Organisation for Economic Co-operation and Development.

Organisation for Economic Co-operation and Development (1997). *Reforming energy and transport subsidies: Environmental and economic implications.* Paris: Organisation for Economic Cooperation and Development.

Organisation for Economic Co-operation and Development (1998). *Improving the environment through reducing subsidies.* Paris: Organisation for Economic Cooperation and Development.

Pagiola, S. (1999). *The global environmental benefits of land degradation control on agricultural land* (World Bank Environment Paper No. 16). Washington, D.C.: World Bank.

Pagiola, S., Agostini, P., Gobbi, J., de Haan, C., Ibrahim, M., Murgueito, E., et al. (2004). *Paying for biodiversity conservation in agricultural landscapes* (World Bank Environment Department Paper 96). Washington, D.C.: World Bank.

Pagiola, S., Arcenas, A., and Platais, G. (2005). Can payments for environmental services help reduce poverty? An exploration of the issues and the evidence to date from Latin America. *World Development* 33 (2): 237–253.

Pagiola, S., Bishop, J., and Landell-Mills, N. (2002). *Selling forest environmental services: Market-based mechanisms for conservation and development.* London: Earthscan.

Paldam, M., and Svendsen, G. (2000). An essay on social capital: Looking for the fire behind the smoke. *European Journal of Political Economy* 16: 339–366.

Panayotou, T. (1997). Demystifying the environmental Kuznets curve: Turning a black box into a policy tool. *Environment and Development Economics* 2: 465–484.

Parry, M., Arnell, N., McMichael, T., Nicholls, R., Martens, P., Kovats, S., et al. (2001). Millions at risk: Defining critical climate change threats and targets. *Global Environmental Change* 11: 181–183.

Pattanayak, S., and Sills, E. (2001). Do tropical forests provide natural insurance? The microeconomics of non-timber forest product collection in the Brazilian Amazon. *Land Economics* 77 (4): 595–612.

Pearce, D. W. (2004). Environmental market creation: Saviour or oversell? *Portuguese Economic Journal* 3 (2): 115–144.

Pearce, D. W. (2005). The social cost of carbon. In D. Helm (Ed.), *Climate change policy*, 99–133. Oxford: Oxford University Press.

Pearce, D. W., and Barbier, E. (2000). *Blueprint for a sustainable economy*. London: Earthscan.

Perotti, R. (1996). Corruption and growth. Growth, income distribution and democracy. What the data say. *Journal of Economic Growth* 1 (2): 149–187.

Porter, G. (2003). Subsidies and the environment: An overview of the state of knowledge. In Organisation for Economic Cooperation and Development (Ed.), *Environmentally harmful subsidies: Policy issues and challenges*, 31–100. Paris: Organisation for Economic Cooperation and Development.

Poulos, C., and Whittington, D. (2000). Time preferences for life-saving programs: Evidence from six less developed countries. *Environmental Science and Technology* 34: 1445–1455.

Pretty, J., and Ward, H. (2001). Social capital and the environment. *World Development* 29: 209–227.

Rausser, G., and Small, A. (2000). Valuing research leads: Bioprospecting and the conservation of genetic resources. *Journal of Political Economy* 108 (1): 173–206.

Ray, D. (1998). *Development economics*. Princeton: Princeton University Press.

Ricker, M., Mendelsohn, R., Daly, D., and Angeles, G. (1999). Enriching the rainforest with native fruit trees: An ecological and economic analysis in Los Tuxtlas (Veracruz, Mexico). *Ecological Economics* 31 (3): 439–448.

Rodrik, D. (2000, January 26). Development strategies for the next century. Paper presented at Institute for Developing Economies/ Japan External Trade Organization Conference on Developing Economies in the 21st Century, Chiba, Japan.

Rojat, D., Rajaosafara, S., and Chaboud, C. (2004). Co-management of the shrimp fishery in Madagascar. In Y. Matsuda and T. Yamamoto (Eds.), *What are Responsible Fisheries? Proceedings of the Twelfth Biennial Conference of the International Institute of Fisheries Economics and Trade (IIFET), July 20–30, 2004, Tokyo, Japan*. Corvallis, Ore.: International Institute of Fisheries Economics and Trade.

Rose-Ackerman, S. (1999). *Corruption and government: Causes, consequences and reform.* Cambridge: Cambridge University Press.

Ruitenbeek, J. (1994). Modelling-economy-ecology linkages in mangroves: Economic evidence for promoting conservation in Bintuni Bay, Indonesia. Ecological Economics 10: 233–247.

Russell, C., and Powell, P. (1996). *Choosing environmental policy tools.* Washington, D.C.: Inter-American Development Bank.

Sachs, J., and Warner, A. (2001). The curse of natural resources. European Economic Review 45: 827–838.

Sagar, A. (2005). Alleviating energy poverty for the world's poor. Energy Policy 33: 1367–1372.

Sander, K., and Zeller, M. (2004). Forest resource management between conservation and poverty alleviation: Experiences from Madagascar. Unpublished document, Institute of Rural Development, University of Göttingen.

Sathirathai, S. (1998). *Economic valuation of mangroves and the roles of local communities in the conservation of natural resources: Case study of Surat Thani, South of Thailand.* Singapore: Economy and Environment Program for Southeast Asia. Retrieved August 1, 2007, from www.idrc.ca/uploads/user-S/10536137110ACF9E.pdf.

Scherr, S., White, A., and Kaimowitz, D. (2004). *A new agenda for forest conservation and poverty reduction: Making markets work for low income producers.* Washington, D.C.: Forest Trends.

Secretariat of the Convention on Biological Diversity (2001). *The value of forest ecosystems* (Convention on Biological Diversity Technical Series No. 4). Montreal: Secretariat of the Convention on Biological

Diversity. Retrieved August 1, 2007, from www.biodiv.org/doc/publications/cbd-ts-04.pdf.

Simpson, D. (2004). *Conserving biodiversity through markets: A better approach* (Property and Environment Research Center Policy Series P32). Missoula, Mont.: Property and Environment Research Center.

Simpson, D., Sedjo, R., and Reid, J. (1996). Valuing biodiversity for use in pharmaceutical research. *Journal of Political Economy* 104 (1): 163–185.

Smith, J., and Scherr, S. (2002). *Forest carbon and local livelihoods: Assessment of opportunities and policy recommendations* (Center for International Forestry Research Occasional Paper 37). Bogor, Indonesia: Center for International Forestry Research.

Tokarick, S. (2005). Who bears the cost of agricultural support in Organisation for Economic Cooperation and Development countries? *World Economy* 28 (4): 573–593.

Tol, R. (2005). The marginal damage costs of carbon-dioxide emissions. In D. Helm (Ed.), *Climate change policy*, 152–166. Oxford: Oxford University Press.

United Nations Development Programme (2002). *Poverty and environment initiative*. New York: United Nations Development Programme.

United Nations Development Programme (2004). *Human development report 2004: Cultural liberty in today's diverse world*. Oxford: Oxford University Press.

United Nations Environment Programme (1991). *The status of desertification and implementation of the United Nations plan of action to combat desertification*. Nairobi: United Nations Environment Programme. Retrieved August 1, 2007, from www.unep.net/des/unced.

United Nations Millennium Project (2005). *Environment and human well-being: A practical strategy—Report of the task force on environmental sustainability*. London: Earthscan.

United Nations Secretary-General. (2000). *We the peoples: The role of the UN in the 21st century* (Millennium Report of the Secretary-General of the United Nations). New York: United Nations Department of Public Information.

van Beers, C., and de Moor, S. (2001). *Public subsidies and policy failures: How subsidies distort the natural environment, equity and trade and how to reform them*. Cheltenham, England: Edward Elgar.

van Beers, C., and van den Bergh, J. (2001). Perseverance of perverse subsidies and their impact on trade and environment. *Ecological Economics* 36: 475–486.

van Beukering, J., Cesar, H., and Janssen, M. (2003). Economic valuation of the Leuser National Park on Sumatra, Indonesia. *Ecological Economics* 44: 43–62.

Vedeld, P., Angelsen, A., Sjaasrad, E., and Berg, G. (2004). *Counting on the environment: Forest income and the rural poor* (World Bank Environmental Economics Series No. 98). Washington, D.C.: World Bank.

von Schirnding, Y., Bruce, N., Smith, K., Ballard-Tremeer, G., Ezzati, M., and Lvovsky, K. (2002). *Addressing the impact of household energy and indoor air pollution on the health of the poor: Implications for policy action and intervention measures* (WHO/HDE/HID/02.9). Geneva: World Health Organization.

Warwick, H., and Doig, A. (2004). *Smoke: The killer in the kitchen—Indoor air pollution in developing countries*. London: ITDG Publishing.

Weladji, R., and Tchamba, M. (2003). Conflict between people and protected areas within the Bénoué Wildlife Conservation Area, North Cameroon. *Oryx* 37 (1): 72–79.

Wells, M., and Brandon, K. (1992). *People and parks: Linking protected area management with local communities*. Washington, D.C.: World Bank.

Winters, P., Murgai, R., Sadoulet, E., de Janvery, A., and Frisvold, G. (1998). Economic and welfare impacts of climate change on developing countries. *Environmental and Resource Economics* 12: 1–24.

World Bank (1994). *Chile: Managing environmental problems: Economic analysis of selected issues* (World Bank Report 13061-CH). Washington, D.C.: World Bank.

World Bank (1996). *Haiti forest and parks protection: Technical assistance project*. (Staff Appraisal Report T-6948-HA). Washington, D.C.: World Bank.

World Bank. (2002). *The environment and the millennium development goals*. Washington, D.C.: World Bank.

World Conservation Union (2003a). *Barotse Floodplain, Zambia: Local economic dependence on wetland resources* (Case Studies in Wetland Valuation No. 2). Gland, Switzerland: World Conservation Union.

World Conservation Union (2003b). *Waza Logone Floodplain, Cameroon. Economic benefits of wetland restoration* (Case Studies in Wetland Valuation No. 4). Gland, Switzerland: World Conservation Union.

World Conservation Union (2005). *Veun Sean Village, Stoeng Treng RAMSAR site, Cambodia: Rapid participatory assessment for wetland valuation* (Case Studies in Wetland Valuation No. 11). Gland, Switzerland: World Conservation Union.

World Energy Council. (1999). *The challenge of rural energy poverty in developing countries.* London: World Energy Council.

Wunder, S. (2005). *Payments for environmental services: Some nuts and bolts* (Occasional Paper 42). Bogor, Indonesia: Center for International Forestry Research.

Yaron, G. (2002). The economic value of Mount Cameroon: Alternative land use options. In D. W. Pearce, C. Pearce, and C. Palmer (Eds.), *Valuing the environment in developing countries: Case studies,* 406–446. Cheltenham, England: Edward Elgar.

Zbinden, S., and Lee, D. (2005). Paying for environmental services: An analysis of participation in Cost Rica's PSA programme. *World Development* 33 (2): 255–272.

3 | Ecoagriculture

Agriculture, Environmental Conservation, and Poverty Reduction at a Landscape Scale

SARA SCHERR, JEFFREY A. MCNEELY, AND SETH SHAMES

The Millennium Ecosystem Assessment (MA) documented the dominant impacts of agriculture on terrestrial land and freshwater use, and the critical importance of agricultural landscapes in providing products for human sustenance, supporting wild species biodiversity and maintaining ecosystem services (MA, 2005). Yet global demand for associated agricultural products is projected to rise at least 50 percent over the next two decades (United Nations Millennium Project, 2005). The need to reconcile agricultural production and production-dependent rural livelihoods with healthy ecosystems has prompted widespread innovation to coordinate landscape and policy action (Acharya, 2006; Jackson and Jackson, 2002; McNeely and Scherr, 2003). However, the dominant national and global institutions—for policy, business, conservation, agriculture, and research—have been shaped largely by mental models that assume and require segregated approaches.

This chapter will discuss a new paradigm, ecoagriculture, which calls for integrated conservation-agriculture landscapes, in which biodiversity conservation is an explicit objective of agriculture, food security, and rural development, and the latter three are explicitly

---

Sara Scherr is the President of Ecoagriculture Partners. Jeffrey A. McNeely is Chief Scientist for the World Conservation Union. Seth Shames is the Policy Project Manager for Ecoagriculture Partners.

considered in shaping conservation strategies. This approach is highly relevant for agricultural systems in environmentally important or threatened areas worldwide, but the focus of this chapter is particularly on low-income farming communities.

To begin, this chapter will explore the web of relationships between food security, poverty alleviation, ecosystem services, biodiversity, and agricultural production. It will also focus on the trends that are creating a need for shifts to an ecoagriculture paradigm; describe the ecoagriculture landscape approach; and present real-world cases of ecoagriculture in low-income farming communities. Furthermore, it will outline strategic actions required to mobilize and scale-up ecoagriculture initiatives in these kinds of communities to a level that would have a meaningful global impact.

The State of Agriculture, Ecosystems, and Rural Poverty

To comprehend the urgent need for a shift toward integrated agriculture-conservation landscapes, it is essential to grasp the intertwined relationships of agriculture, ecosystems, and rural poverty. Of the estimated 800 million people who do not have access to sufficient food, half are smallholder farmers, a fifth are rural landless, and a tenth are principally dependent on rangelands, forests, and fisheries. For most of these people, reducing poverty and hunger will depend centrally on their ability to sustain and increase crop, livestock, forest, and fishery production. Yet widespread land and water degradation affects production on at least half of all croplands and threatens resource-dependent livelihoods. Moreover, the MA has confirmed that agriculture is now the dominant terrestrial land use (2005). Agricultural expansion and intensification have become the main drivers of biodiversity loss and ecosystem degradation.

Subsistence farmers throughout the world depend directly on the ecosystem services and biodiversity of agricultural landscapes. Ecosystem services—ecological processes and functions that sustain and improve human well-being (Daily, 1997)—can be divided into four categories: (1) *provisioning services* or ecosystems that provide food, timber, medicines, and other useful products; (2) *regulating services* such as flood control and climate stabilization; (3) *supporting services* such as pollination, soil formation, and water purification; and (4) *cultural services*, which are aesthetic, spiritual, or recreational assets that provide both intangible benefits and tangible ones such as ecotourism attractions (Kremen and Ostfeld, 2005). Provisioning has often been considered the highest priority service for poor farmers provided by agricultural landscapes, but it is now recognized that even the breadbaskets and rice bowls of the world also provide other ecosystem services, such as water supply and quality, or pest and disease control (Wood, Sebastian and Scherr, 2000).

Although wild biodiversity and ecosystem services are closely linked, they are not synonymous. A landscape with relatively intact wild biodiversity is likely to provide a full complement of ecosystem services; however, many ecosystem services can also be provided by nonnative species, or by combinations of native and nonnative species in heavily managed settings, such as permanent farms. The implication is that, even where wild biodiversity has been significantly reduced to make way for food and fiber production, high levels of ecosystem services can often still be provided through intentional land-management practices.

Agricultural biodiversity is an ecosystem service of particular relevance to rural livelihoods. As defined by the 1992 Convention on Biological Diversity (CBD), agricultural biodiversity includes genetic diversity of domesticated crops, animals, fish, and trees; diversity of wild species on which agricultural production depends, such as wild pollinators, soil microorganisms, and predators of agricultural pests; and diversity of wild species and ecological communities that use agricultural landscapes as their habitat. Unfortunately, like the other ecosystem services critical to poor farmers, agricultural biodiversity is deteriorating throughout world.

Ecosystem Service Degradation

The rural poor in developing countries need to increase agricultural production to feed their families and supply growing markets. However, widespread ecological degradation in many areas is already either reducing yields or increasing the cost of production. Up to 50 percent of the world's agricultural land and 60 percent of ecosystem services are now affected by some degree of degradation, with agricultural land-use the chief cause of land degradation (Bossio et al., 2004; MA, 2005). Half the world's rivers are seriously depleted and polluted, and 60 percent of the world's 227 largest rivers have been significantly fragmented by large dams, many built to supply irrigation water. Estimates are that 20 percent of irrigated land suffers from secondary salinization and waterlogging, induced by the buildup of salts in irrigation water (Wood et al., 2000). The food system will also have to address or adapt to the collapse in harvests of wild game and wild fisheries in many regions around the world, due to overexploitation and habitat loss or pollution (Hassan, Scholes, and Ash, 2005). Climate change will exacerbate these stresses for poor farmers in the tropics. Farmers who depend directly on natural resources are the first to feel the pressure of this degradation and the least able to adapt.

These resource constraints also build on themselves as farmers become more desperate. Forests and natural vegetation are cleared for agricultural use, timber, and wood fuels, as formerly fertile land degrades. These land-use shifts fragment landscapes, breaking formerly contiguous wild species populations into smaller units that are more vulnerable to extirpation. Farmers generally have sought to eliminate wild species from their lands in order to reduce the negative effects of pests, predators, and weeds; however, these practices often harm beneficial wild species like pollinators (Buchmann and Nabhan, 1996), insect-eating birds, and other species that prey on agricultural pests.

The threats agriculture poses to ecosystems have been a key motivator for conservationists to develop protected areas where agricultural activity is officially excluded or largely circumscribed. Nonetheless, the MA calculated that more than 45 percent of 100,000 protected areas had more than 30 percent of their land area in crop production (2005). These realities are requiring farmers, agricultural planners, and conservationists to reconsider the relationship between production agriculture and conservation of ecosystems and biodiversity.

Impacts of Degradation and Ecological Poverty

Ecological degradation threatens the livelihoods and food security of vulnerable, poor farmers. Increases in global aggregate agricultural production could, theoretically, ensure food security to the world's entire population. However, food access is determined not only by the available supply, but also by having the arable land and inputs needed to produce it or the cash income to purchase it. The world's poorest have none of these, even if food is available elsewhere in the world. Poor transportation, food storage, and marketing systems throughout large areas of the developing world mean that even food that can be imported cheaply to port cities becomes too expensive for people in the hinterland to buy, even if it reaches such remote areas.

Thus, even in the agricultural breadbaskets of the tropics, where food is amply available, poor people with low cash incomes are unable to buy enough. Some 1.2 billion people worldwide earn less than one dollar a day, a poverty line that adjusts for differences across countries and times in purchasing power. Some 75 percent of them live in rural areas. Projections indicate that more than 60 percent of the poor will continue to be rural people in 2025 (International Fund for Agricultural Development, 2000). For them to climb out of poverty, they must first secure their food supply, a process in which healthy ecosystems are essential.

Much of this agriculturally linked ecological degradation and food insecurity feeds into a cycle of ecological poverty for the rural poor. The term *ecological poverty* describes the type of widespread poverty that arises from degradation or loss of natural capital (Coward, Oliver, and Conroy, 1999). But ecological poverty both leads to economic poverty and results from it. As discussed earlier, when poor people have trouble finding food because of insufficient agricultural production or income, they may become even more dependent on the products of wild biodiversity, clearing new fields from natural habitat, and poaching and encroaching on protected areas. Such measures may provide emergency relief, but they are not sustainable and may reduce the area's natural capital or even destroy natural food supplies.

Ecological poverty is concentrated in many of the areas where wild biodiversity is richest or most threatened. An estimated 325 million poor people live on favored agricultural lands in developing countries, while 630 million live on marginal agricultural, forested, and arid lands (Consultative Group on International Agricultural Research, 1997). Some 300 million people, most of them poor, live in forested areas and another 200 million live around them (Panayotou and Ashton, 1992). Many indigenous ethnic groups, among the most impoverished and marginalized peoples, live in lands where extensive wild biodiversity remains.

Trends of Agriculture-Environment-Livelihoods Interaction

Any system that aims to reverse the cycle of degradation, food insecurity, and ecological poverty among the rural poor while meeting increasing demand for agricultural products will require the integration of production increases, ecosystem service conservation, and livelihood development. In addition to understanding underlying causes of poverty, efforts to achieve this integration must take into account the following trends.

Increasing Demand for Agricultural Products in Ecologically Sensitive Areas

Human population is expected to grow from a bit more than six billion people today to more than eight billion by 2030, an increase of about a third, with another two to four billion added in the subsequent fifty years (Cohen, 2003). But food demand is expected to grow even faster as a result of growing urbanization and rising incomes and, if hunger is to be reduced among the 800 million people currently undernourished (UN Millennium Project, 2005), more land will surely be required to grow crops, even more so if biofuels become a greater contributor to energy needs.

Tilman et al. (2001) predicts that feeding a population of nine billion using current methods could result in the conversion of another billion hectares of natural habitat to agricultural production, primarily in the developing world, together with a doubling or tripling of nitrogen and phosphorous inputs, a twofold increase in water consumption, and a threefold increase in pesticide use. A serious limiting factor is expected to be water, as 70 percent of the freshwater used by people is already devoted to agriculture (Rosegrant, Cai, and Clein, 2002). Scenarios prepared by the MA, thus, suggest that agricultural production in the future will have to focus more explicitly on ecologically sensitive management systems (Carpenter et al., 2005).

Most Increased Food Production Will Be on Marginal Lands

An estimated 90 percent of food products consumed in most countries will be produced domestically. Total export levels increased sharply between 1961 and 2000, but agricultural exports still accounted for only about 10 percent of production (McCalla, 2000). This pattern seems unlikely to change over the next few decades, even though continuing globalization of agriculture will influence product mix and prices. Large and growing interior populations in large countries will continue to be fed mainly by local and national producers.

Lower-productivity lands such as drylands, hillsides, and rainforests now account for more than two-thirds of total agricultural land in developing countries (Consultative Group on International Agricultural Research, 1997). Because current yields are relatively low, existing technologies can double or even triple yields, with adequate investment, market developments and attention to good ecosystem husbandry (UN Millennium Project, 2005). Extensive grain monocultures are not likely to be sustainable in such areas, calling for more diversified land use approaches. Although the bulk of new production will come mainly from existing croplands, the most promising areas with significant new land for agriculture are in places like the forest and savanna zones of Brazil and Mozambique, which are the main remaining large reservoirs of natural habitat. These habitats will be seriously damaged by highly simplified, high-external-input production systems, but an integrated production-conservation approach could significantly reduce the damage.

Agricultural Sustainability Will Require Investment in Ecosystem Management

The conservation community is moving towards an ecosystem approach to conserving biodiversity, in light of the dependence of protected areas (PAs) on a supportive matrix of land and water use and the creation of biological corridors (Convention on Biological Diversity, 2000). The international community has set a goal of having at least 10 percent of every habitat type under effective protection by 2015 (The Nature Conservancy, 2004). This strategy—if successful—will protect many species and ecological communities. But some estimates suggest that more than half of all species exist principally outside PAs, mostly in agricultural landscapes (Blann, 2006). For example, conservation of wetlands within agricultural landscapes is critical for wild bird populations (Heimlich et al., 1998). Such species will be conserved only through initiatives by and with farmers. The concept of agriculture as ecological sacrifice areas is no longer valid in many regions because agricultural lands both perform many ecosystem services and provide essential habitat to many species.

Wild Products Will Continue to Be Important

People in low-income developing countries and subregions will continue to rely on harvesting wild species. Wild greens, spices, and flavorings enhance local diets, and many tree fruits and root crops serve to assuage preharvest hunger and provide famine foods when crops or the economy fails. Frogs, rodents, snails, edible insects, and other small creatures have long been an important part of the rural diet in virtually all parts of the world (Paoletti, 2005). Bushmeat is the principal source of animal protein in humid West Africa and other forest regions, and efforts to replace these with domestic livestock have been disappointing. Fisheries are the main animal protein source of the poor worldwide. In Africa and many parts of Asia, more than 80 percent of medicines still come from wild sources. Gathered wood remains the main fuel for hundreds of millions of people, while forests and savannas provide critical inputs for farming in the form of fodder, soil nutrients, and fencing. Achieving food security will, therefore, require the conservation of the ecosystems providing these foods and other products.

Agricultural Systems Will Need to Diversify to Adapt to Climate Change

Strategic planning for agricultural development has begun to focus on adaptation of systems to climate change, anticipating rising temperatures and more extreme weather events. The U.S. Department of Agriculture and the International Rice Research Institute have both concluded that with each one degree centigrade increase in temperature during the growing season, the yields of rice, wheat, and maize drop by 10 percent (Brown, 2004; Tan and Shibasaki, 2003). Cash crops, such as coffee and tea, that require cooler environments will also be affected, forcing farmers to move higher up the hills, clearing new lands as they climb. Montane forests important for biodiversity are likely to come under increasing threat. Effective responses to climate change will require changing varieties and modified management of soils and water, as well as new strategies for pest management as species of wild pests, their natural predators, and their life cycles change in response to climates. Increasing landscape and farm-scale diversity are likely to be an important response for risk reduction (Jackson et al., 2005).

The challenges described above are unlikely to be met by the solutions of industrial agriculture, the original Green Revolution, sustainable agriculture and natural resource management (with its primary focus on sustaining the resources underpinning production), or even the ecotechnology approach of Swaminathan (1994) with its focus on the farmer's field, although all of these have major elements to contribute. Approaches to biodiversity conservation also need to move beyond the wild biodiversity focus of strictly protected areas and the modest goals of integrated conservation and development projects. Ecoagriculture—a fully integrated approach to agriculture, conservation, and rural livelihoods, within a landscape or ecosystem context—is needed in many regions.

Ecoagriculture: Integrating Production and Conservation at a Landscape Scale

Ecoagriculture explicitly recognizes the mutual interdependence among agriculture, biodiversity and ecosystem services. Ecoagriculture landscapes are mosaics of areas in natural/native habitat and areas under agricultural production. Effective ecoagriculture systems rely on maximizing the ecological, economic, and social synergies among them and minimizing the conflicts.

The term *landscape* itself is functionally defined, depending upon the spatial units needed or actually managed by the group of stakeholders working together to achieve biodiversity, production and livelihood goals. Ecoagriculture landscapes are land-use mosaics with:

- *Natural areas* with high-habitat quality and niches to ensure critical elements for habitat or ecosystem services that cannot be provided in areas under production, which are also managed to benefit agricultural livelihoods either through positive synergies with production or other livelihood benefits;

- *Agricultural production areas* that are productive and profitable and meet food security, market, and livelihood needs, which are also configured and managed to provide a matrix with benign or positive ecological qualities for wild biodiversity and ecosystem services; and

- *Institutional mechanisms* to coordinate initiatives to achieve production, conservation and livelihood objectives at landscape, farm and community scales, by exploiting synergies and managing trade-offs among them.

The concept of ecoagriculture further recognizes that agriculture-dependent rural communities are critical, and sometimes the principal, stewards of biodiversity and ecosystem services. While protected natural areas are essential in ecoagriculture landscapes, to ensure critical habitat for vulnerable species, maintain water sources, and provide cultural resource, these resources often may be owned or managed by local communities and farmers.

Ecoagriculture Approaches

Broadly, ecoagriculture landscapes rely on six basic strategies of resource management, three focused on the agricultural part of the landscape and three on the conservation areas. In production areas, farmers sustainably increase agricultural output and reduce costs in several ways that enhance the habitat quality and ecosystem services, including:

- minimize agricultural wastes and pollution;

- manage resources in ways that conserve water, soils, and wild flora and fauna; and

- use crop, grass, and tree combinations to mimic the ecological structure and function of natural habitats.

Farmers and other conservation managers protect and expand natural areas in ways that also provide benefits for adjacent farmers and communities, such as:

- minimize or reverse conversion of natural areas;

- protect and expand larger patches of high-quality natural habitat; and

- develop effective ecological networks and corridors. (McNeely and Scherr, 2003)

The relative area and spatial configuration of agricultural and natural components, as well as other elements, such as physical infrastructure and human settlements, are key landscape-design issues (Forman, 1995). The conservation of wild species that are highly sensitive to habitat disturbance—as are some of the most endangered and rare globally—requires large, well-connected patches of natural habitat. But some wild species, including many that are threatened and endangered, can coexist in compatibly managed agricultural landscapes, even in high-yielding systems.

Numerous approaches to agriculture, conservation, and rural development contribute management practices and planning frameworks that can be applied in ecoagriculture landscapes. The outcomes of planning and negotiations among the multiple stakeholders in any particular landscape will take diverse forms, depending on the context of local cultures and philosophies of land management.

Ecoagriculture Cases

There are numerous documented cases of small farmers managing ecoagriculture landscapes jointly for agricultural production and conservation. The following examples provide illustrations of how ecoagriculture works in the real world.

TRANSBOUNDARY COMANAGEMENT IN COSTA RICA AND PANAMA

The Grandoca-Manzanillo National Wildlife Refuge on Costa Rica's Caribbean coast connects with Panama's San Pondsak National Wildlife Refuge. This 10,000-hectare refuge is comanaged by local communities, nongovernmental organizations (NGOs), and government agencies. Small farm agro-ecosystems are integral to regional biodiversity conservation. More than three hundred farmers hold secure land titles in the refuge's buffer zone. A regional small farmers' cooperative, Smallholder Association of Talamanca (APPTA), supports more than 1,500 small farmers and has become Central America's largest volume organic producer and exporter, generating 15–60 percent increases in small-farmer revenue. Conservation-based carbon offset schemes are being developed to provide additional revenue for stewardship-focused farming.

INDIGENOUS HONEYBEE CONSERVATION IN THE HINDU KUSH HIMALAYAS

Agriculture in the mountainous Hindu Kush–Himalayan region is in a stage of transition from traditional cereal crop farming to high-value cash crops, such as fruits and vegetables. This ongoing transformation from subsistence to cash-crop farming poses a number of new challenges, including low production and crop failures due to inadequate pollination. The decline in pollinator intensity presents a serious threat to agricultural production and livelihoods. The International Centre for Integrated Mountain Development (ICIMOD), an international development and research organization, is working with local people to address this issue through a program focused on the conservation and sustainable management of wild bees (Ahmed et al., 2004).

DRYLAND RESTORATION:
COMMUNITY WATER HARVESTING IN RAJASTHAN, INDIA

Until recently, drought and environmental degradation severely impaired the livelihood security of local communities within Rajasthan's Arvari Basin. Crop failure, soil erosion, and watershed degradation were widespread, with communities facing a continual challenge to meet water needs. Twenty years ago, the Tarun Bharat Sangh, a voluntary organization based in Jaipur, India, initiated a community-led watershed restoration program. The response was based upon reinstating *johads*, a traditional indigenous technology. Johads are simple concave mud barriers, built across small, uphill river tributaries to collect water. As water drains through the catchment area, johads encourage groundwater recharge and improved hillside forest growth, while providing water for irrigation, wildlife, livestock, and domestic use. More than five thousand johads now serve close to 1,050 villages in the region. Community leadership over watershed management is coordinated through purposefully established village councils. The transformation in Rajasthan's social, economic, and biophysical landscape is evident, most notably in the restoration of the Avari River, which had not flowed since the 1940s. In turn, enhanced water availability has resulted in more sustainable agricultural practices, improved livelihood security and, overall, strengthened emphasis on community-led natural resources management within the region.

Although many examples of successful approaches linking biodiversity conservation with sustainable agriculture and rural development have been documented, current knowledge, institutional arrangements, and policies are inadequate to meet ecoagriculture objectives at a globally significant scale.

Scaling Up Ecoagriculture Landscapes for the Rural Poor

To climb out of poverty through ecoagriculture development, low-income farmers need effective community organizations to manage resources and advocate for rights; institutional arrangements that support multi-stakeholder planning processes and knowledge sharing; policies that enable ecoagriculture planning; and a research agenda that informs this emerging approach.

Empowering Communities

A core feature of ecoagriculture landscapes is the role of local farming or pastoral communities as key stewards, decision makers, and managers of biodiversity. Public agencies may operate forests and protected areas, but their viability and sustainability depend on the matrix of private land uses in the landscape. Economic and social incentives can motivate collective action among local communities. Hundreds of community-based organizations have been documented to mobilize or engage in landscape-scale ecoagriculture initiatives (Campbell, 1994; Isely and Scherr, 2003; McNeely and Scherr, 2003; Rhodes and Scherr, 2005). The institutions leading these initiatives are hybrids linking conventional farmer cooperatives, rural development committees, and community-based conservation organizations (Buck et al., 2004). In the Philippines, for example, local farmer-based Landcare groups work with conservation organizations, municipal governments and research organizations to revegetate hillsides, conserve biodiversity in populated protected areas, and improve water quality (Cramb and Culasero, 2003).

An important implication of the central role of communities in conservation, especially outside protected areas, is that conservation organizations need to embrace and re-orient their role explicitly to support local community stewardship in ways that respect and realistically address the central role of agriculture and livelihoods in planning and implementation methodologies (Bumacas et al., 2007). For communities to function in this way, they will also need stronger rights over their natural resources. This will require that the poor and disenfranchised groups within the landscape organize themselves for political strength, that they join coalitions with other stakeholders, and that they are supported strategically in their negotiations with more powerful groups.

Support Landscape-Scale, Multi-Stakeholder Planning Platforms

To achieve objectives at the landscape scale requires not only processes of collective action to support communities of producers, but also mechanisms to coordinate action among all key stakeholders in the landscape. These links must often bridge sectors with historical legacies of distrust. Development or adaptation of institutions for engagement, coordination and governance of ecoagriculture becomes the critical challenge. Scaling up and sustaining ecoagriculture landscapes that involve multiple stakeholders requires a process, and usually an institution, that enables multi-stakeholder assessment, planning, implementation, and monitoring for adaptive management. Currently, ecoagriculture initiatives take numerous forms, mobilized by community organizations, public agencies, NGOs, or national/international projects. Methodologies that have been developed to assist the planning and governance process include landscape visioning and scenario-building processes; participatory landscape modeling; community biodiversity assessments; and guidelines for adaptive collaborative management (Buck et al., 2001; Edmunds and Wollenberg,

2001). Multi-stakeholder trust-building processes and negotiation platforms are being adapted to the specific context of agriculture-biodiversity conflict situations. Diversity of approaches is expected and desirable, but more systematic and comparative evaluation of effectiveness in achieving sustainable processes and outcomes is needed.

Knowledge-sharing systems need to be reshaped to provide services to rural resource stewards, and to accelerate exchange of practical knowledge among them and across sectors. Rural communities must be acknowledged as key stewards of biodiversity and ecosystem services, and professional conservation and development organizations, public agencies and others need to re-orient their activities to reflect this reality and help link innovators to each other, as well as to other stakeholder groups, to enhance information exchange on ecoagriculture stewardship.

Promote Policies that Support Ecoagriculture Landscapes

Ecoagriculture innovators around the world highlight the need for a more supportive policy environment for ecoagriculture, or simply the removal of major policy barriers (Rhodes and Scherr, 2005). Core policy needs that could spur ecoagriculture among low-income farmers include compatibility and coordination of agricultural development and biodiversity conservation policies, and increased rights of farming communities to natural resources.

Consumers, policymakers, and investors are beginning to focus on the link between agriculture and conservation, and responding with new demands on the agricultural system, through systems of voluntary certification, industry standards, and government regulation (Secretariat of the CBD, 2005; CBD, Agricultural Biological Diversity, 2002; Agricultural Biological Diversity, 2000; Conservation and Sustainable Use, 1996). Ecosystem/landscape-scale programs and projects are being initiated by government agencies and NGOs, often in multi-stakeholder partnerships, and financed through public budgets, as in the case of India and China, and international development loans (Fernandes, 2004). New political coalitions are being formed to promote integrated cross-sectoral policies, bringing in voices and sectors not traditionally involved in either agricultural or conservation policy, such as municipal governments, in the context of political decentralization; urban consumer groups; international financial organizations concerned with screening investments for environmental sustainability; parts of the food industry; public health advocates; and good governance proponents seeking to reduce wasteful spending on subsidies.

At the international policy level, ecoagriculture strategies are being integrated into the work programs of the relevant international conventions. For example, the CBD has adopted a new biodiversity goal of 30 percent of agricultural areas under biodiversity-friendly management by 2010 (CBD, Global Strategy for Plant Conservation, 2002), and will focus on agriculture in meetings during 2008, as will the Commission on Sustainable Development. Rules developing under the World Trade Organization will need to be carefully scrutinized to ensure that they do not disadvantage producers in ecoagriculture landscapes.

As noted earlier, secure rights to local natural resources are a foundational element to ecoagriculture development. Some countries, notably Australia, Brazil, and India, have adopted legislation that explicitly recognizes the rights of indigenous and other local communities to manage and conserve forests and natural habitats (Ellsworth, 2004; Molnar, Scherr, and Khare, 2004). The Convention on Biological Diversity and other international bodies are beginning to focus on opportunities for community-led conservation, although many elements of the conservation community are still uncomfortable directly addressing and supporting agricultural development. Consequently, farming communities are often underrepresented in many conservation policy processes.

Produce and Share Knowledge for Ecoagriculture

The challenge of shaping agricultural landscapes to meet joint production, conservation and livelihood goals for low-income farmers will require a dramatic scaling up and refocusing of research in national research systems, at the Future Harvest Centers supported by the Consultative Group for International Agricultural Research, and in centers of conservation science, national academies of science, and universities. The priorities are:

- to understand the interaction and dynamics of conservation and production areas;

- to develop production systems, including improved varieties of more diverse domesticated species, that explicitly meet biodiversity objectives and mimic natural ecosystems; and

- to make more elements of farming systems ecologically sustainable, including industrial processing, packaging, and transport.

Ecoagriculture systems that appear to be successful need to be fully documented, both in terms of landscape-scale outcomes and specific interventions. Mapping of spatial overlays between important agricultural areas, in terms of national product supply and local livelihoods, and important biodiversity will be essential.

Conclusion

For poor farmers throughout the world, ecoagriculture is an approach whose time has come. The transformation of agricultural production from being one of the greatest threats to global biodiversity and ecosystem services to becoming a major contributor to ecosystem integrity is unquestionably a key challenge of the twenty-first century and of particular importance to the rural poor. An ecoagriculture landscape approach could help to achieve the critical goals of agricultural sustainability, resilience of food systems, adaptation to climate change, food security, and poverty alleviation.

It is simply no longer possible to consider agricultural systems outside of their ecological contexts. Small farmers in degrading ecosystems are reminded of this daily as yields become increasingly scarce and unreliable. They need institutional, policy, and research support to manage ecoagriculture systems that can provide for ecosystem services to agricultural and surrounding ecosystems. There must also be a buy-in from all relevant stakeholders in a landscape, including conservationists. For these multi-stakeholder processes, and the ecoagriculture landscapes management systems that emerge from them, to be truly sustainable, the actors must have the courage to move beyond historical compartmentalization and distrust and to recognize that ultimately their goals are often the same.

References

Acharya, K. P. (2006). Linking trees on farms with biodiversity conservation in subsistence farming systems in Nepal. *Biodiversity and Conservation* 15: 631–646.

Ahmed, F., Partap, U., Joshi, S. R., and Gurung, M. B. (2004). Indigenous honeybees: Allies for mountain farmers. *LEISA* 20 (4): 12–13.

Blann, K. (2006). *Habitat in agricultural landscapes: How much is enough? A state-of-the-science literature review.* West Linn, Ore.: Defenders of Wildlife.

Bossio, D., Noble, A., Pretty, J., and Penning de Vries, F. (2004, August). Reversing land and water degradation: Trends and "bright spot" opportunities. Paper presented at the Stockholm International Water Institute/ Comprehensive Assessment on Water Management in Agriculture Seminar, Stockholm.

Brown, L. R. (2004). *Outgrowing the earth: The food security challenge in an age of falling water tables and rising temperatures.* New York: Norton.

Buchmann, S. L., and Nabhan, G. P. (1996). *The forgotten pollinators.* Washington, D.C.: Island Press.

Buck, L., Gavin, T., Lee, D., and Uphoff, N. (2004). *Ecoagriculture: A review and assessment of its scientific foundations* (Ecoagriculture Discussion Paper Series, No. 1). Washington, D.C.: Ecoagriculture Partners.

Buck, L. E., Geissler, C. C., Schelhas, J., and Wollenberg, E. (2001). *Biological diversity: Balancing interests through adaptive collaborative management.* Boca Raton, Fla.: CRC Press.

Bumacas, D., Catacutan, D., Chibememe, G., and Rhodes, C. (2007). Community leadership in ecoagriculture. In S. J. Scherr and J. A. McNeely (Eds.), *Farming with nature: The science and practice of ecoagriculture*. Washington, D.C.: Island Press.

Campbell, A. (1994). *Landcare: Communities shaping the land and the future*. St. Leonards, Australia: Allen and Unwin.

Carpenter, S. R., Pingali, P., Bennett, E., and Zurek, M. (Eds.) (2005). *Ecosystems and human well-being: Scenarios*. Washington, D.C.: Island Press.

Cohen, J. E. (2003). Human population: The next half century. *Science* 302: 1172–1175.

Consultative Group on International Agricultural Research, Technical Advisory Committee, Food and Agriculture Organization of the United Nations (1997). *Report of the study on CGIAR research priorities for marginal lands*. Retrieved July 6, 2007, from www.fao.org/Wairdocs/TAC/X5784E/x5784e02.htm#TopOfPage.

Convention on Biological Diversity, Agricultural biological diversity (COP 6, Decision VI/5) (2002). Retrieved July 6, 2007, from www.cbd.int/decisions/default.shtml?m = COP-06&id = 7179&lg = 0.

Convention on Biological Diversity, Agricultural biological diversity: Review of phase I of the programme of work and adoption of a multi-year work programme (COP 5, Decision V/5) (2000). Retrieved July 6, 2007, from www.cbd.int/decisions/default.shtml?m = COP-05&id = 7147&lg = 0.

Convention on Biological Diversity, Article 2: Use of terms (1992). Retrieved July 6, 2007, from www.cbd.int/convention/articles.shtml?a = cbd-02.

Convention on Biological Diversity, Conservation and sustainable use of agricultural biological diversity (COP 3, Decision III/11) (1996). Retrieved July 6, 2007, from www.cbd.int/decisions/default.shtml?m = COP-03&id = 7107&lg = 0.

Convention on Biological Diversity, Ecosystem approach (COP 5, Decision V/6) (2000). Retrieved July 6, 2007, from www.cbd.int/decisions/default.asp?lg = 0&m = cop-05&d = 06.

Convention on Biological Diversity, Global strategy for plant conservation (COP 6, Decision VI/9) (2002). Retrieved July 6, 2007, from www.biodiv.org/decisions/default.asp?dec = VI/9.

Coward, E. W., Jr., Oliver, M. L., and Conroy, M. E. (1999, September). Building natural assets: Rethinking the [Future Harvest] Centers' natural resources agenda and its links to poverty alleviation. Paper presented to Assessing the Impact of Agricultural Research on Poverty Alleviation workshop, San Jose, Costa Rica.

Cramb, R. A., and Culasero, Z. (2003). Landcare and livelihoods: The promotion and adoption of conservation farming systems in the Philippine uplands. *International Journal of Agricultural Sustainability* 1 (2): 141–154.

Daily, G. (1997). *Nature's services: Societal dependence on natural ecosystems.* Washington, D.C.: Island Press.

Edmunds, D., and Wollenberg, E. (2001). A strategic approach to multi-stakeholder negotiations. *Development and Change* 32 (2): 231–253.

Ellsworth, L. (2004). *A place in the world: A review of the global debate on tenure security.* New York: Ford Foundation.

Fernandes, E. (2004, September). Ecoagriculture investment: Lessons from the World Bank. Paper presented to the International Ecoagriculture Conference and Practioners' Fair, Nairobi.

Forman, R. T. (1995). *Land mosaic: The ecology of landscapes and regions.* Cambridge: Cambridge University Press.

Hassan, R., Scholes, R., and Ash, N. (Eds.) (2005). *Ecosystems and human well-being: Current state and trends,* vol. 1. Washington, D.C.: Island Press.

Heimlich, R. E., Wiebe, K. D., Klassen, R., and Gadsby, D. (1998). *Wetlands and agriculture: Private interests and public benefits.* Washington, D.C.: U.S. Department of Agriculture.

International Fund for Agricultural Development. (2000). *Issues and options in rangelands development: IFAD's Experience* (Final Draft Report of the Thematic Group on Community-Based Management of Natural Resources (I-Rangelands). Rome: International Fund for Agricultural Development.

Isely, C., and Scherr, S. (2003). Community based ecoagriculture initiatives: Findings from the 2002 UNDP equator prize nominations, a joint research project conducted by the Equator Initiative and Ecoagriculture Partners. Retrieved July 6, 2007, from www.ecoagriculturepartners.org/documents/reports/CIreportdraft12-23%5B1%5D.pdf.

Jackson, D. L., Bawa, K., Pascual, U., and Perrings, C. (Eds.) (2005). *Agrobiodiversity: A new science agenda for biodiversity in support of sustainable agroecosystems* (DIVERSITAS Report, No. 4.) Davis, CA: DIVERSITAS.

Jackson, D. L., and Jackson, L. L. (Eds.) (2002). *The farm as natural habitat: Reconnecting food systems with ecosystems.* Washington, D.C.: Island Press.

Kremen, C., and Ostfeld, S. (2005). A call to ecologists. Measuring, analyzing, and managing ecosystem services. *Frontiers in Ecology and the Environment* 3 (10): 540–548.

McCalla, A. F. (2000, March). *Agriculture in the 21st century* (International Maize and Wheat Improvement Center Economics Program Distinguished Economist Lecture, No. 4). Mexico City: International Maize and Wheat Improvement Center.

McNeely, J. A., and Scherr, S. J. (2003). *Ecoagriculture: Strategies for feeding the world and conserving wild biodiversity.* Washington, D.C.: Island Press.

Millennium Ecosystem Assessment (2005). *Ecosystems and human well-being: Synthesis.* Washington, D.C.: World Resources Institute.

Molnar, A., Scherr, S., and Khare, A. (2004). *Who conserves the world's forests? A new assessment of conservation and investment trends.* Washington, D.C.: Forest Trends.

Nature Conservancy (2004). *The Nature Conservancy's 2015 goal.* Retrieved July 6, 2007 from http://sites-conserveonline.org/gpg/projects/tnc2015goal.html.

Panayotou, T., and Ashton, P. (1992). *Not by timber alone: Economics and ecology for sustaining tropical forests.* Washington, D.C.: Island Press.

Paoletti, M. G. (Ed.) (2005.) *Ecological implications of minilivestock: The role of insects, frogs, and snails for sustainable development.* Enfield, N.H.: Science Publishers.

Rhodes, C., and Scherr, S. (Eds.) (2005). *Developing ecoagriculture to improve livelihoods, biodiversity conservation and sustainable production at a landscape scale: Assessment and recommendations from the first international ecoagriculture conference and practitioners' fair, Sept. 25–Oct. 1, 2004.* Washington, D.C.: Ecoagriculture Partners.

Rosegrant, M. W., Cai, X., and Clein, S. A. (2002). *World water and food to 2025: Dealing with scarcity.* Washington, D.C.: International Food Policy Research Institute.

Secretariat of the Convention on Biological Diversity (2005). *Handbook of the convention on biological diversity.* Montreal: Secretariat of the Convention on Biological Diversity.

Swaminathan, M. S. (Ed.) (1994). *Ecotechnology and rural employment: A dialogue.* Madras: Macmillan India.

Tan, G., and Shibasaki, R. (2003). Global estimation of crop productivity and the impacts of global warming by GIS and EPIC integration. *Ecological Modeling* 168: 357–370.

Tilman, D., Fargione, J., Wolff, B., D'Antonio, C., Dobson, A., Howarth, R., et al. (2001). Forecasting agriculturally driven global environmental change. *Science* 292 (13): 281–284.

United Nations Millennium Project, Task Force on Hunger. (2005). *Halving hunger: It can be done.* London: Earthscan.

Wood, S., Sebastian, K., and Scherr, S. J. (2000). *Pilot analysis of global ecosystems: Agroecosystems.* Washington D.C.: International Food Policy Research Institute and World Resources Institute.

4 | Conserving Biodiversity and Ensuring Sustainable Livelihoods

Sustainable Production and Consumption in Global Supply Chains

TENSIE WHELAN

The Rainforest Alliance works in more than sixty countries with people whose livelihoods depend on the land. We help people transform the way they grow food, harvest wood, and host travelers. From large multinational corporations to small, community-based cooperatives, the Rainforest Alliance involves businesses and consumers worldwide in our efforts to bring responsibly produced goods and services to a global marketplace where the demand for sustainability is growing steadily.

We set standards for sustainability that conserve wildlife and wildlands and promote the well-being of workers and their communities. Farms and forestry enterprises that meet our comprehensive criteria receive the Rainforest Alliance and Forest Stewardship Council certification seal. We also work with tourism businesses to help them succeed while leaving a small footprint on the environment and providing a boost to local economies.

Since the organization was founded in 1987, we have made great strides in advancing our mission. Juan Carlos Menses, an earnest young Colombian coffee farmer in his twenties, speaks to the effectiveness of the Rainforest Alliance's approach to sustainable production and forging linkages in the global supply chain. Juan Carlos says

Tensie Whelan is the Executive Director of the Rainforest Alliance.

that while coffee farming might be perceived by his generation as old-fashioned—the vocation of parents and grandparents—he has found a new way to conceive of this traditional occupation (Juan Carlos Menses, personal communication, October 2007). He now believes that coffee cultivation means protecting birds, drinking water, and trees and providing a living for people in the community (ibid.). He feels part of a larger cause (ibid.). If more people shared this holistic understanding, he believes his generation could be enticed to come back to the farms to maintain their livelihoods and land in a sustainable way (ibid.).

Juan Valdez's Commitment to Sustainability

Juan Carlos is from one of the more than 170 farms in Colombia that are certified by the Rainforest Alliance under the standards of the Sustainable Agriculture Network. The Sustainable Agriculture Network is a coalition of locally based leading conservation groups that, in collaboration with scientists, farmers, government officials, and businesspeople, has devised and implemented a comprehensive set of standards that protect wildlands and wildlife as well as the rights and welfare of workers and their families and communities. Bogotá-based Sustainable Agriculture Network member Fundación Natura audits the Colombian farms annually to guarantee their continued compliance with the Sustainable Agriculture Network standards.

The Federación Nacional de Cafeteros de Colombia (the National Federation of Coffee Growers of Colombia, or FNC), which was founded in 1927, supports all 631,000 coffee growers in Colombia. Some 94 percent of these coffee growers have farms with less than 12 acres (5 hectares) in coffee. There are 2,158,900 acres (873,600 hectares) of coffee in the country, and coffee is the top export crop (M. Salazar, Colombian Coffee Federation, personal communication, June 2007). Colombia is one of the largest producers of coffee in the world and the largest producer of quality coffee. Unlike most countries, Colombia consists of a diversity of microclimates that facilitate that year-round harvest of coffee. Much of the coffee is still produced under shade, providing unique habitat for migratory birds and other wildlife as well as protecting watersheds.

Some years ago, the FNC developed the brilliant brand icon Juan Valdez, a Colombian coffee grower who represents authenticity and quality. The brand captured the imagination of coffee drinkers worldwide. In 2007, Juan Valdez earned a new decoration—a Rainforest Alliance Certified seal. Coffee in the new 100 percent Rainforest Alliance Certified line travels from the steep, coffee-covered Colombian mountainsides to the aisles of stores such as retail giant Wal-Mart. The success of certified coffee sales reinforces farmers like Juan Carlos' commitment to social and environmental responsibility in coffee growing and encourages others to join him in good land-stewardship practices.

Large and small companies working with Rainforest Alliance are driving the producers' commitment with their own commitment to sourcing Rainforest Alliance Certified coffee. Kraft, a multinational food company that has coffee brands such as Yuban, Kenco, and Maxwell House, is buying a significant amount of Rainforest Alliance Certified coffee from Colombia and other countries (57,000 bags in 2006). Caribou Coffee, an American chain of more than 400 shops (making it the second largest specialty coffee chain in the United States, after Starbucks), has made a commitment that 50 percent of its coffee will be Rainforest Alliance Certified by 2008. It buys from Colombia. Nespresso, which is also working with the Rainforest Alliance on improving the practices of producers, is another big buyer of coffee from Colombia.

Colombian coffee farmers have great resources to protect through sustainable management. For example, the Sierra Nevada region is a particularly good coffee-growing region with a wealth of ecosystem services and cultural heritage. The mountain range is the headwaters for thirty-five rivers that supply most of the water for Colombia. It is home to a large indigenous population, representing many different tribes, who say that the region has mystical resonance for them. The government has provided them with a reserve in the heart of the mountains. These mountains have been the site of heavy fighting between the Revolutionary Armed Forces of Colombia (FARC) and the government forces, and many *cafeteros* were forced to leave their farms on and off over the years. Fortunately, the government has been firmly in control for the last year, allowing the producers to return.

Edgar Ramirez, lead FNC extension agent in the Sierra Nevada, educates coffee growers about the benefits of Rainforest Alliance certification. Ramirez is a de-facto historian of the *cafeteros* of the region, and he tells visitors how coffee was introduced to the Sierra Nevada in the 1950s. To this day, it is grown in traditional fashion, under the shade of the rainforest. The unique climate—intense rain eight months of the year, with a four-month harvest period—produces very large beans. Large beans command a higher price than small ones. Coffee prices plummeted here ten years ago during the coffee crisis but have bounced back in the last two years. In response to the crisis, the communities in the Sierra Nevada got together to discuss if they should convert from shade-grown coffee to sun-grown in order to increase productivity. (Growing beans under full sun with select chemicals enables the grower to produce more beans.) After much deliberation, these poor and in many cases illiterate farmers decided to continue with traditional coffee production because they felt they had a responsibility to their neighbors living in the valley below. They knew that if they cut down their trees to grow sun-grown coffee, the soil on the steep slopes would erode and chemicals would leach into the water of their fellow citizens.

Over the last year, some of the Sierra Nevada farmers have been working with the FNC and extension agents to obtain Rainforest Alliance certification. The premium that they receive for Rainforest Alliance Certified coffee helps them maintain their commitment to protecting the environment of the entire region. In addition, the producers find that the requirements of certification help them to manage their farms better, increase productivity, and improve quality.

La Floresta, a farm in the Sierra Nevada region, is owned by the Hortua family. The Hortuas had to leave their land during the conflict, but returned to the farm in 2005. Says German Suarez Hortua, the nephew and farm manager, "We have managed our farm for 66 years; we are three generations. We managed as artisans without technical assistance or guidance. With the [Rainforest Alliance] program, we began to understand the weaknesses we had in our systems, in cultivation, waste, water; that we shouldn't use so many chemicals and that the people who work with us deserve a better life. Rainforest Alliance showed us how to do things correctly" (personal communication, October 20, 2006).

German's aunt, Daissy Hortua, doesn't hesitate to tell visitors about the impact of improvements they made to achieve Rainforest Alliance certification: "We have learned from Rainforest Alliance how to take care of the birds, the *beneficio* [profits], the management of the workers—their kitchen, food and bathrooms" (personal communication, October 20, 2006). The most difficult change required by certification, according to her, is the expense of the water treatment technology (it costs about USD 500) and the need to provide social-security benefits to the workers (ibid.).

Gabriel Silva, president of the FNC, spent a fair amount of time in the Sierra Nevada region as a boy. When he visits farms like la Floresta, he listens intently to the *cafeteros* and talks with the children. In Colombia, the FNC is like the Red Cross. They are seen as a neutral organization and they work hard to maintain this status so they can help people across the country, even in war-torn districts. The FNC has a large research facility, a thousand extension agents, a coffee university, Juan Valdez marketing, brand management, coffee shops, and extensive technical-assistance programs (which include reforestation and water conservation). They have developed a sophisticated database and geographic information system that has data on producers, farms, coffee, soils, temperatures, and other matters.

In 1940, the FNC formed the National Coffee Fund. This fund enables them to buy coffee at a decent price from any coffee grower who wants to sell it. The producers can also sell through importers/exports and direct to buyers.

Gabriel Silva and the FNC have been working with the Rainforest Alliance for several years now to extend Rainforest Alliance certification throughout the country, including providing technical assistance to farmers to prepare them for certification. In 2003, there were 650 sixty-kilo bags of Rainforest Alliance Certified coffee grown in Colombia, while in 2006 there were some 192,000. The Rainforest Alliance is present in nine of the country's nineteen states; in 2006 there were 11,000 hectares (27,000 acres) of Rainforest Alliance Certified coffee-producing land in Colombia.

The benefits of Rainforest Alliance certification documented by the FNC include:

- Producers receiving a USD 0.25 per pound premium for Rainforest Alliance Certified coffee;

- Increases of productivity—as measured in bags harvested per hectare—of up to 20 percent;

- Installation of water-treatment technology;

- Social security and good housing for workers; and

- Increased producer consciousness regarding their role in creating a sustainable future for their children and country (A. Gil, Colombian Coffee Federation, personal communication, October 2006).

Santander, a Colombian state with mostly shade coffee, has firmly committed to Rainforest Alliance certification and aims for every farm in the state to achieve certification within the next five years. Alvaro Bautista, a Santander farmer, has been organizing the resourceful, intelligent, and committed producers in his cooperative to get certified. These farmers must be managers, marketers, accountants, physical laborers, and students of nature at the same time. They must embody virtually every skill available to humans, if they are to excel. "Every *cafetero* is an entrepreneur, a protector of biodiversity, a provider of jobs," says Bautista proudly (personal communication, October 18, 2006).

He describes the work they have done: "We have inventoried every tree species, its size and age, and are working on diversification. We have stopped the contamination by coffee pulp of the water. We are now better protecting the microfauna that are important for the fertility of our land. We have created buffer zones around water sources. On other farms people throw their garbage on the ground or into the streams. Our farms are clean and we are recycling paper and aluminum and glass. Our workers' housing used to have dirt floors. We have fixed that and built kitchens and bathrooms. Our permanent workers have health benefits and we are making sure that everyone's kids go to school. We have put in place financial management; 70 percent of our farms have computers and we are tracking cost and prices" (personal communication, October 18, 2006).

Farmers all around Colombia testify to the positive impact of Rainforest Alliance certification. Jorge Julian Santos and his family own Finca Morros, the first farm in Colombia to achieve Rainforest Alliance certification. "We have always produced coffee under shade," he says, "but now we do it in an organized way, with the environment, managing chemicals, social security for workers and taking care of their basic needs. On the farm, before we used chemicals to take of the borer [broca beetle], now we use no chemicals. . . . Cenicafe [the FNC research facility] found twelve species of migratory birds on our farm. We are proud that we are helping the environment. All the farmers in the region feel that way. For example, hunting for export (not for food) used to be part of the culture, now it has completely stopped" (personal communication, October 18, 2006).

In 2006, new housing for seasonal workers—required for Rainforest Alliance certification by the standards of the SAN—was built at Finca Morros. Santos and others hope that the investment will help them attract good workers, because there are never enough at harvest time. People have moved to the cities or to other pastimes, seeing coffee work as arduous and badly paid.

The FNC is committed to tripling the production of Rainforest Alliance Certified coffee. "The farmers clearly understand that this is a process of continuous improvement, meaning that every day they must contribute somehow toward sustainability and certification," says Silva.

The Rainforest Alliance, Sustainability Certification, and Corporate Sourcing

Through sustainability certification, farmers, foresters, and tour operators worldwide are rewarded—as the certified Colombian farmers who work with the FNC and the Rainforest Alliance have been—for conserving biodiversity while ensuring that their livelihood will be profitable for years to come. At the Rainforest Alliance, we certify not only to the standards of the Sustainable Agriculture Network for agricultural crops—coffee, bananas, cacao, citrus, pineapple, passion fruit, ferns, cut flowers, and more—but also to those of the Forest Stewardship Council, which we helped to establish in 1993. The Forest Stewardship Council, with its multi-stakeholder approach, is the global standard setter for responsible forestry.

For tourism, we take a slightly different approach. There are almost seventy existing sustainable tourism certification initiatives worldwide. The Rainforest Alliance supports local certification programs, and it works to increase the international recognition of sustainable tourism certification and to get regional networks of certification programs to share resources and information and create standards for certification criteria. We also provide marketing support, training, and technical assistance to certified businesses and businesses in the process of becoming certified.

We increasingly see linkages in the sustainable supply chain as the key to ensuring and increasing sustainable production. For example, preferential purchasing by Caribou Coffee and Kraft—both large players in the international coffee market—is in part responsible for the premium that Colombian coffee growers receive for their Rainforest Alliance Certified coffee. Strong demand for certified products encourages producers to increase supply. In order to increase the supply, more farmers and more land must be engaged in conservation and improving conditions for workers and communities near farms and forestry operations. The recently completed U.S. Agency for International Development (USAID)–funded Certified Sustainable Products Alliance, a three-year partnership to increase the amount, value and sales of Rainforest Alliance Certified timber, bananas and coffee from

Mexico and Central America, proved the success of targeting supply chains in order to ensure sustainable production and consumption patterns. Private sector investments in the project totaled USD 44,000,000. As Glenn Anders, USAID Mission Director for Guatemala and Central American Programs states, "By linking responsible buyers for certified products with responsible suppliers in these global markets, the Alliance constructs and seals a circuit in which all players—producers, purchasers, distributors and consumers—are winners" (personal communication, October 2005).

Companies Work with the Rainforest Alliance to Make the Market System More Sustainable

The following studies are examples of connections made in the agriculture, forestry and tourism industries that have resulted in biodiversity conservation and ensured sustainable livelihoods for producers engaging in Rainforest Alliance certification to Sustainable Agriculture Network or Forest Stewardship Council standards, or for operators who utilize best management practices for sustainable tourism.

Kraft Takes the Lead in Supporting Sustainable Coffee Production

For four generations, Diego Llach's family has farmed coffee amid the rainforests of the steep and shaded slopes of El Salvador's volcanoes. The family has nine farms in the Los Nogales group, scattered across these mountains. When Kraft Foods former CEO Roger Deromedi visited one of the farms, Llach told him that the price Kraft paid for quality Rainforest Alliance Certified coffee allowed him to invest in improvements such as better housing for workers and a doctor to care for their children.

Kraft's unprecedented commitment to buying Rainforest Alliance Certified sustainably produced coffee has commanded the attention of farmers everywhere. In 2004, Kraft purchased five million pounds of coffee from certified farms in Brazil, Colombia, Costa Rica, Guatemala, El Salvador, Honduras, Mexico, and Peru. In 2006, the company purchased thirteen million pounds, and it has plans for continuing growth.

"This is the most extensive commitment to sustainable coffee ever made; Kraft is setting the bar for the rest of the coffee sector and proving that sustainability can be integrated into the way a company does business," explains Rainforest Alliance Chief of Agriculture Chris Wille. Torrential rains from Hurricane Stan devastated rural communities in Mexico and Central America in October 2005. Thousands of people were left homeless, and crops and lives were ruined. As a demonstration of Kraft's commitment to farmers in the region, the company bought 5,100 bags of coffee from affected farms—Kraft's first coffee purchase from Mexico. At the same time, Kraft continued to purchase beans at a premium from farms in Guatemala, despite their reduced supply. The premiums helped farmers to recover from the disaster.

In addition to buying record levels of certified coffee, the global food company is also partnering with the Sustainable Agriculture Network to train agronomists and help farmers toward meeting the comprehensive certification standards.

Given Kraft's global prominence in coffee, this partnership is the first indisputable evidence that the concept of sustainability, once limited to niche markets, is entering the mainstream. With this unprecedented commitment from Kraft, the Rainforest Alliance was able to demonstrate that coffee farming can be environmentally friendly, equitable, and profitable.

Growing and harvesting certified coffee for Kraft in 2005 brought higher incomes and improved working conditions for more than 100,000 farmers and farmworkers and their families. It also brought a combined total of 21,437 hectares (53,000 acres) of farmland into compliance with the most rigorous environmental standards. In addition to making their coffee holdings ecofriendly, farmers in the program are conserving thousands of hectares of natural habitats and valuable ecosystems. Beyond its own program, Kraft is supporting efforts of the coffee industry to adopt industrywide standards of sustainability. Producers such as Diego Llach appreciate the higher prices for certified coffee, and recognize that the benefits of the program go far beyond the sales price.

By following the Rainforest Alliance guidelines, farmers are able to cut costs, improve quality, protect the environment, and upgrade worker health and safety. For example, Llach has planted 25,000 trees, replaced pesticides with a marigold flower extract, and rebuilt worker housing. "We made some investments in improved housing, sanitation and biological pest control, and many of the changes paid back in efficiency and worker morale," notes Llach, who learned how to manage a coffee farm from his grandfather and father. "The Rainforest Alliance taught us how to protect natural resources like water supplies; it's a philosophy that goes hand in hand with our four generations of farming experience" (personal communication, January 2005). These changes in the coffee farmlands are important to Kraft. As Roger Deromedi noted, the company has been in the coffee business for a hundred years and wants consumers a century from now to enjoy its great coffee also (Diego Llach, personal communication, January 2005). That means that coffee farmers and the environments that support them must also last; they must be sustainable (ibid.).

Kraft is one of the world's leading coffee companies. In 2005, Kraft introduced 100 percent certified lines of popular brands such as Kenco, Jacques Vabre, Gevalia, and Splendid in the United Kingdom, France, Sweden, and Italy. Kraft buys more certified coffee than it needs for the packages sporting the Rainforest Alliance seal and blends the extra beans into other well-known brands such as Maxwell House, Jacobs, and Carte Noire. By integrating certified coffee into mainstream brands, Kraft is helping to grow overall demand and expand the global market for sustainably produced coffee, benefiting an ever increasing number of farm communities and wild areas. In 2007, McDonald's UK announced that all 1,200 McDonald's restaurants in the United Kingdom and Ireland would sell exclusively Kenco coffee, a Kraft Foods high-quality Arabica coffee containing 100 percent Rainforest Alliance Certified beans. McDonald's move is the latest example of a recent trend toward sustainable certified products breaking out of specialty niches and into mass markets, gaining popularity with large companies, mainstream outlets, and consumers.

Chiquita Reaps a Better Banana

Since 1992, Chiquita Brands International has been working with the Rainforest Alliance to reinvent the banana industry one farm at a time. Poor farming practices were a major source of pollution and deforestation on the Caribbean slope of Central America and Colombia fifteen years ago. Pesticide-impregnated plastic bags—used to protect bananas while they grow—littered riverbanks near farms, agrochemical runoff killed fish and other aquatic life, and sediment choked rivers and coral reefs. Workers often endured long hours and unsafe conditions and suffered health problems caused by agrochemicals used on the farms. In order to establish economically viable solutions to these problems, the Rainforest Alliance and partner organizations spent two years conferring with scientists and industry representatives and visiting farms. That effort resulted in nine guiding principles to promote environmental sustainability and social equity on banana farms. Those standards include zero tolerance for deforestation, prohibition of "dirty dozen" pesticides, protection of wildlife, conservation of water and soils, better pay, safe and pleasant working conditions, and the right of unions to organize at farms.

In 1992, Chiquita began applying the Rainforest Alliance's social and economic standards on two farms in Costa Rica, which took two years to get certified. A systematic transformation of Chiquita's other farms followed that experiment. At significant expense, the company has planted buffer zones along streams, installed systems for filtering wastewater from packing plants, managed garbage and recycled all plastics, instituted occupational safety programs, protected forest patches, improved worker housing, sanitary facilities, storage facilities, and other infrastructure, stopped using agrochemicals that posed a risk to workers and aquatic life, and switched to low-toxicity "protectors" to decrease the need for fungicides.

By 2000, all of Chiquita's company-owned farms in Latin America were Rainforest Alliance Certified. Chiquita has since concentrated on helping the independent farmers who supply more than two thirds of the company's bananas to adopt those same standards. Since Rainforest Alliance certification demands steady improvement and criteria are made increasingly strict to take advantage of new technologies and practices, the situation on certified farms gets a little better every year.

According to Raúl Gómez, a farm manager in Costa Rica who has worked for Chiquita for fifteen years, the institution of the Rainforest Alliance's standards has been the equivalent of a "social revolution" (personal communication, July 2005). During the past fifteen years, Gómez has helped prepare several Chiquita farms for certification, so he has experienced its effectiveness first hand. "Everything has changed thanks to the Rainforest Alliance program. We've cut agrochemical use. We've planted hundreds of trees along roads and streams. We're promoting environmental education," says Gómez (ibid.). "And it's all for the good of humanity, so that we can leave something for our children" (ibid.).

Bananas are big business—the world's number one export fruit and the fourth most important food crop after rice, wheat, and maize (Global Crop Diversity Trust, 2006)—and Chiquita is a giant in the industry, supplying nearly 25 percent of the bananas consumed by North Americans and Europeans (D. McLaughlin, personal communication, July 2005). Improvement of the company's farms has consequently had a tremendously positive impact on vast areas of land and more than a hundred communities, whereas certification of the independent farms that supply the company with bananas is steadily increasing the acreage and population that benefits from the Rainforest Alliance's standards. Chiquita protects patches of rainforest, recycles or reuses nearly 80 percent of the plastic bags and twine used on company farms—about 3,200 metric tons per year—and has reforested more than 1,000 hectares (2,500 acres) with nearly one million trees and bushes to establish buffer zones along rivers and roadways and around housing and other facilities. Pesticide use is strictly controlled, workers who apply them are required to wear protective gear, and the company has planted groundcover on more than half of its farmland, which reduces soil erosion and eliminates the need for herbicides. Workers have clean and safe conditions, showers, bathrooms, and eating areas, and their families have access to health care, education, and recreational facilities.

Chiquita recently began tackling community development, creating conservation projects, and installing filtering and recycling systems in its packing plants that reduce water use by 80 percent. The company is also investigating biological controls and new fungicides that could significantly cut the toxicity of agrochemicals used on farms. Chiquita has demonstrated that environmental and social conditions can be improved without sacrificing production. Rainforest Alliance Certified farms, whether managed by Chiquita or other producers, are among the most productive farms in the world. Although Chiquita has invested more than USD 20 million to make required capital improvements, it has reduced its production costs by more than USD 100 million (Esty and Winston, 2006). "In addition to gaining improved morale and productivity in our farms, we have saved money in the process. Everybody wins—the workers, the company and the environment, not to mention the Rainforest Alliance, which deserves enormous credit for showing us a better way," said Bob Kistinger, president and chief operating officer of the Chiquita Fresh Group (personal communication, July 2005).

Motivated by its experience with the Rainforest Alliance, Chiquita implemented a companywide code of conduct and began publishing corporate responsibility reports, which have been widely hailed as straightforward and honest. All of the company's farms are now certified according to Social Accountability SA8000 criteria, the most rigorous and verifiable social standards currently available. Chiquita also signed a historic labor rights framework agreement with regional and international unions in 2001.

In the fall of 2005, Chiquita added the Rainforest Alliance seal to 50 million bananas a week in nine countries: Austria, Belgium, Denmark, Finland, Germany, Holland, Norway, Switzerland, and Sweden. That is 500,000 boxes a week, or about half of Chiquita's supply in Europe. The launch was accompanied by large print ads in newspapers across Europe and memorable and amusing television ads, all featuring the new label that combines the company's famous Miss

Chiquita logo with the Rainforest Alliance frog. "Adopting the Rainforest Alliance standard has been one of the smartest decisions Chiquita has ever made. Not only have we helped the environment and our workers—through better training and equipment, less use of agrochemicals and better systems for recycling and managing waste, for example—but we also learned that profound cultural change was possible. . . . We are deeply committed to corporate responsibility, not simply as an element of our strategy, but because it's the right thing to do," said Cyrus F. Freidheim Jr., former Chiquita Brands board chair (personal communication, July 2005).

Conserving the Lands, Culture, and Economy of Europe's Cork-producing Regions

Since the 1600s, when Benedictine Monk Dom Perignon first thought to seal a bottle of sparkling wine with cork instead of oiled rags, cork has been the wine-bottle stopper of choice. Bowing to pressure by supermarkets to protect against possible tainting, oxidation, and leakage, vintners throughout the United States, Europe, South America, and South Africa have been replacing their natural cork with synthetics. In doing so, they are endangering one of the last natural forest ecosystems in Western Europe, and along with it an economy and culture that has grown up around cork farming over thousands of years.

In an effort to preserve the environmental integrity of these regions, to protect an industry that supports some 100,000 cork workers, and to promote sustainable forest management in one of the Mediterranean's biodiversity "hotpots," the Rainforest Alliance has been working with cork producers across Spain and Portugal to conserve their endangered forests through Forest Stewardship Council certification. So far, more than 15,000 hectares (37,000 acres) of cork producing lands have received Forest Stewardship Council/Rainforest Alliance certification, with more certifications underway.

"Without the demand for cork, economic pressures could force farmers to abandon the active management of cork forests, which may lead to rural exodus as well as unbalance the ecosystems that preserve the biodiversity of these regions," explains Rainforest Alliance European forestry division manager Jamie Lawrence. Synthetic stoppers already make up 8 percent of the more than 13 billion stoppers made each year to supply the international wine market (Cork Quality Council, 2007a). If more wine-bottle stoppers used worldwide are made from plastic, Europe's cork industry may suffer, and along with it, the cork forests of Portugal, Spain, Morocco, Italy, Tunisia, and France.

"While some supermarkets and others have claimed that the use of plastic in lieu of cork will contribute to the forests' environmental protection, this is absolutely false," explains Richard Donovan, the Rainforest Alliance's chief of forestry. Unlike its synthetic counterparts, cork is an inherently sustainable resource, both renewable and biodegradable. The cork oak tree (*Quercus suber*) is unique in that its thick bark can be stripped off every decade to extract the cork without damaging the trees, which live 170 to 250 years on average (Cork Quality Council, 2007b). Carried out by skilled craftsmen each summer, the stripping process has remained virtually unchanged for nearly three thousand years. In fact, regular cork stripping is necessary to prevent the bark from aging and to maintain the health of the tree and the forest ecosystem that provides habitat to endangered species, including the imperial eagle, the Iberian lynx, and the Barbary deer. The forestlands where the cork oaks thrive are largely open swaths made up of grassland and scrub vegetation interspersed by trees, where farmers have practiced a low-intensity mix of agriculture and forestry for millennia. Careful forest management not only provides for the continued extraction of the cork oak but also helps to create the conditions for a diverse range of other products that are harvested from the woodlands. A harmonious balance is maintained, where local people can provide for their needs without damaging the ecosystem or threatening the long-term sustainability of their most important natural resource. Villagers gather edible fungi for their own consumption, use rockrose bushes for firewood in their traditional stone bread ovens, and tap local beehives for honey flavored with native lavender and rosemary. On even a small patch of cork land, a farmer can raise a herd of goats, a few cows, and some pigs, which forage for acorns and graze beneath the trees. Income from cork can represent anywhere from 30 to 100 percent of a farmer's income (Amorim, personal communication, June 2006).

The recent increase in demand for synthetic cork threatens to undermine the economic basis of cork farming and thereby the cork-producing areas of the Iberian Peninsula, where cork oak forests represent around 21 percent of the forest area and are responsible for the production of more than 50 percent of the cork consumed worldwide (Cork Quality Council 2007a). Farmers there have already been converting their forests to fast growing species like eucalyptus, which can be harvested quickly and at a greater profit than cork. The Forest Stewardship Council/Rainforest Alliance certification of cork-producing lands allows manufacturers like Amorim, which owns two certified processing plants and is a leading cork producer and processor, to supply certified cork to a market that increasingly demands that forest and other goods be responsibly produced. "The Forest Stewardship Council/Rainforest Alliance certification for both cork forests and industrial cork companies clearly meets the growing market demand for sustainable natural products, especially from major international distribution chains," explains Carlos de Jesus, marketing and communications director for Amorim (personal communication, June 2006). "But, as importantly, it also validates the unique ability of cork to contribute to the advance of crucial environmental, economic, cultural, and social aspects relevant to the entire Western Mediterranean Basin. We hope the Forest Stewardship Council–certified status recently granted to Portuguese industrial and forestry companies inspires organizations in other countries to realize that it is distinctly possible to create wealth while protecting the environment" (ibid.). According to Jamie Lawrence, "The economy, culture, and environmental sustainability of some of the last natural areas of the Iberian Peninsula rests on increased demand for sustainably produced cork. This certification is a step toward conserving some of the last natural landscapes of Western Europe along with key species of animals, plants, and birds."

Helping Communities Conserve the Maya Biosphere Reserve

To Carlos Crasborn, president of the Carmelita forestry cooperative, it is clear that his community's future rests on the conservation of the surrounding forest. By implementing best management practices for sustainable forestry, Crasborn and his neighbors in Guatemala's Maya Biosphere Reserve have steadily increased the profits from their forestry businesses, and have invested more than a third of their earnings in community development, improved technology, and sustainable management methods (personal communication, October 2006). In so doing, the communities have reduced forest fires on the land they manage to twelve times lower than in the core protected area of the Reserve, and the average annual deforestation rate to twenty times lower than that of the protected areas where harvesting of wood and nontimber forest products is prohibited (Hughell, 2007). "Our parents protected this forest for our benefit, and it is our responsibility to protect it for future generations," reflects Crasborn (personal communication, October 2006).

According to the twenty-three-year-old cooperative leader, investments in a community sawmill, carpentry shop and training have allowed coop members to increase their earnings while reducing logging to less than one percent of their 57,000 hectare (141,000 acre) forest concession (personal communication, October 2006). Crasborn explains that the concession is divided into logging blocks, each of which will be allowed to recuperate for forty years following timber extraction, whereas more than half of it is reserved for ecotourism and the sustainable harvest of decorative palm leaves and chicle tree sap—the traditional base of chewing gum (ibid.).

Carmelita is one of twelve communities managing tracts of forest within the Maya Biosphere Reserve, a mosaic of concessions, national parks, and other protected areas covering more than 1.5 million hectares (3.7 million acres) of wilderness in northern Guatemala. The reserve holds over a dozen important archaeological sites and such rare wildlife as jaguars, brocket deer, scarlet macaws, and ocellated turkeys. The Rainforest Alliance has certified twelve community and two private forestry operations in the reserve, and it assesses them annually to ensure that they follow strict standards for protection of the environment and people.

Those efforts have significantly improved community-based forestry in the Maya Biosphere, where communities have hired professional foresters to design and administer forest management plans, improved their administrative capacities, ensured safe working conditions, and made various other changes in order to earn their certification. Once certified, the Rainforest Alliance helps these communities gain access to preferential international markets and provides the training necessary to meet buyers' demands. As a result, certified operations are making weekly shipments of palm leaves to the United States and exporting milled wood and finished products, mostly manufactured from little-known woods such as *pucté* and *manchiche*, for which there were no markets just a few years ago.

According to José Román Carrera, Rainforest Alliance forestry manager for Central America, the export of new products—and the access to buyers willing to pay higher prices for value-added certified wood—has provided much-needed additional income to the more than six thousand people involved in managing the biosphere reserve's forest concessions. He notes that this has not only led to new jobs and improved household incomes, but that part of the profits have also been invested in community works such as a potable water system, new schools, clinics, and an emergency medical fund for poor families. "The increased earnings not only raise living standards, they also raise people's awareness of the need to manage the forest in a sustainable manner," says Carrera.

The community of Uaxactún, set in the rainforest north of Tikal National Park, has sold nontraditional wood species to several companies, produced special cuts of mahogany for Gibson Guitars, and exports weekly shipments of jade palm leaves to U.S. floral supplier Continental Floral Greens. According to Floridalma Ax, a member of the organization that administers Uaxactún's forest concession, community members have invested part of the money from those sales to hire teachers for the town's understaffed school and provide scholarships for older students to take computer courses in the nearest city (personal communication, October 2006). "We invest in education because we want the next generation to be well trained and capable of defending our interests," says Ax (ibid.).

The success of this strategy for conserving the area's natural resources is immediately apparent, especially when contrasted with the conditions found in nearby national parks. For example, Laguna del Tigre National Park, the reserve's largest protected area, has already lost more than 40 percent of its forests to illegal loggers and slash-and-burn farmers, whereas the concessions have lost less than 4 percent of their forest cover.

According to Benedín García, a former forest ranger and one of the founders of the organization that administers Uaxactún's forest concession, the reason the concessions are better conserved than the parks is that they are protected by the people who rely on them for their livelihoods. He explains that part of the money earned from the sale of certified wood is used to pay local forest guards who patrol the concession every day, but all of the town's residents also contribute to that vigilance. "Our secret is that we have more than 150 people working in this forest, collecting palm leaves, chicle and allspice, and if one of them sees something happening that shouldn't be, they report it to us, and we send a delegation to that area immediately."

For Carrera, this community approach to conservation is not only the best means of protecting the Maya Biosphere Reserve, but could be the key to saving the region's other large wilderness tracts, all of which are threatened. A career conservationist, Carrera was the regional director of Guatemala's National Council of Protected Areas, having joined the organization when the biosphere reserve was created, and spent years battling illegal loggers and squatters before going to work for the Rainforest Alliance. "I used to think that the way to protect the forest was to say, 'Stop, don't touch.' We put people in jail and confiscated the illegal wood, but the forest just kept getting smaller and smaller," explains Carrera. "I realize now that a more effective way to conserve the rainforest is to show the people who live there that they can make a better living by managing the forest sustainably than they would it if they cut it down. This is something we are accomplishing in Guatemala and that we would like to repeat in and around Central America's other biosphere reserves, in order to ensure the survival of this region's endangered wilderness."

Indonesia's Teak Farms: A Sulawesi Cooperative Supplies Certified Teak to the World Market

What started as a quiet rebellion, of sorts, by a group of disenfranchised villagers in Indonesia has sprouted into an economically viable, environmentally sustainable teak business. In 1970, the Indonesian government appropriated large chunks of land from villages in South Konawe District in Southwest Sulawesi, and then hired local villagers to establish teak plantations on the very land that had just been taken from them. In response, the villagers stashed a few teak seeds in their pockets and brought them home to plant in their fields and gardens. Today, those homegrown trees, sprouted in private farm plots, are proving a highly effective tool to combat illegal logging on state lands while providing villagers with a reliable source of income.

Over the past thirty-five years, global demand for teak has surged. Teak has always been highly valued for its unique properties. Its high oil content makes it extremely dense and virtually impervious to water. It holds up under heat, cold, wind, and rain. It resists disease, and it has the extraordinary ability, when in contact with metal, to prevent rust. The virtually maintenance-free qualities of teak make it one of the most prized, high-quality building materials in the world, and a favorite of outdoor furniture makers and boat builders alike. This demand for teak has put enormous pressure on the governments' plantations, tempting many of South Sulawesi's poor to venture into the plantations to log illegally. The gains have been few. Villagers who harvest and sell illegal teak find themselves at the mercy of middlemen, who pay notoriously low prices. Illegal logging depletes the teak resource, removing long-term income potential. Without careful management, teak groves can quickly be degraded and the resource loses its value. When a tree is cut, many shoots grow up from the stump, but these may be crooked, less vigorous, full of low branches and not suitable for sale as timber. In the meantime, communities around the forest are further impacted by erosion and subsequent siltation and the depletion of water resources as unsustainable numbers of trees are removed. Conversely, on a well-managed teak plantation, the stumps are removed after harvest and the soil is replanted with seed. Because teak grows on twenty- to thirty-year cycles (R. Barr, Tropical Forest Trust, personal communication, October, 2005), replanting should be a continuous process, which does not happen when trees are taken illegally.

Thanks to the farmers in South Konawe District, the trend of illegal teak harvesting is beginning to be reversed. The farmers have taken to managing their home teak plots so responsibly that they have merited Forest Stewardship Council certification by the Rainforest Alliance. Whereas once these farmers might have earned supplemental income by risky illegal logging, Rainforest Alliance certification means they can now command premiums high enough from their own teak plots to survive financially. The premiums for certified teak may be significantly higher than illegally logged teak because the demand is high. "Businesses on the tail end of the supply chain are wary of procuring illegal teak; and they want traceability of this resource," explains Jeff Hayward, former Asia Pacific regional manager of the Rainforest Alliance's SmartWood program. "Now farmers living in this region can access markets previously unavailable to them."

The road to certification involves a rigorous process. In South Konawe District, people from forty-six villages started by creating a cooperative called Koperasi Hutan Jaya Lestari (KHJL). Nearly two hundred farmers joined the cooperative. In 2004 they began working with Tropical Forest Trust, a nonprofit organization based in Switzerland, to close the gap between existing management practices and those that the Forest Stewardship Council deems as responsible forestry. KHJL applied for the certification assessment at the end of 2004. After on-site evaluations of forest areas in a sample of twelve of the active villages involved in the cooperative, the auditing team compiled a full assessment report, which was reviewed by SmartWood. In May 2005, KHJL farmers received their certification.

Abdul Harris Tamburaka, head of KJHL, points out, "The Forest Stewardship Council/Rainforest Alliance certificate gives credibility to

KJHL because the world believes and agrees that we are managing the forest sustainably." Farmers like this model because it allows them to maintain traditional land-use practices, while servicing the global teak supply chain. Teak plots are usually one to ten hectares, intercropped with other cash crops such as cocoa, coffee, sago, cashews, pepper, and candlenut trees. This ensures a diversity of income sources without overtaxing the soil's fertility. Word is spreading throughout the communities that certification is the way to go. Robin Barr, social forestry officer for Tropical Forest Trust, says, "One beneficial (and unexpected outcome) has been that the cooperative has attracted farmers away from illegal logging. By managing their own teak, farmers now see that they can make more money than through illicit means" (personal communication, October 2005).

KHJL's certification was funded through Tropical Forest Trust with the support of retailers from Europe and Indonesia who all wanted to secure a reliable supply of FSC-certified teak. Franck Moreno of Castorama, a French furniture retailer, explains, "When we joined Tropical Forest Trust in 2003, there was no high quality FSC-certified teak available on the market" (personal communication, October 2005). Thanks to Castorama's financing, certified teak is now available. "It's a win-win situation," says Moreno. "The communities get access to the global market and we secure raw materials" (ibid.).

Comforted by higher premiums and better markets for certified teak, these farmers can take a long-term approach to land use, rather than exhausting the surrounding natural resources for immediate day-to-day survival. "By rewarding a local, community-based enterprise for managing its land sustainably, the Forest Stewardship Council certificate affirms to the people of South Konawe that their teak harvest is beneficial socially, environmentally and economically," explains Hayward. "The self-reliant teak farmers in these communities have been planting and replanting furniture-quality teak for decades. These trees are an investment in the future of farmers' children and grandchildren." Now there is extra income for school fees, building and repairing of houses, medical expenses, and marriage ceremonies.

EarthChoice: Domtar's Declaration of Environmental Support

While paper production can have a devastating impact on the environment, forests can also be managed sustainably to support a healthy range of biodiversity, communities, and businesses. So, in November 2003, when the Canadian-based paper giant Domtar Inc. publicly committed to certifying all 7.28 million hectares (18 million acres) of its forests, eleven pulp and paper mills, and fourteen sawmills to Forest Stewardship Council standards, the announcement had broad ramifications for paper manufacturers and forests everywhere.

Why did North America's third largest paper producer go to all of the trouble and expense of companywide certification? The answer was simple: As Lewis Fix, director of business development for Domtar's line of EarthChoice papers, explains, "It was the right thing to do" (personal communication, May 2006). The Domtar commitment to sustainability did not stop with certification. "The more forests we certified, the more certified fiber we had available," continues Fix. "So we created a line of papers called EarthChoice. It started with doing the right thing in the forest, which we then turned into an opportunity to put new products into the marketplace" (ibid.). Introduced in April 2005 with a campaign designed to overcome corporate end-users' common concerns about the quality, selection, availability, and price of ethical papers, Domtar EarthChoice is now the most comprehensive line of socially and environmentally responsible paper products on the market, with over 700 available options. Domtar EarthChoice papers are intended for business applications, including corporate identity packages, brochures, promotional materials, annual reports, business forms, training manuals, direct mail, newsletters, catalogs, magazines, book covers, invitations, presentation folders, calendars, envelopes, labels, and business cards. Each paper in the line has been certified for meeting the environmental and social standards of the Forest Stewardship Council. Many of the papers have both Forest Stewardship Council/Rainforest Alliance Certified virgin pulp and post consumer waste pulp, offering a comprehensive solution to reducing the environmental impact of paper production.

Not only did Domtar develop an innovative family of products with their EarthChoice papers, but the company also embarked on a groundbreaking promotional effort that began with the "No More Excuses" ad campaign in early 2005, and targeted corporate end users with a direct mailing to corporate executives at Fortune 1000 companies. Included in the mailing was a coffee-table book on boreal (northern) forests and EarthChoice samples. The campaign included ads in *The Wall Street Journal, Forbes, Inc.,* and *Business Week*. Domtar also included a 10 percent rebate offer for companies willing to print their entire annual report on EarthChoice paper, which would then earn them the right to display the FSC, Rainforest Alliance, and EarthChoice logos on the back cover.

Along with "No More Excuses," Domtar launched "Share the Vision," a promotional event showcasing Grammy nominated artist Gary Burden's album cover designs for musicians including Joni Mitchell, the Eagles, The Doors, Crosby, Stills and Nash, and others, reimaged on Domtar EarthChoice papers. Domtar took the exhibit on the road, displaying it in venues across the United States and Canada, targeting the design community, merchants, and printers, touring seven cities in six months. "We wanted to blow away the myth that environmental papers are high priced, low quality, and not widely available," explains Fix. In addition to external communications and marketing, Domtar conducted internal training within the organization to ensure that staff could actively promote these products. They have created EarthChoice specialists within Domtar who go out within their localities and train merchants and printers throughout North America on the virtues of Forest Stewardship Council certification. As a result, Domtar has had merchants become Forest Stewardship Council Chain-of-Custody certified.

What Is Good for the Environment Is Good for Business: Pictorial Offset

About fifteen years ago, Donald Samuels watched a documentary about a company whose maintenance personnel routinely dumped used solvent out the back door, a practice that inadvertently resulted in the contamination of the town's water supply. Samuels, a managing partner of Pictorial Offset Corporation, realized with some dismay that anyone's mistake could result in an environmental mishap, if not a full-blown catastrophe. His epiphany led to a companywide policy of good corporate stewardship that includes recycling, reduced water consumption, reduced atmospheric gas emissions and most recently, the use of paper from Forest Stewardship Council/Rainforest Alliance Certified forests.

Since 1980, the Samuels brothers—Donald, Gary, and Lester—have been at the helm of Pictorial Offset, the largest family-owned single-facility commercial printer in the United States. The New Jersey printing firm was founded by their grandfather in 1938. They have increased its size from twenty-eight employees and USD 1 million in annual sales to three hundred employees and USD 78 million in annual sales. Pictorial's clients represent a wide-range of consumer and service industries. Pictorial Offset is also one of the nation's corporate environmental leaders and was the first printer in the world certified to an integrated International Standards Organization ISO 9002 (quality) and ISO 14001 (environment) management system.

Pictorial Offset's decision to earn Rainforest Alliance certification came at the urging of another Rainforest Alliance partner company, the Mohawk Paper Company. Mohawk, which was certified in 2002, was interested in increasing its business with Pictorial Offset. "Pictorial Offset's commitment to sustainable paper use has enormous implications for the industry and the environment," remarks Rainforest Alliance chief of forestry Richard Donovan. "By using paper that comes from well managed forests, the company is providing an incentive for conservation to landowners and sending a strong message to the paper and printing community that sustainability is a viable and successful business strategy."

Pictorial Offset has seen that as their commitment to corporate environmental and social responsibility has increased, more corporations, designers, and print buyers are specifying the use of products that reduce consumption of natural resources. While Donald Samuels' initial impulse toward good corporate stewardship grew from his commitment to the environment, and out of concern for his children and future generations, he has also been keenly aware that what is good for the environment is good for business. Pictorial Offset has doubled the size of its business since it became ISO 14001 certified in 1998. After his environmental awakening, Samuels and his brothers went "dumpster diving" to see exactly what they were throwing in the landfills. Their thorough assessment of the company's waste stream led to a comprehensive recycling and water-reduction program.

Today, Pictorial Offset recycles roughly 12 million pounds of paper each year, which means savings of between USD 800,000 and USD 1 million per year. The companies reduced water usage translates to half a million gallons per month in savings. Just in 2005, Pictorial achieved an additional reduction in volatile organic compound emissions—gases that are a byproduct of printing and lead to air pollution—of over 5 percent. "Our conservation measures are so deeply embedded in our corporate culture at this point that we no longer itemize all of the cost benefits," says Samuels (personal communication, March 2006).

Pioneering Horizontes Nature Tours
Plays Key Role in
Costa Rica's Conservation

In 1984, when Tamara Budowski started Horizontes Nature Tours with partner Margarita Forero, it was highly unusual for a woman to own a business in Costa Rica, much less a woman in her twenties. But Budowski, who had studied biology, and Forero, who had studied dentistry, found they were able to explore the country's natural riches through their newfound career choice. Taken with the sheer beauty of Costa Rica's national parks, they chose to make them the focus of their nature-based tourism business, and soon realized that this particular approach to travel could also be an incentive for park conservation.

To raise consciousness and awareness in Costa Rica about the potential of nature-based tourism as a conservation and development tool, Budowski and Forero teamed up with two newly formed companies, Rios Tropicales and Marenco Lodge. Together, they held a large conference attended by NGOs, university professors and presidents, and soon garnered the full support of their country's scientific and educational communities. Two years later, when the World Conservation Union chose Costa Rica as the location for its general assembly, Horizontes was asked to organize the logistics of the conference. Young entrepreneur Tamara Budowski was tapped to give a speech on the emerging phenomenon of ecotourism in Costa Rica. "I was only twenty-eight, and it was one of the most frightening things," recalls Budowski (personal communication, April 2006). "I wanted to make a good impression, and I also wanted to represent my country to the conservation world as a model, to show what we were doing and inspire others to follow in our steps" (ibid.).

While the Rainforest Alliance does not offer certification to tourism businesses that operate sustainably, it does work with hotel owners and other tourism operations, giving them the training to meet the standards of reputable certification systems. As a means of advancing sustainable tourism in Latin America, the Rainforest Alliance has signed cooperative agreements with major tour operators, encouraging them to motivate their affiliated hotels and business enterprises to conserve biodiversity and reduce tourism's negative impacts.

In this regard, Horizontes has been one of the Rainforest Alliance's most ardent supporters. Budowski and Forero have actively encouraged hotels in Costa Rica to participate in workshops run by the Rainforest Alliance, where participants learn the nuts and bolts of best practices including methods for recycling, reducing waste and ensuring that local communities benefit from tourist dollars. "In 2004, we decided it was time to make a statement promoting sustainable business practices to the tourism industry," recalls Budowski (personal communication, April 2006). "That year we built a booth at our National Tourism Fair (EXPOTUR) using Rainforest Alliance Certified wood. We served Rainforest Alliance Certified coffee and organic snacks, and talked about sustainable tourism to everyone who would listen" (ibid.). But Budowski and her team found that most of their colleagues had little understanding of what was meant by sustainability, and that relatively few hotels in the country were affiliated with a certifier, since standards were not geared for small and medium sized businesses. "The Rainforest Alliance approached us with a solution in which we could help create awareness among our suppliers and put them on a path toward more sustainable business practices based on commitment, not profits, both in the short and over the long term," Budowski explains (ibid.).

In collaboration with Horizontes, the Rainforest Alliance has sponsored a number of workshops throughout Costa Rica. "People leave these workshops very inspired, clear on the need to improve their practices and reduce their impact on our planet," Budowksi points out (personal communication, April 2006). Based on a signed agreement with the Rainforest Alliance, Horizontes directs clients to hotels that have participated in the Rainforest Alliance workshops.

In addition to its work with the Rainforest Alliance to promote sustainable tourism, over the years Horizontes has instituted a number of initiatives aimed at "greening" its own business and industry. Besides favoring certified hotels and lodges, Horizontes promotes low-impact travel. So, for example, when Monteverde Cloud Forest Reserve was suffering from the environmental effects of too many visitors, Horizontes voluntarily curtailed its number of visits and actively sought alternative destinations to share with its partners. As a policy, the company sponsors only small group travel with highly trained guides, most of whom are university graduates and are required to take a comprehensive course on conservation.

Since 1995, Horizontes has organized and funded workshops aimed at raising the level of professionalism of Costa Rica's naturalist guides. In tourism, guides are the bridges between nature and travelers. They can help tourists understand not only the many wonders of the natural world but also its fragility, its connection to our own lives and the need to protect it. The company's weeklong seminars bring together local and international experts in conservation, thematic interpretation, and biological sciences. Horizontes also sponsors guide training workshops with Costa Rica's National Biodiversity Institute (INBio) and Costa Rica's Association for Tourism Professionals (ACOPROT). "We've set the standard for well-trained guides in Costa Rica," reports Budowski (personal communication, April 2006). "Many of our colleagues have followed our example and as a result have better, more professionally trained staff with which they develop long-lasting relationships" (ibid.). Once they go through the Horizontes workshops, guides can be the "watchdogs" of tourism businesses' interactions with the natural and cultural environment. Well-trained tour guides can educate tourism businesses about the importance of implementing sustainability practices. They can also teach travelers about behaving responsibly in natural and cultural destinations and making responsible decisions about which tourism businesses to patronize during their travels and vacations.

Through a joint effort with RARE, Costa Rica's Neotropical Foundation, and the Harvard University–affiliated World Teach, the company has also helped fund scholarships for young people living close to the national parks so that they can become guides and help protect their own communities and lands. Over the years, Horizontes has extended philanthropic efforts to community development and park projects, including a fund for national park guards to prevent poaching and support of trail building, research stations, and conservation organizations. Currently the travel firm is supporting projects aimed at curbing deforestation of the wild Osa Peninsula, as well as preventing the pollution of Costa Rica's aquifers, rivers, and streams. Its commitment to conservation goes beyond the country's wildlands to its own San José office, which was one of the first in Costa Rica to institute a strict recycling and reducing policy. Staff regularly gather for talks on current conservation projects and they have organized a committee responsible for "greening" office practices and procedures. A weekly recycling drive recently gained the support of neighboring homes and businesses. Not least of all, Horizontes supplies all of its clients with welcome packages that include maps and information on responsible travel, further fostering awareness of and concern for conservation in one of the world's leading nature travel destinations.

Raising Awareness and Reducing the Environmental Impact of Tourism on the Gulf of Honduras

As owner of the Guatemalan tour operator Explore, one of Carlos Jiménez's most popular trips is down the Rio Dulce, which flows through a jungle-draped gorge in eastern Guatemala into the Gulf of

Honduras. From palm-lined beaches to idyllic cays, rainforests, rivers, coral reefs, colonial fortresses, and traditional cultures, the Gulf of Honduras region holds a fantastic combination of resources. Though the area—which comprises Guatemala's Caribbean coast, southern Belize, and northwest Honduras—still receives relatively few visitors, the extensive development of nearby Ambergris Caye, in northern Belize, and the Bay Islands of Honduras suggest that more tourism is on the way.

The Rainforest Alliance is providing the gulf's inhabitants with the information they need to take advantage of their area's tourism potential in a sustainable manner. For Jiménez, the Rainforest Alliance is strengthening a process that will not only improve tourism's impact on the gulf's resources and people, but should make the region a more attractive destination. "We aren't going to be able to improve the image of our region unless we take care of the environment, the social aspects and quality. They are three essential elements," Jiménez says (personal communication, March 2006).

In order to provide the gulf's inhabitants with the tools they need to conserve their cultural and natural resources while benefiting their communities, the Alliance has trained more than a hundred tourism entrepreneurs, government officials, and representatives of rural communities at a series of workshops held in the Guatemalan town of Puerto Barrios. Those workshops have introduced participants from Guatemala, Honduras, and Belize to best-management practices. These practices not only improve the environmental and social impact of tourism businesses but can also lower their operating expenses and make them more marketable. The workshops include use of a guide to best practices published by the Rainforest Alliance, examples of businesses that have adopted those practices, and an introduction to the evaluation process required for certification.

Francisco Enríquez, technical advisor for Asociación Ak' Tenamit, a Mayan development organization that coordinates an ecotourism project in the Río Dulce area, attended a workshop in 2003 that he called accessible and practical, even though he was familiar with most of the concepts covered in it. "I believe these workshops are very important, because most of the people who participated in the one I went to were unfamiliar with many details of the issues covered," he says (personal communication, March 2006).

Nirma Méndez, administrator of the Hotel MarBrissa in Puerto Barrios, where the workshops were held, was one of those people. "I've learned a lot, such as ways to save energy and water, and do things for the environment," she says (personal communication, March 2006). Méndez explains that she has made a number of changes in the hotel since attending the first workshop, such as switching to energy-efficient light bulbs and air conditioners, and asking guests to keep their air conditioners on the low setting and to hang their towels to dry if they want to reuse them. "They're small changes, but they help a lot," she says, adding that the hotel plans to replace all its air conditioners and build a sewage-treatment plant (ibid.).

The Rainforest Alliance works with an array of tourism operations in the region, from modern hotels to community projects. Ak' Tenamit's ecotourism program includes a lodge that is owned and operated by the Q'eqchi Maya village of Plan Grande Quehueche, at the edge of Rio Dulce National Park, where visitors can hike through the forest, and learn about the Q'eqchi culture. "Ecotourism is an increasingly important segment of the tourism industry, and a way for communities to improve their standard of living, while conserving their values and culture," notes Enriquez. He explains that even though the village's lodge applied many best practices from the start, the workshops gave them ideas for improvements, such as better water management methods for the bathrooms they were about to rebuild.

Another enterprise participating in the workshops is the Hacienda Tijax Jungle Lodge, where two-thirds of the property is covered with rainforest that guests can explore on low-impact trails. "The workshop gave us up-to-date information about different ways that people in the hotel business can improve our service and ensure that our operations have the least possible impact on the environment," observes Esvin Chacón, Hacienda Tijax's sales manager (personal communication, March 2006). Méndez says the Rainforest Alliance's workshops have not just helped hotels improve, but have also had a positive impact on Puerto Barrios as a whole. She notes that the mayor and representatives of the regional government have participated in workshops, adding that the town's beaches are cleaner than they were a year ago. "In the workshops, they give advice for improving the entire port," she says. "It's for the whole port, so that more tourists come here, and so that those who come leave content."

Conclusion

Partnerships between civil society and the private sector have been effective in strengthening the supply chain for certified sustainable products. By the end of March 2008, the Rainforest Alliance had certified more than 108 million acres of sustainable forest and agriculture land. Companies large and small have committed to sourcing certified products, encouraging even more sustainable production. The companies that work with the Rainforest Alliance and the producers and tour operators who supply these companies are learning that sustainable practices benefit not only the environment and workers but also their bottom line. Whether they realize cost savings through reduced chemical use or increased productivity or obtain new markets or price premiums, entities at all levels of the supply chain profit financially (in the short or long term) from sustainable production. Producers, businesspeople, and consumers are learning together that development and environmental conservation are not opposing goals—in fact, together they can transform the global economy into one that can sustain itself for centuries to come.

References

Cork Quality Council (2007a). Cork production industry statistics. Retrieved April 23, 2007, from www.corkqc.com/production/production.htm.
Cork Quality Council (2007b). Renewable harvesting: The cork growing cycle. Retrieved April 23, 2007, from www.corkqc.com/harvesting/harvest2.htm.

Esty, D. C., and Winston, A. S. (2006). *Green to gold: How smart companies use environmental strategy to innovate, create value, and build competitive advantage.* New Haven: Yale University Press.

Global Crop Diversity Trust (2006). Bananas in peril. Retrieved April 23, 2007, from www.croptrust.org/main/threats.php?itemid=123.

Hughell, D., and Butterfield, R. (2008). *Impact of FSC Certification on Deforestation and the Incidence of Wildfires in the Maya Biosphere Reserve.* New York: Rainforest Alliance.

5 | Restoring Ecosystems and Renewing Lives

The Story of Mhaswandi, a Once-Poor Village in India

MARCELLA D'SOUZA AND CRISPINO LOBO

With India clocking an enviable economic growth rate of over 9 percent per annum, it is easy to forget that as much as 70 percent of its people depend on agriculture and allied activities (including forests), a sector that today contributes only 24 percent of the country's GDP and grew a mere 2.3 percent during the period 2002–2007 (Chidambaram, 2007). Despite impressive gains in manufacturing and services-related capacities, India is still a predominantly agrarian country: as many as 115 million families till the land (Chidambaram, 2007). A failure during the monsoons can have severe impacts on the agricultural and forest sectors, which, in turn, affect the overall economic scenario: decreases in demand and production slow the pace of growth.[1] For the vast majority of Indians living in rural India engaged in subsistence farming, it is the difference between hunger and well-being.

A crucial factor that determines the productivity of agriculture, forests, and biotic resources is the overall vitality and robustness of the

---

Marcella D'Souza is the Executive Director of the Watershed Organization Trust (WOTR). Crispino Lobo is the Executive Director of the Sampada Trust and cofounder of WOTR.

1. The Finance Minister estimated that an annual growth rate of 4 percent in agriculture would be required to push overall growth into the double digits (Chidambaram, 2007).

local ecology and environment. One of the major challenges India faces is how to speed up economic growth and food security for its more than one billion people without exhausting the resources on which this growth depends. Environmental health is of utmost importance for India's future, inasmuch as 65 percent of its population ekes out a living from soil, water, forests, and biomass. Healthy soil and land resources are critical to providing the ever-increasing human population with food and livestock with fodder. Forests support many forest dwellers, help in water retention in river catchments, provide fuel for heating and cooking, and supply fodder.

An audit of the basic resources and developmental indicators in India shows a grim picture. Even after repeated and sustained efforts at achieving food security, more than 330 million of India's people live below the poverty line. The infant mortality rate is high. Tuberculosis, malaria, waterborne diseases, and maternal mortality during childbirth still take a huge toll. High rates of illiteracy are prevalent, especially in rural areas and particularly among women. Some 38 percent of the households still do not have access to clean drinking water. Only 23.7 percent of households have access to toilet facilities, of which only one-tenth lie in the rural areas.

The most important environmental problems in rural areas are land degradation, the loss of soil nutrients, forest and groundwater depletion, and diminishing biodiversity (Ministry of Environment and Forests, 2001). In urban areas, pollution and related problems are more serious (ibid.). The total geographical area of India is 329 million hectares, of which 141 million hectares (43 percent) are severely affected by water and wind erosion, while desertification alone impacts 33 percent of the land (ibid.). The net sown area is 143 million hectares, nearly two-thirds (59 percent) of which is affected by some kind of soil degradation (ibid.). Another 8.53 million hectares are waterlogged (ibid.).

Over 70 percent of the agricultural land in India is rainfed and produces more than 50 percent of grains, coarse cereals, and oil seeds. Productivity in this area is completely dependent upon the availability of rain, but the rainfall pattern is erratic in terms of frequency, intensity, and spatial distribution. Generally, the rainy days do not exceed forty-five days per year and are concentrated within a maximum period of three months. The general rainfall pattern ranges from scant to medium (100 mm to 1,200 mm), with some regions experiencing heavy rainfall (2,000–4,500 mm). The summer monsoon season (June to September) is the major time of agricultural activity, with severe moisture stress from January to May. The vegetative growth period in these areas is less than 150 days.

The pressure on the land is extremely high, both from the human and animal populations, and has resulted in fragile environments and disrupted ecological relationships. Overall, depending upon the topography, rainfall, and vegetative cover, all these areas are prone to severe erosion, heavy runoff from flash floods and cloudbursts, and low infiltration. Soil erosion is quite severe. It is estimated that an average of sixteen tons of soil is washed away from each hectare of land annually, along with essential micronutrients.

Among the most noticeable and defining characteristics of the majority of people living in rain-dependent, ecologically fragile regions are poverty and marginalization. This is because of the close and symbiotic relationship between the environment and the local communities residing therein. As much as 80 percent of a rural household's basket of consumption comes from natural and biotic resources obtained from the local environment, its primary source of survival. A robust and vibrant ecology leads to a healthy and resource abundant environment, which for the inhabitants means better provisioning, as well as secure and stable sources of food, water, fuel, fodder, and livelihood. An ecologically fragile and resource-poor environment, on the other hand, means a harsher and leaner level of subsistence, frequently punctuated by periods of heightened stress and insecurity. This puts immense pressure on existing resources and creates unequal and exploitative social relationships that lead to marginalization, alienation, pauperization, and migration on a seasonal or permanent basis.

The story of Mhaswandi, a remote village in western India, vividly underscores the symbiotic and direct relationship between the environment and the well-being of the community living therein. It also powerfully illustrates the transformations that occur when people come together to regenerate and restore their area of survival, namely, the watersheds in which they live.

Mhaswandi: From Resource Exploitation to Resource Mobilization

Until 1994, Mhaswandi was a remote nondescript village, one of the thousands dotting the foothills of the Sahyadri range in the northeastern part of the Ahmednagar district, a rainshadow and drought-prone region of Maharashtra, a state straddling the western seaboard of India.[2] Nestled in the upper reaches of a large watershed, the Mhaswandi watershed is 1,145 hectares (11.45 sq. km) spread across undulating and steep hills, most of which, until recently, were barren and degraded. It is a well-defined watershed that receives a highly variable quantum of rainfall that averaged 302 mm from 1984 to 1992, but then 735 mm in 2006. Temperatures range from 38–40 C in the summer to 16–22 C in the winter.

Because of the hilly nature of the terrain, arable lands constitute only a little over half of the watershed (598 ha). Until 1994, only 10 percent of cultivated lands had access to any irrigation facilities, with the bulk being dependent on the vagaries of the monsoons. With agriculture barely able to support the 287 families (1,579 people) living in Mhaswandi for three or four months a year, most of the able-bodied people would migrate for the better part of the year to nearby towns and resource-endowed areas in search of livelihoods and employment. Those who remained behind had to contend with a harsh environment where fuelwood, fodder, and water were scare commodities. Women would trek for kilometers in the blazing sun in search of potable water and firewood. Cattle were let loose to fend for themselves.

Mhaswandi was archetypal of the many remote, hilly, dusty villages dotting vast tracts of India that offered only a hardscrabble, subsistence-level existence to the majority of its inhabitants. All this

---

2. Unless otherwise noted, this section is adapted from information available at the Watershed Organization Trust Web site, www.wotr.org.

changed in 1994 when some people in Mhaswandi came to hear of watershed projects being undertaken in nearby villages. The projects had made a huge difference in these villages and created quite some buzz in the neighborhood. The Mhaswandi residents decided to approach Fr. Hermann Bacher, who was the inspiration and mover behind these efforts, with the request to undertake a similar effort in Mhaswandi itself. Fr. Bacher, Chairman of the Watershed Organization Trust and Promoter-Founder of the Indo-German Watershed Development Program (IGWDP) in Maharashtra, decided to take the villagers up on their offer (Lobo, 2006).[3]

Fr. Bacher visited Mhaswandi and laid down a number of conditions that the people would have to agree to and adopt in order to be considered for inclusion in the IGWDP. The principal requirements were as follows: free grazing would have to be regulated, clear felling of trees banned, and spudding of bore wells and planting of water intensive crops proscribed; the community would have to contribute at least 16 percent of project costs, either through cash payments or voluntary labor; the villagers would have to implement the project themselves, and establish a representative executive body called the Village Watershed Committee (VWC), which would plan, organize, and supervise the entire project, including funds disbursement; and, finally, the women would have to be mainstreamed into the project, organized into self-help groups, and federated at the village level.

The ban on free grazing and tree cutting was particularly difficult because it affected significant aspects of livelihood and household maintenance.[4] After much debate, hesitation, and not a little anxiety,

---

3. The IGWDP is a large-scale, bilaterally assisted project funded by the German government, involving the German Agency for Technical Cooperation (GTZ), the German Bank for Development (KfW), the National Bank for Agriculture and Rural Development (NABARD), and the Watershed Organization Trust (WOTR).

4. Cattle were largely let loose for free grazing; controlled grazing meant that either a person had to watch over the cattle, thus imposing a cost for an asset that was nondescript and having low productivity, or sell them off, which was an emotional issue. Trees and shrubs provided firewood for cooking; however, trimming, lopping, and pruning were permitted. Improved and more efficient stoves were introduced to compensate for the ban.

the villagers decided to take up the challenge. Developing the watershed they lived in became an overriding priority, and this effort engaged the entire village for the next seven years.[5] The Sangamner Cooperative Sugar Factory (SSCF), which accepted Fr. Bacher's request to provide the villagers with institutional and technical support, assisted them in this effort. The Watershed Organization Trust (WOTR), a nongovernmental organization that developed and manages the capacity building phase of the IGWDP, provided capacity building support to both the SSCF and the villagers themselves (Watershed Organization Trust, n.d.).

The people organized themselves and began planning the various treatments that would need to be implemented across the watershed, beginning from the hillsides and reaching into and across the valleys. On the hilltops, along the ridgeline, they dug broad water absorption trenches (WATs). Across the barren hill slopes, they dug continuous contour trenches (CCTs) and, after refilling these, planted trees interspersed with grasses on the earthen mounds and between the trenches. On wastelands of gentler slope, they dug CCTs or boundary trenches, using the excavated soil to form earthen bunds behind the trenches. On these, too, they planted trees and grasses, as well as on the lands in between. On farmlands, they raised compartment bunds. Along the drainage channels, from the hilltops down into the valleys, across every rill and stream, they built a series of water harvesting and erosion control structures, including stone bunds, gully plugs, gabion structures, earthen nullah bunds, and check dams. On the rivers, they built medium-sized masonry check weirs and, at appropriate points, percolation tanks.

Thus, when it rained, as water gushed down the hills, its velocity and momentum were broken. The WATs and CCTs made rushing water slow down to a canter; cantering water was made to walk; and walking water was brought to a halt. The standing water was then allowed to percolate into the ground. The farm bunds and contour bunds held the water in situ, on the land itself. Along the rills and rivulets, as surplus water from the trenches and the adjoining areas

---

5. The project, which began in 1994, officially ended in 2001.

poured in, the stone and gully plugs acted as miniature breakwaters, slowing down the torrents and trapping silt and soil; this slowed but still cascading water was further obstructed and contained by the gabion structures, earthen nullah bunds, check dams, and masonry weirs. By the time the water reached the valley, it was slow-moving and contained only a minimum silt load, the rest having been deposited in the various structures across the hillsides and along the drainage channels. This life-giving water was now impounded in percolation tanks, the surplus of which, when fully released, fed the downstream areas.

This comprehensive and integrated way of harvesting rainwater and mitigating soil erosion not only increased the availability of water and improved the soil moisture regime but also, equally important, recharged the depleted groundwater aquifers. These aquifers feed the wells, which are the primary source of water for agriculture, livestock, and domestic purposes in the long, hot summer months, when surface water bodies have long evaporated and dried up.

These conservation measures resulted in significant impacts on the health and vitality of the local ecology and environment. Springs that had run dry sprang to life, and others fed the streams and rivulets with water for several months longer than they had in the past. Owing to controlled grazing and restrictions on tree cutting, as well as the massive afforestation effort undertaken, the hillsides began to green.[6] Grasses that were hardly seen before reappeared and proliferated; shrubs, bushes, and trees that had been stunted by severe grazing grew taller and denser; planted trees took root and reached upward, covering the barren soil. Biotic life and activities increased manifold. Wildlife and migratory birds reappeared, as the many water bodies that had been created became centers of life and congregation.[7] The significant impact on grass cover can be gauged from the fact that fodder collection, which was only 950 tons in 1994, grew to 2,500 tons in 2006, an increase of 163 percent.

6. More than 600,000 trees have been planted.
7. There are now sightings of deer, foxes, jackals, hares, and even the occasional leopard.

The impact on the village community was no less dramatic. Before their eyes, the villagers began to see their barren hills and degraded environs come to life. A new spring filled the air. There was the promise of a new dawn, a better tomorrow. Seeing that water was now available for a longer period, people began to invest effort and resources in developing their farms. They dug wells, bought electric motor pumps, adopted modern cultural practices and methods,[8] and began to diversify the cropping pattern. While still growing the traditional cereal crops of sorghum and millet, which are the mainstay of their food, they began to grow new crops for the market: fruits and vegetables such as tomatoes, onions, cabbage, garlic, okra, cucumbers, squashes, watermelons, and pumpkins, as well as flower cash crops of marigolds, chrysanthemums, and roses.

By 2006, the number of wells had grown from 33 to 70 (+112 percent). Most of these wells now have water throughout the year, with a standing water column ranging from twenty to sixty feet. Perennially irrigated area that was only 29 hectares in 1994 expanded to 175 hectares (+503 percent); seasonally irrigated area has also increased from 35 hectares to 423 hectares (+1,109 percent).

More important, from the social and developmental perspective, the agricultural employment season grew from three months to year-round by 2006, thus resulting in a complete stopping of migration in search of employment. In fact, reverse migration has been observed. Some people who had left the village earlier for towns or irrigated regions in search of a livelihood, have now returned to the village.

The benefits for the landless have been particularly noteworthy. Because of a shortage of labor, many farmers have begun to focus on the well-irrigated portions of their farms and have leased out the rest to the landless on a sharecropping basis. These sharecroppers have also diversified into rearing crossbred cows and small ruminants

---

8. While this resulted in a significant increase in chemical fertilizer, pesticide, and herbicide use, the people have now realized the severe adverse long-term effects of these inputs and are now increasingly using organic farming practices. This has become possible because compost and manure are now more easily available and the market for organic foods is growing.

(goats, sheep). This transformation from assetless, wage-dependent agricultural labor to land cultivating and investing farmers is a life-changing milestone—a social and psychological watershed, indeed, in both the personal and social lives of this otherwise disempowered group of people.

With increased water availability and fodder, the people then ventured into rearing crossbred milch cattle. They established a Milk Producers Cooperative Society in the village and raised loans from the nearby banks. Milk production, which was 190 liters per day before the project, has increased to 464 liters per day. With the income from the sale of milk becoming a major contributory to family income, the villagers established a bulk cooling facility, which allowed for milk collection throughout the day and reduced spoilage. Milch cattle have grown from 8 in 1994 to 130 (1,525 percent). Scrub cattle, and sheep and goats, have reduced by 81 percent and 63 percent respectively.

The quality of life of the people has seen a noticeable and marked improvement. Dependency on firewood for cooking has greatly reduced: there are now eleven biogas plants, and 150 households have purchased liquefied petroleum gas (LPG) cylinders. None existed previously. Similarly, more than 130 households have built individual toilets; before there were none. From one TV set in the village before the project, there are now 125; bicycles have gone up from 13 to 30 (+131 percent) and motorcycles from one to 85.

The price of land is often a fairly accurate barometer of the vitality of the local economy, the productivity of land, and people's perception of the future, as well as of what they see as their future main source of income, risk insurance and security. Irrigated farmland that cost Rs. 15,000 (USD 349) per hectare in 1994 now commands Rs. 750,000 (USD 17,442), an increase of 4,900 percent.[9] Rainfed lands cost Rs. 5,000/ ha (USD 116) then; now, Rs. 125,000 (USD 2,907), an increase of 2,400 percent.

The village has also become a kind of demonstration and showcase. People from far and wide visit regularly to seek inspiration, see what has been done, and learn how to do the same in their villages.

9. We assume an exchange rate of USD 1 = Rs. 43.

Villagers who were diffident, shy, and loath to speak to outsiders now confidently and with ease show visitors around and share with them their experiences and stories. Visiting groups are given a memento of their visit and charged a fee for the time invested in them.[10] The villagers have a trained pool of accomplished guides who take turns to welcome guests. Fees thus earned are credited to the village account pool (VWC account), and guides are paid an honorarium. What is particularly impressive is the key role women play in these events: they are articulate, have many achievements to their credit, which they proudly show, and are active members of the guide pool. The village has been the recipient of several public awards and has been extensively covered in the print and electronic media.

Drivers of Change: The "How" and "Why" of Getting There

Successful outcomes that last are usually the result of a happy coming together of opportunity, the right people, representative, effective, and accountable institutions, and inclusive practices and processes.[11] These result in mutual trust and good relationships developing that result in buy-in and ownership. We outline some of the key institutions and processes adopted by the villagers of Mhaswandi as they implemented their project.

The Village Watershed Committee and Forest Protection Committee

When the people decided to implement the watershed project, they nominated a Village Watershed Committee (VWC), as well as a Forest

---

10. It is estimated that, as of December 2006, as many as 400 groups and a total of 25,000 people have visited the village.

11. Unless otherwise noted, the information in this section comes from innumerable conversations and informal learning opportunities. In the absence of substantive formal references, we would like to give credit to those individuals who have made significant contributions to the ideas captured within.

Protection Committee (FPC). The VWC is a representative body with executive powers and consisted of representatives from each social group and geographical area of the village. Initially, there were twenty-one members, of whom four were women. This body met on a monthly basis during project implementation and was accountable to the Gram Sabha, the gathering of all adult and voting members of the village that met quarterly. The mandate, role, and responsibilities of the VWC were as follows: motivate the community and individual landowners to accept and observe the project disciplines; plan the nature, location, and phasing of physical measures; receive project funds, and manage and account for them; organize implementation of measures; supervise the same and make payments; resolve disputes, solve problems, and enforce social sanctions and penalties; and collect the statutory contributions, and liaison with project authorities, as well as government functionaries.

The FPC is a body registered under the regulations of the Forest Department and has permission to enter and treat forest lands, share the usufructs thereof, and, in collaboration with the Forest Department personnel, ensure protection of the trees planted and growing on these lands. Every household of the village is a member of the FPC, and, while it has an executive committee of its own, for all practical purposes, it is synonymous with the VWC.

Women's Self-help Groups and the Apex Joint Women's Committee

Women, though comprising about 50 percent of the village and doing over 70 percent of the farm work, have traditionally had no say in village matters. Yet, any effort at regenerating the environment could be sustainable only if women had a stake in it as decision makers, not merely as laborers. Hence, one of the conditions of inclusion in the watershed program was that the men would have to consent to women becoming part of the effort as members of the VWC and by organizing themselves into self-help groups (SHGs) or solidarity groups. In the beginning, the men were rather uneasy, but eventually they came to terms with the situation and even supported the women (WOTR, n.d.).[12] In the beginning, 160 women came together into eight SHGs focused on savings and credit activities. They then formed a village-level federation of these groups called the Apex Joint Women's Committee (Samyuktya Mahila Samittee, SMS) with representation from each of the SHGs. The purpose of the SMS is to present a united women's voice in the village and representation in the various village bodies. At first, many women, especially those belonging to the marginalized communities, had reservations about the agenda and usefulness of these groups, since membership involved monthly contributions of money (between Rs. 20 and 50 per month) and intra-group lending. Their previous experiences of dealing with members of the dominant groups and of financial transactions had not been encouraging. Today, however, there are 22 SHGs with 440 members, and the SMS consists of 19 members.

The SMS and the SHGs have played and continue to play a vital role in empowering women and mainstreaming them into the institutional and political life of the village. Previously, they were hardly seen and would attend village meetings only as spectators. Today they not only actively participate in the meetings of the VWC and the Village Panchayat (village council, the center of local self-government), but they also attend to visitors and proudly explain what they and their village have achieved. While previously they would leave the village only to go to nearby areas, they now venture to far-off areas in search of new knowledge and business opportunities.

Participatory Net Planning

At first the farmers and landowners were greatly hesitant about developing their lands with exogenous funds, since their previous

---

12. The Women's Promotion Team in WOTR developed a successful strategy of empowering women called the Gender Oriented Participatory Operational Pedagogy (GO-POP), which actively sought and secured the men's support and involvement in the development of their women. GO-POP is detailed in WOTR's Operations Manual—Guidelines for Operations (Capacity Building) in the Indo-German Watershed Development Program, Maharashtra (n.d.).

experience in this regard had been negative.[13] In order to overcome this fear and obtain the necessary contribution from the farmer/landowner, the accompanying nongovernmental organizations (NGOs), as well as the VWC, had to take into confidence each family and landowner.[14]

In addition to several public meetings and intense discussions, they used a method called the Participatory Net Planning Method (PNP).[15] The PNP is a unique approach that engages the owner-couple in an on-site dialogue on their land holding with knowledgeable local farmers and agricultural professionals. A resource inventory is undertaken and measures for enhancing conservation and productivity are planned. The planned measures, inputs and expected outcomes are in keeping with the farm couple's decision, knowledge, and aspirations, as they are deepened and informed with insights from modern land-husbandry practices. This results in a sense of ownership, as well as site-specific conservation and productivity-enhancing interventions.

This approach was repeated with each farming and landholding household in the village and resulted in wholehearted support and cooperation. This was very important, since watershed treatments, such as contour bunds and water surplusing arrangements, sometimes cut across farm boundaries and have to be undertaken irrespective of land holding patterns.

Social Audits, Qualitative Assessment Matrix, and Participatory Impact Monitoring

Maintaining the confidence of a contributing public requires public disclosure and accountability, especially when substantial exogenous

---

13. Previously, works undertaken by the government on private lands, for the purpose of soil and water conservation, were later converted into loans, and a lien was placed on the lands. Moreover, the landowners and farmers were not consulted or involved in the works undertaken on their lands. These efforts were, therefore, viewed as having been imposed and implemented by the government, and largely not maintained.

14. Farmers were required to contribute 16 percent of the labor costs of physical measures undertaken on their lands.

15. The PNP was developed by WOTR and is now widely adopted across India, in both public and donor-funded projects.

funds are involved. Project details, including the activities to be undertaken and costs and funds received and spent, were displayed in a public place and regularly updated. The VWC would meet at least fortnightly in the early days of the project, and later, monthly. The entire village would meet once every three months and receive an accounting of works undertaken, plans to be executed, and costs incurred. Problems encountered were discussed and solutions worked out.

In addition to an audit of financial and operational issues, a Quality and Impact audit was also conducted using the Qualitative Assessment Matrix (QAM) and Participatory Impact Monitoring (PIM) formats on a quarterly and annual basis, respectively.[16] The trends were regularly monitored by the accompanying NGOs.

The QAM is a set of key indicators that measures key events and milestones of the various aspects and processes involved in successfully implementing, in this case, a watershed project. Its use increases awareness in the community of what is taking place, helps participants undertake steps and measures to address shortcomings, and strengthens dynamics for achieving project outcomes.

In the PIM, the villagers identify a set of indicators for assessing impacts, both positive and negative, using various tools. This is followed by a detailed collection of data in a participatory manner at various points in time; the group then uses this information to make assessments. Such an approach injects a fair degree of objectivity in perceptions, as well as being a process for learning and change.

Peer Group Review

Competition and public approbation are powerful incentives to doing better and excelling. The Peer Group Review (PGR) is an annual event, which begins from the second year of project implementation, wherein those desirous of entering for a prize, at either the regional and/or state level, submit their entries to WOTR, which then assembles a team of representatives from participating watershed project

---

16. Both the QAM and the PIM were developed by WOTR and have now been widely adopted and adapted for a variety of developmental interventions.

villages, augmented with external persons of repute, to visit the projects and assess progress and achievements as well as grade the project against criteria that they themselves have predetermined.[17] This is a time of great excitement, discussion, and learning as the assessment is publicly undertaken in each of the villages. Winners are publicly congratulated at well-publicized events attended by high-ranking politicians and governmental and social leaders. Though voluntary, all projects doing reasonably well participate, since being publicly acknowledged has its advantages: it confers a respectability to the feted villages and attracts developmental funds and resources from public and private agencies.

Maintenance Fund

In order to ensure that funds are available for maintenance of project-created common assets and infrastructure, as well as to undertake developmental activities as needed, the project required the setting up of a maintenance fund (MF) from the very beginning. This fund was to be developed and managed by the VWC and an account of its utilization given to the entire village assembly on a quarterly basis. Of the 16 percent contribution toward labor costs that villagers had to make, the project returned 50 percent of this amount to the MF. Additional resources came from donations received, fines imposed on those who violated the ban on free grazing and tree cutting, fees charged to visitors for services rendered, and proceeds obtained from the leasing of common property resources.[18] The money had to be invested in secure deposits in a bank, and the interest used for maintenance of common assets and financing, on a returnable or loan basis, social development and income generating activities. By the end of 2006, the MF had

---

17. This tool and method was developed by WOTR to primarily serve as an instrument for cross-project sharing and learning, as well as to stimulate healthy competition for promoting excellence. Three prizes each are given for those participating in the regional- or state-level competitions.

18. For instance, the VWC annually auctions off three check weirs, the waters of which are used for raising fish, for Rs. 45,000 (USD 1,047), on average. Grass, on common lands, is auctioned off for Rs. 40,000 (USD 930), on average, annually.

grown to Rs. 913,000 (USD 21,233) (Jangal, 2007), which, by village economy standards, is very substantial and quite unique in the locale.

The Issue of Sustainability

The issue of what happens after the project is completed and the facilitating agencies have withdrawn is crucial, not only because of the substantial investments that have been made but also because, if the hope and expectations created during the project implementation are lost, disappointment and frustration can develop. This dissatisfaction, in turn, leads to fatalism and a weakening, if not unraveling, of the social and institutional fabric that brings and holds people together to undertake common activities. Such failings have potentially serious developmental consequences, as change and adopting new ways of thinking and acting, key requirements for the poor and marginalized to benefit from growth and development, become that much more difficult on the community level.

Generally, nothing succeeds like success, especially if those enjoying it have had a stake in making it possible. Using this criterion, there is no reason why the benefits that have accrued to date should not generally continue, barring natural or human-caused calamities.

The project ended in 2001. Since then, as indicated, not only have the benefits that were accruing in the early years continued and in several aspects increased, but they have also extended to an increasing number of beneficiaries.

In recent years, a number of formal and informal associations and groups have been formed, each of which has developed significant capital funds from membership contributions, attesting to growing levels of income and the resultant cultural and recreational needs that arise.[19]

---

19. Most of these groups and associations are for undertaking celebration of religious, social, or cultural events. Funds collected range from Rs. 20,000 (USD 465) for recently formed groups to Rs. 600,000 (USD 13,954) for older groups.

Furthermore, the institutions developed continue to exist and function. They have undergone leadership changes, remained accountable, and continue to enjoy the confidence of the majority of villagers. The VWC also has the financial means to maintain the common structures that have been created, as well as to undertake developmental activities as needed.

Of particular importance are the status and role of women, who, in watershed projects, hold the key to sustainability. Not only do they draw heavily upon locally available natural resources to meet household needs, but they are also the transmitters of culture; not for nothing is it said that "the hand that rocks the cradle, rules the world." The women of Mhaswandi have proven themselves an entrepreneurial lot.

In order to mitigate the impact of the restrictions on tree cutting, as well as reduce health hazards, the project provided women with improved stoves. The project also encouraged and financially supported them to convert to LPG cooking stoves. Instead of merely becoming customers, the women approached the company itself and obtained for their village a gas dealership, thus securing substantial cost savings for their members. The SMS now manages this activity and has provided loans on generous terms for the purchase of gas stoves and cylinders.

In order to reduce their dependence on the moneylenders and middlemen, the women used the funds allocated to them to purchase seeds and fertilizers in bulk directly from the wholesalers, thus cutting out the traders. This effort was so successful that they found it difficult to manage the demand, given their myriad other household and farm responsibilities, and eventually handed this activity over to the VWC. In return, the VWC annually gives the SMS an amount of at least Rs. 10,000 from its profits.

Women have also undertaken several income-generating activities, such as setting up a unit that makes chili powder and masala (a mixture of spices used for making curries), which is now self-financing. They have provided each homemaker with a large shopping bag on which the logo and activities of the SMS and the services provided are printed. These bags are used by the village women in the local market and serve as promotional items.

The children are remarkably knowledgeable about what has been done in their village and why. Their understanding of the relationship between a regenerated environment and their well-being is truly impressive, a clear pointer to the fact that the women have developed a strong sense of ownership of the watershed effort and are now conveying it to their children, the generation that will have to sustain the created assets.

Conclusions

The Mhaswandi experience and story offer several pointers on what could make a developmental initiative successful, sustainable, and inclusive. We shall focus on the key lessons.

People must have a strong and urgent need for a particular intervention and be willing to pay the price for realizing it. They must "walk the walk," accept the consequences that follow, and enforce the discipline required to realize their hopes and aspirations. In other words, there are no free lunches. Interventions that require community-wide cooperation, as in the case of natural-resource-management projects, will only be successful if there is a minimum amount of social capital existing; that is, a working level of trust among different groups or at least the willingness to discuss and sort out issues. A key requirement for project success and sustainability is a sense of ownership, which is engendered only when communities and individuals indicate their commitment through real contributions. Freebies only undermine the spirit of self-help. People will only truly commit themselves when they are involved in all the aspects of project implementation through negotiated and agreed-upon outcomes, in terms of inputs, conditionalities, obligations, activities, and phasing.

Capacity building is also central to sustainability, equity, and empowerment. A project is only as good as the people involved in it. It is important that sufficient time and effort be devoted to building up capacities in the areas of project management, organization, accountability, and documentation in order to ensure transparency, equity, and ownership. Capacity building should be systematic; sequenced; premised on the principles of learning-by-doing, regular on-site accompaniment, and participatory review and monitoring; and geared toward achieving desired outcomes (Lobo and D'Souza, 1999).[20] This requires a willingness to accept mistakes being made, sometimes costly ones, because they can become valuable learning occasions that will bear on the success and long-term sustainability of the project.

Transparency, accountability, and the free flow of information to all concerned are necessary to secure commitment, foster ownership, sustain the process and momentum, and enable the establishment of effective management and maintenance mechanisms (Lobo, 2005). Indigenous technology should be adopted or adapted to the greatest extent possible wherever feasible and given a comparative advantage over modern approaches. This validates the wisdom, heritage, and technical prowess of the community and instills a sense of pride, self-confidence, and ownership of the project. Regular and participatory monitoring of jointly agreed-upon indicators should also be undertaken. It not only promotes accountability but also provides occasions for learning. Effective monitoring, however, requires the establishing of an appropriate MIS that captures and tracks the essential elements of the project on a continuous basis so as to provide inputs for corrective actions, information dissemination, and learning (Lobo, 2005).

Unless the needs and interests of the vulnerable and marginalized sections of a community—women in general and women-headed households in particular, as well as the landless, tribal people, artisans, and pastoralists—are factored in as a specific objective of a developmental effort, they could well end up worse off than before.

---

20. An example of such a capacity building approach is the Participatory Operational Pedagogy (POP), developed by the WOTR, that is implemented in the IGWDP, Maharashtra, and of which Mhaswandi was an early beneficiary. The POP systematically and sequentially builds up the capabilities of community-based organizations (CBOs) and NGOs to undertake participatory watershed development in a transparent and accountable manner while meeting the expected quality levels and benchmarks.

Hence, care must be taken to address gender-related issues, as well as equitable distribution of benefits for disadvantaged groups.

Sustainability is possible only if development with equity is ensured. It is necessary to ensure that all members of the target group, especially the poor and marginalized, see themselves as benefiting from the project, if a sense of ownership is to be created, social harmony strengthened, and created assets maintained after the project period.

Women hold the key to sustainability of natural resources, treated watersheds, regenerated forests, pastures, and plantations. Since women undertake the bulk of agricultural and sedentary pastoral operations, involving them as decision makers and not merely as laborers, with preference given to their priorities, choice of technology, and planting material, ensures their stake in the maintenance of created assets. More important, they are then given incentives to encourage their children, the next generation of resource users, to become careful husbanders of the resources they inherit from their parents. Conscious and sustained effort is required to enhance women's participation and position in the community by ensuring their effective integration into village institutions and into decision-making processes. This requires putting emphasis on raising gender awareness among men and building confidence among women on a sustained basis (D'Souza, 1997 and 1998).

Mhaswandi had a number of enabling and favorable initial conditions; however, the Mhaswandi experience has been repeated in hundreds of villagers in Maharashtra and other parts of the country, in varying degrees, depending upon local conditions. These successes validate the thesis that a regenerated environment, or a watershed developed, has not only significant positive externalities, such as contributing to mitigating the adverse effects of climate change (Lobo, 2003), but also profound impacts on the community involved in its transformation. The effort and its benefits catalyze a developmental dynamic and set loose social processes of cohesion and inclusion that strengthen the community foundations of development, progress, and good participatory governance (Lobo and D'Souza, 2004).

References

Chidambaram, P. (2007, February). Budget speech 2007–2008. Speech of Minister of Finance to the Parliament of India, New Delhi. Retrieved November 6, 2007, from http://indiabudget.nic.in/ub2007–08/bs/speecha.htm.

D'Souza, M. (1997). *Gender and watershed development.* Retrieved May 5, 2007, from www.odi.org.uk/agren/papers/agrenpaper_88.pdf.

D'Souza, M. (1998). Watershed development—Creating space for women. Retrieved May 5, 2007, from www.wotr.org/Articles/creating_space.pdf.

Jangal, J. (2007). Indo-German watershed development project at Mhaswandi watershed, Sangamner Taluka, Ahmednagar district, Maharashtra (Watershed Organization Trust Internal Evaluation Report). On file with author.

Lobo, C. (2003). Watershed management: A sustainable strategy for augmenting water resources and mitigating climate changes. *Annals of Arid Zone* 41: 359–364.

Lobo, C. (2005). Reducing rent seeking and dissipative payments: Introducing accountability mechanisms in watershed development programs in India. Retrieved May 5, 2007, from www.wotr.org/Articles/reducing_rent.pdf.

Lobo, C. (2006). Facilitating change—Creating institutional and policy spaces: A case study of the Indo German Watershed Development Program (IGWDP). Retrieved May 5, 2007, from www.wotr.org/articles/facilitating_change.pdf.

Lobo, C., and D'Souza, M. (1999). Qualification and capacity building of NGOs and village self-help groups for large scale implementation of watershed projects. *Journal of Rural Development*, Special Issue on Watershed Development, part 2, 18 (4).

Lobo, C., and D'Souza, M. (2004, November). Watershed development, water management and the millennium development goals. Paper presented at World Bank and Government of Uttar Pradesh Watershed Summit, Chandigarh, India. Retrieved May 5, 2007, from http://wotr.org/Articles/wsd_and_mdg.pdf.

Ministry of Environment and Forests, Government of India (2001). India national action programme (NAP) to combat desertification in the context of United Nations Convention to Combat Desertification (UNCCD). Retrieved June 11, 2007, from www.envfor.nic.in/unccd/desert1.html.

Watershed Organization Trust (n.d). Operations manual—Guidelines for operations (capacity building) in the Indo German Watershed Development Programme. On file with author.

Watershed Organization Trust (2007). Watershed voices: Mhaswandi watershed project. Retrieved May 5, 2007, from www.wotr.org/wsvoices.htm.

Part II INVESTING IN THE ENVIRONMENT, PROTECTING COMMUNITIES FROM NATURAL DISASTERS

6 | The International Disaster Risk Reduction Regime

Escaping the Cycle of Poverty and Tragedy

ALENA HERKLOTZ

Since the second half of the twentieth century, natural disasters have been increasing in frequency, severity, and intensity (UN Inter-Agency Task Force, 2001). Today, disasters affect more than 200 million people every year (UN World Conference on Disaster Reduction Secretariat [UNWCDR], 2004). In 2005, for example, the global community suffered 150 disasters with a combined toll of 90,000 human lives and USD 220 billion in economic losses (UN International Strategy for Disaster Reduction [UNISDR], 2006b). The October 2005 earthquake in Asia injured 70,000 people and claimed the lives of another 73,000 (Braine, 2006). The 2005 hurricane season in Central and North America was the most active on record for the Atlantic, shattering numerous records, most notably with Katrina, which was by far the costliest hurricane in U.S. history and the most deadly since 1928 (National Oceanic and Atmospheric Administration, 2006a; 2006b).

Natural disasters present a major global concern. They strike worldwide and can overwhelm the capacity and resources of even the most developed nations. In the past fifteen years alone, the world has witnessed well over five thousand natural disasters and lost nearly one

---

Alena Herklotz is the Adam and Brittany Levinson Fellow in International Law of Sustainable Development within the Sustainable Development Legal Initiative (SDLI) of the Leitner Center for International Law and Justice at Fordham University School of Law.

million people and close to USD 1,200 billion as a result (UNISDR, n.d.a; n.d.d; n.d.e). The 2004 Indian Ocean earthquake triggered a series of the most deadly tsunamis in recorded history, which battered ten South Asian and East African countries, killing 283,100 people and displacing 1,126,900 (U.S. Geological Survey, 2007). Since the 1990s, two-thirds of disasters have been floods and storms across all five continents, while earthquakes and tsunamis claimed more than 400,000 lives despite accounting for only 8 percent of recorded events (UNISDR, n.d.b; n.d.e). The period also saw the three warmest years on record, in which heat waves killed thousands in Europe and Asia (UNWCDR, 2004). The El Niño phenomenon alone has caused 20,000 fatalities and USD 35 billion in material damages (UNWCDR, 2004). Another 40,000 lives have been lost to severe landslides, debris- and mudflows (UNWCDR, 2004). When examined in terms of lives affected, that is, the number of people requiring immediate emergency assistance for basic survival needs, the impacts are still more staggering: the number left homeless, hungry, or injured is nearly 3.5 billion (UNISDR, n.d.c). Floods, storms, and drought account for nearly all human impacts, affecting 2 billion, 435 million, and 950 million lives, respectively (UNISDR, n.d.c). In an average year, natural disasters will claim more than 60,000 lives and affect more than 250 million people (UK Department for International Development, 2004).

Though better known for the human impacts of lives lost and affected, natural disasters also present a formidable economic threat. Together, floods, storms, and drought also have the greatest impact in terms of economic losses, which encompass damage to infrastructure, crops, housing, revenues, employment, and market stability (UNISDR, 2006a). Over the past fifteen years, storms have been the most costly at USD 443 billion, but are followed closely by floods, which have cost over USD 360 billion (UNISDR, n.d.d). The lesser financial impact of drought, over USD 106 billion, is more than made up for by the disproportionate USD 271 billion incurred from earthquakes and tsunamis (UNISDR, n.d.d).

The rise in the frequency and severity of natural disasters has been accompanied by a shift in their global perception, from a conception of catastrophes beyond human control to which people can only struggle to respond, to recognition that natural disasters are, in fact, the product of risk accumulated through years of vulnerability and underlying hazards (Schipper and Pelling, 2006). Natural hazards, such as droughts, earthquakes and tsunamis, epidemics, floods, landslides, tropical storms, volcanic eruptions, and wildfires are affecting human populations more because people are simply more vulnerable (International Institute for Sustainable Development [IISD], 2005). Poor planning, population growth, urbanization, and increasing environmental degradation create the high-risk conditions that invite and enable disaster (IISD, 2005). Most significantly, however, natural disasters are inextricably linked to poverty. Poverty and exclusion increase the vulnerability that enables natural hazards to become disasters, which, in turn, swiftly eliminate development gains made in the fight against poverty: "disasters triggered by natural hazards are a consequence of development failure as much as failed development is a product of disasters" (Schipper and Pelling, 2006).

The Hyogo Framework for Action (HFA) 2005–2015, adopted at the January 2005 World Conference on Disaster Reduction (WCDR), is the global community's plan of action for assisting countries and communities to escape this cycle. The WCDR, its outcome documents, and its arrangements set forth the principles, priorities, and measures by which the world will seek to reduce the risk of natural disasters. This chapter will examine the response of the international community to the formidable issue of natural disasters. It reviews the mounting evidence of the strong interrelation between natural disasters and poverty, analyzing both the ways in which poverty produces vulnerability and the resulting disasters' adverse effects on development. The chapter then examines the international regime leading up to the WCDR, before turning to the current system, the progress made to date, and the challenges that remain.

Disasters and Deprivation: Inextricable Links

Far from being great equalizers that devastate wealthy and poor nations and people alike, natural disasters, in reality, prey on the disadvantaged. While the figures presented earlier attest to the universal nature of the problem, examination taken just one layer further reveals the gross disparities that poverty creates: over 90 percent of deaths caused by natural disasters occur in developing countries (UNISDR, 2004). While the citizens of least developed countries make up only 11 percent of those exposed to natural disasters, they suffer more than half of the fatalities that result (Bureau for Crisis Prevention and Recovery, United Nations Development Programme [UNDP], 2004). Likewise, although the absolute dollar losses are greater in the developed world, economic losses are proportionately much higher in the developing world, in terms of percentage of gross domestic product (GDP): as much as twenty times the relative loss sustained in the industrialized world (UNISDR, 2004; World Bank, 2005). The USD 125 billion loss that the United States incurred from Hurricane Katrina amounted to a mere 0.1 percent of GDP, whereas the USD 418 million in economic damages that Tajikistan suffered from devastating flooding in 1992 totaled in excess of 378 percent of its GDP (UNISDR, n.d.g). The countries that suffer the greatest relative loss are those that have the least: small island states and the poorest nations worldwide (UNISDR, n.d.g).

Poor people fare no better: Poverty places people at greater risk. Three factors, in particular, converge to make the poor more vulnerable to adverse impacts from natural disasters: risk awareness; location; and low disaster resilience (de Ville de Goyet and Griekspoor, 2007). Awareness of risk is directly related to levels of formal education and access to information and, therefore, as disparately denied the poor. Just as education, information access and risk awareness are typically a function of wealth, so too is the choice of a safe location for one's residence. The poor can seldom afford the more secure locations that, by virtue of their greater security, come at a greater price (ibid.). Lack of choice is not limited to the selection of areas that are less prone to natural hazards, but extends as well to the means for escaping when disaster strikes. Even when the government manages to warn people in time for safe evacuation, many simply cannot afford to act (ibid.). Poverty similarly limits the information and resources available for disaster resilient construction materials and techniques. Lack of education and access to information leave the poor unfamiliar with the best products and practices, while their poverty often means that they could not afford them even if informed (ibid.). These disadvantages plague local and national governments in the developing world as well, as they are, likewise, unaware and/or financially incapable of achieving disaster resilience in public infrastructure, such as hospitals and schools. A 1999 earthquake in Pereira, Colombia damaged 74 percent of the schools because the buildings were not built according to hazard resistant standards, with devastating impacts on primary education (UNISRD, n.d.f).

Not only does poverty render the poor more vulnerable to the adverse affects of natural disasters, but natural disasters can also eliminate years of development gains in a matter of minutes. Disasters commonly reduce, and even reverse, economic growth by destroying productive infrastructure and industries, such as agriculture and tourism (UNISDR, 2004). In addition, immediate, emergency disaster relief diverts already scarce funds from national and international development programs (Schipper and Pelling, 2006). As with the relative impacts on GDP, the impact on household income is also felt the most by those who have the least. The sudden loss of half an annual income of USD 50,000 is a hardship, whereas the same loss for a family living on USD 2/day is often a matter of life and death (de Ville de Goyet and Griekspoor, 2007). The poor also tend to hold what scant assets they possess in tangible resources that may be lost in disasters, while the wealthy hold a greater percentage of their wealth in intangible assets and can afford to insure all of their tangible assets (de Ville de Goyet and Griekspoor, 2007). Natural disasters stall and undermine economic development and harm individual households.

Disaster response, when available, drains development resources and threatens to overwhelm global capacity to recover and resume poverty alleviation efforts worldwide. The international community recognized the significance of natural disasters in the development context at the 2000 United Nations Millennium Summit, the largest gathering of world leaders in history,[1] at which a new global partnership to reduce extreme global poverty was negotiated. The Summit's

---

1. Some 103 heads of state and 89 heads of government attended the summit. For a full list of participants, see United Nations (2000b).

outcome document, the UN Millennium Declaration (2000), noted the importance of reducing the number and impacts of natural disasters, while its implementation road map calls for the development of early warning systems and vulnerability maps, support of international partnerships and cooperation, action to address man-made determinants of disasters, and incorporation of risk reduction into national planning processes (UN Road Map, 2001).

The primary outcome of the Summit, however, is a series of timebound targets for achieving the objectives set out by the Declaration that have become known as Millennium Development Goals (MDGs) (UN, 2002). The MDGs are quantified benchmarks to be achieved by 2015 that address poverty in its many dimensions, from tackling hunger, disease, and child mortality to promoting gender equality, education, and environmental sustainability (UN, 2002). Natural disasters, however, cut across these varied dimensions to threaten both the achievement and lasting impact of the MDGs.

Disasters threaten MDG1, the eradication of extreme poverty and hunger, by destroying individual livelihoods and the national infrastructure necessary for their promotion and security. MDG2's target of primary universal education is, likewise, compromised by the damage disasters do to public infrastructure, specifically schools. Furthermore, disasters displace not only households and communities but also family priorities regarding education, as children are kept home to supplement lost assets and conserve income. Natural disasters undermine the empowerment of women pursued under MDG3, as scarce resources are reserved for male family members, and women are forced to take on greater workloads to assist in the family's recovery. Outside the home, disasters and the havoc they wreak leave women more vulnerable to sexual violence and insensitive relief programs that are merely superimposed over preexisting inequities. (Schipper and Pelling, 2006).

MDG4 targets child mortality; however, disasters injure countless children, while those who survive are among the most vulnerable to postdisaster disease by virtue of their undeveloped immune systems. Disasters also threaten infants and young children with the loss or injury of caregivers and income earners (UNISDR, n.d.f). Disasters stress expectant mothers and the household and public-health resources on which they depend, in direct contravention of MDG5's aims to improve maternal health. The chronic diseases taken on by MDG6, more generally, also prey on the vulnerabilities that disasters create by attacking people's health and ability to generate the income needed to care properly for themselves and their families (Schipper and Pelling, 2006).

By their very definition, natural disasters threaten the environmental resources and sustainability protected and pursued under MDG7. In addition, damaged ecosystems fail the communities they formerly sustained, while weakened drainage, drinking water, and sanitation infrastructure compound environmental damage and attendant human impacts (Schipper and Pelling, 2006). Finally, natural disasters undermine MDG8, which calls for a global partnership for development, by drawing funds away from traditional development assistance and destroying hard won gains achieved through past international cooperation and investment. In light of the inextricable links between natural disasters and poverty, and, in particular, the threat they pose to the MDGs, it is crucial that natural disasters be taken into consideration in national and international development planning. Disaster-risk reduction must be integrated into development, so as to minimize the diversion of resources from development to disaster response, and the loss of hard-won development gains.

The International Disaster Reduction Regime

As noted, human and economic impacts from natural disasters have been increasing since the second half of the twentieth century. In 1989, responding to a significant increase in disasters in the 1980s, the UN General Assembly (GA) declared the 1990s the International Decade for Natural Disaster Reduction (IDNDR), with the objective of addressing disaster reduction across the wide range of natural hazards (UN, 1989). In the resolution convening the IDNDR, the GA emphasized the rising human and economic harms caused by disasters, but also the growing scientific and technical knowledge regarding their mitigation, stating that "the international community as a whole has now improved its capacity to confront this problem and that fatalism about natural disasters is no longer justified" (UN, 1989, para. 5).

The objective of the IDNDR, as set forth in the Framework of Action annexed to the resolution proclaiming the Decade, was to reduce loss of life and property, as well as social and economic disruption, caused by natural disasters. Underlying this objective, the IDNDR had the following goals: improving national capacity to quickly and effectively mitigate natural disaster impacts; developing guidelines and strategies for applying current science and technology with cultural and economic sensitivity; fostering research to close existing knowledge gaps; disseminating current and new assessment, prediction and mitigation information; and developing and monitoring measures for assessment, prediction, prevention, and mitigation, through locally tailored technical assistance and technology transfers, demonstration projects, and education and training. To these ends, the Framework called on member states to establish national mitigation programs and integrate disaster prevention policies into their national development programs, as well as to launch national committees to cooperate with the scientific and technological communities in pursuit of the Decade's objectives (UN, 1989).

The Framework also requested that the Secretary-General establish, *inter alia*, a scientific and technical committee of experts to identify priorities and gaps in knowledge and assess and make recommendations for the Decade's programs (UN, 1989). The Scientific and Technical Committee adopted three targets in pursuit of the Decade's goals and objectives, which the GA swiftly endorsed (UN, 1991; 1999a). Thereunder, the IDNDR was to ensure that all countries would have comprehensive national risk assessments; local and national mitigation plans of practical measures for prevention, preparedness and community awareness; and ready access to early warning systems in place, as part of their sustainable development plans, by 2000 (UN, 1999a). The wide range of work areas identified to achieve these targets included:

- Comprehensive research activities;

- Knowledge and technology transfer and expanded access to relevant data;

- Structural measures based on risk assessment and mapping to strengthen human settlements and public infrastructure;

- Public information campaigns and formal education and training programs;

- Adoption of local and national administrative legislation on prevention, preparedness, and mitigation;

- Integration of disaster prevention into national planning, including risk management and relief capacities;

- Inclusion of hazard awareness, vulnerability analysis and risk assessment in land-use planning; and

- Decentralization of resources and responsibilities to empower local communities with greater self-reliance and resilience. (UN, 1999a)

To measure the progress made toward meeting the Decade's goals and targets, the UN convened the 1994 World Conference on Natural Disaster Reduction (WCNDR), in Yokohama, Japan, for a midterm review. Most significantly, the conference marked a significant shift in strategy from science and technology to social sciences and economics, and from emergency preparedness to the reduction of vulnerability and risks (UN, 1999c). In addition, the WCNDR generated discussion of the activities to be pursued for the remainder of the decade and beyond, facilitated further exchange of information, and raised international awareness of the importance of the issues under consideration (UN, 1994b).

The WCNDR adopted the Yokohama Strategy for a Safer World and a Plan of Action, two important policy statements that call for a "new spirit of partnership to build a safer world, based on common interest, sovereign equality and shared responsibility to save human lives, protect human and natural resources, the ecosystem and cultural heritage" (UN, 1994b, chap. 1, annex 1, para. 9). The Yokohama Strategy establishes guidelines for incorporating disaster prevention, mitigation, preparation, and relief into development plans and ensuring adequate implementation mechanisms at the community, national, regional and international levels (UN, 1994b). Based on the principles of risk assessment; disaster prevention and preparedness, reduction and mitigation; early warning; public participation; technical, financial, and scientific cooperation; and environmental protection, the Yokohama Strategy presents an increased emphasis on human action and signifies an important shift from response after the fact to reduction through prior action:

> Disaster prevention, mitigation, and preparedness are better than disaster response in achieving the goals of the Decade. Disaster response alone is not sufficient, as it yields only temporary results at very high cost. We have followed this limited approach for too long. . . . Prevention contributes to lasting improvement in safety and is essential to integrated disaster management. (UN, 1994b, chap. 1, annex 2, para. 4)

The Strategy aims to promote and strengthen national capacities and capabilities, as well as subregional, regional, and international cooperation, with special attention being paid to the least developed and small island states (UN, 1994b). The GA endorsed both the Strategy and the Action Plan in December of that year, noting the primary responsibility of each country to protect its people, infrastructure and other national assets, as well as the Strategy's call for the promotion and strengthening of cooperation within and across regions (UN, 1994a). In addition, the GA recommended that donor countries accord greater priority to natural disasters in their aid programs and budgets, and called on all countries to integrate disaster reduction in their development planning and pursue regional cooperation in accordance with the conference recommendations (UN, 1994a).

Reporting on the achievements of the IDNDR program at the close of the decade, the Secretary-General found that it had accomplished a great deal. The Decade had succeeded in establishing synergies between its disaster reduction strategy and several other UN agencies and activities; advancing progress in early warning capacities; integrating its preventative approach through UN reform; and matching advances at the international level with widespread regional and national achievements. The Secretary-General's report also highlighted several areas in which further efforts were required, including innovative approaches to reduce the vulnerability of the poor and the ecosystems on which they depend; improve land-planning and urban infrastructure resilience; emphasize and support community-based actions; increase public awareness, education and training, the use of new communications technology, and the accuracy and speed of hazard warnings; and promote public private partnerships and the further integration of risk management in development and environmental planning (UN, 1999a).

The international community welcomed the Secretary-General's report, as well as his recommendations for successor arrangements to carry on the Decade's activities (UN, 2000a). The Secretary-General had proposed an interagency framework and mechanism with a dedicated interagency task force and secretariat to serve as the focal point for natural-disaster reduction within the UN system, and both the UN Economic and Social Council (ECOSOC) and the GA, keen to build upon the vital links the IDNDR had forged between the political, scientific, and technological communities, agreed that this was the best approach for continuing efforts under the Yokohama Strategy and Plan of Action (UN, 1999b; 2000a). The resulting International Strategy for Disaster Reduction (ISDR), launched in 2000, now serves as the framework for the disaster reduction activities of the United Nations system.

The two main objectives of the Strategy are to enable communities to become resilient to the effects of disasters, thereby reducing the compound threat to modern social and economic vulnerabilities; and to evolve "from protection against hazards to the management of risk, by integrating risk prevention strategies into sustainable development activities" (UN, 1999b). These objectives, in turn, inform the four main goals around which the Strategy is structured: increasing public awareness of the risks posed by disasters; securing commitments by public authorities to reduce those risks; engaging public participation at all levels of implementation to create disaster-resistant communities; and reducing economic and social losses from disasters (UN, 1999b).

Four years after the launch of the ISDR, the GA convened a conference to measure the progress made in fulfilling these goals and objectives, as well as to adopt a new plan of action for the new decade and century (UN, 2004). In addition to emphasizing disaster risk reduction as an important element for sustainable development and reiterating the disparate impact of disasters on developing nations, the resolution convening the 2005 World Conference on Disaster Reduction (WCDR) specified five objectives for the event: to review the Yokohama Strategy and Action Plan, to update the guiding framework; to identify specific activities aimed at ensuring implementation of existing commitments; to share good practices and lessons learned, and identify remaining gaps and challenges; to increase awareness and, thereby, implementation of disaster reduction policies; and to increase the reliability and availability of public and agency access to information (UN, 2004a).

The first objective, the review of the Yokohama Strategy and Action Plan, was carried out by the ISDR Secretariat, by reference to the accounts and documentation collected by the ISDR and its predecessor, the IDNDR. The review was further supplemented by inputs from states, organizations, and individual experts engaged in disaster and risk management, sustainable development and poverty eradication. Although it found evidence of a growing awareness of the interrelations between poverty, sustainable environmental practices, natural-resource management, and disaster risk, the Secretariat concluded that this understanding had yet to be fully implemented in practice. Specifically, the review identified five remaining challenges:

- Governance: organizational, legal, and policy frameworks;

- Risk identification, assessment, monitoring, and early warning;

- Knowledge management and education;

- Underlying risk factor reduction; and

- Preparedness for effective response and recovery. (UN, 2004b)

Guided by these findings, as well as the enduring Principles of the Yokohama Strategy, the international community adopted a renewed policy framework for disaster reduction.

The WCDR produced two major resolutions: the Hyogo Declaration and the Hyogo Framework for Action 2005–2015: Building the Resilience of Nations and Communities to Disasters (UN, 2005). While both documents are soft-law instruments, lacking any legally binding force, they are powerful in their aspirations and leadership. The WCDR message emphasizes the relationship between disaster reduction, sustainable development, and poverty alleviation:

> We are convinced that disasters seriously undermine the results of development investments in a very short time, and therefore, remain a major impediment to sustainable development and poverty eradication. We are also cognizant that developments that fail to appropriately consider disaster risks could increase vulnerability. Coping with and reducing disasters so as to enable and strengthen nations' sustainable development is, therefore, one of the most critical challenges facing the international community. (UN, 2005: 3)

The approach advocated is also highly human-centered, focusing on the generation of a culture of prevention and resilience that builds from the individual level on up to the international, through education, risk assessment, early warning systems, and other proactive approaches and activities that defy powerlessness and reduce vulnerability. Maintaining the primary responsibility of states to protect their people and property, the Declaration also stresses the urgent need to enhance the capacity of the more disaster-prone and least developing countries to do so, calling for international cooperation, including financial and technical assistance. The Declaration proclaims the HFA as a guiding framework for disaster reduction over the following decade, and calls on all stakeholders to translate this shared vision into concrete action (UN, 2005).

The negotiation of the HFA required numerous formal and informal meetings and several late nights. The first issue of contention was the definition and scope of the "hazards" the Framework would cover: delegates debated whether "hazards" should include natural hazards only, or those induced by human activities as well, such as environmental degradation and technological hazards (IISD, 2005). The final agreement refers to both. Debate also ensued over the inclusion of references to climate change, which the United States wished to remove, provoking strong opposition from the European Union among many others (ibid.). Agreement to retain several references was reached late on the eve before the final plenary (ibid.). A European Union proposal on developing targets and indicators, of which many countries were in favor, was defeated by resistance from the United States (ibid.). Instead, the ISDR and Inter-Agency Task Force were directed to develop indicators to assist states in measuring their progress toward successful implementation of the HFA, while the task of developing targets was maintained at the national level. Reference to the principle of common but differentiated responsibility, championed by Brazil in particular, was ultimately replaced by less specific language referring to the importance of international cooperation and partnerships, at the insistence of the United States, European Union, and Japan (ibid.). Mention of a new funding mechanism for assisting developing country development of national disaster risk reduction strategies and action plans was, likewise, discarded in favor of a specification that resource mobilization would be conditioned on donor financial capacity (ibid.).

The Hyogo Framework, as ultimately adopted, resolves to pursue "the substantial reduction of disaster losses, in lives and in the social, economic and environmental assets of communities and countries" by 2015 (UN, 2005: 8). To that end, the HFA outlines the following three strategic goals:

- The more effective integration of disaster risk into sustainable development policies, planning and programming at all levels, with a special emphasis on disaster prevention, mitigation, preparedness and vulnerability reduction;

- The development and strengthening of institutions, mechanisms and capacities at all levels, in particular at the community level, that can contribute to building hazard resilience; and

- The systematic incorporation of risk reduction approaches into emergency preparedness, response and recovery programs in the reconstruction of affected communities. (UN, 2005: 9)

In addition, the HFA outlines five priorities for action: ensuring that disaster-risk reduction is a national and local priority with a strong institutional basis for implementation; identifying, assessing, and monitoring disaster risks and enhancing early warning; using knowledge, innovation, and education to build a culture of safety and resilience at all levels; reducing underlying risk factors; and strengthening disaster preparedness for effective response at all levels. Environmental, social, and development practices figure prominently in the key activities to be undertaken under the fourth priority, reducing risk, which include environmental and natural resource management; promotion of food security and health-sector and public-infrastructure resilience; strengthening social safety nets; income diversification for vulnerable populations; financial risk-sharing; public-private partnerships; alternative and innovative financing; and land-use planning and other technical measures, such as risk assessments and revision of building standards and practices (UN, 2005).

In its implementation and follow-up provisions, the HFA calls for a multisectoral approach, in which states and regional and international organizations integrate disaster risk reduction into all levels of their sustainable development policies, planning, and programming. The Framework also encourages voluntary partnerships for its implementation and emphasizes the need for support for the least developed countries, and for Africa in particular, as a matter of priority, including financial and technical assistance. All states are called upon to designate a national coordination mechanism, prepare and publish national baseline assessments and program summaries, review national progress, and take part in the international regimes for disaster reduction, sustainable development, and climate change. Similar tasks are requested of regional and international bodies; however, while the ISDR is assigned the development of indicators for assessing progress in the implementation of the HFA, the document does not contain any clear donor commitments or timebound substantive targets. Attempts to introduce clear-cut obligations were soundly defeated (UN, 2005).

Subsection F, on resource mobilization, lists several important state actions, including provision and support for implementation in disaster-prone developing countries, such as financial and technical assistance, debt sustainability, mutually agreeable technology transfers, public-private partnerships, and both North-South and South-South cooperation; the mainstreaming of risk reduction into development assistance; adequate contributions to the UN Trust Fund for Disaster Reduction; and risk spreading and insurance partnerships (UN, 2005). The operative language is, however, weak, limiting state action to "the bounds of their financial capabilities" and using only the aspirational "should" (UN, 2005). The nonbinding character of the framework and its noncommittal language confer a great deal of flexibility on the state parties, allowing for the selective implementation of its technical and organizational recommendations according to their needs and capacities. At the same time, however, the lack of any legally binding commitments exposes its objectives to frustration and failure. Limited by its legally nonbinding status and reserved language, the HFA, arguably, cannot measure up to the formidable threats it is intended to address.

Hyogo Outcomes: Accomplishments and Remaining Challenges

In the absence of any firm targets or commitments, the test of the conference decisions will lie in the extent to which states adhere to their vision and follow-through with concrete action. For its part, the

ISDR system has diligently supported national efforts; facilitated regional and international coordination; stimulated the sharing of good practices; documented and assessed the progress made; and produced practical tools to assist in the promotion and implementation of disaster risk reduction measures in various countries and regions. Among the most important tools for decision makers is its publication, *Words into action: A guide for implementing the Hyogo Framework* (2007).

Developed through extensive consultation with key actors in disaster risk reduction, including partner agencies and experts, national platforms and regional agencies, the guide distills a wealth of risk management and reduction experience from around the world. Because the HFA places primary responsibility for disaster risk reduction on states, the guide is geared toward government actors and toward assisting states in measuring their progress and identifying opportunities for improvement. The guide comprises five chapters of specific recommended tasks, one for each of the HFA's five priorities for action. Together they contain a set of twenty-two suggested tasks for addressing key implementation efforts, accompanied by illustrative examples, practical instructions, and supplemental references (UN-ISDR, 2007).

In addition, the ISDR system has been strengthened to expand its representation and outreach. In 2006, the UN Under-Secretary-General for Humanitarian Affairs, in his capacity as chairperson of the ISDR, launched a consultative process to consider ways to strengthen the ISDR system and better support states in their efforts to implement the HFA. Among the outcome proposals was a plan to create a multi-stakeholder Global Platform for Disaster Risk Reduction to serve as the principal global forum of the ISDR system for advocacy and awareness raising, experience sharing, progress reporting, and for identifying gaps and challenges for the ISDR system (Global Platform for Disaster Risk Reduction, n.d.).

The Global Platform met in Geneva for the first time in June 2007. In preparation, ISDR partners across the world contributed to the drafting of a global progress report, entitled *Disaster risk reduction: 2007 global review* (UN, 2007). Under the HFA, the ISDR is to prepare periodic reviews of progress made and challenges identified, and the 2007 review also previewed several of the elements that will be included in these biennial reports, the first of which is expected in 2009. The review draws on implementation reports from seventy member states, as well as on regional reviews conducted by the ISDR Secretariat in cooperation with the World Bank and partners from all over the world. The authors conclude that international priorities will need to be radically realigned to address the current challenges faced in implementing the HFA and delivering on its promises of enhanced safety and security (UN, 2007).

The review distinguishes between two classes of risk: intensive risk, in which concentrated populations and economic activities suffer catastrophic disasters from large-scale hazards; and extensive risk, in which dispersed populations and activities suffer localized, cumulative impacts from small-scale, primarily climactic hazards (UN, 2007). In this context, it focuses primarily on earthquakes and climate change, with regard to both mortality and economic losses (ibid.). The review finds that those risk reduction fields in which the most progress has been made—namely, early warning, preparedness, and response—have little impact on intensive earthquake risk in low and middle income countries undergoing rapid urban growth (ibid.). Unless current trends change, a substantial and growing proportion of disaster deaths will be caused by large-scale earthquake catastrophes (ibid.). Similarly, while early warning, preparedness, and response can lower mortality in areas of intensive climate risk, they do little to prevent economic losses in these zones, which are, therefore, predicted to rise in both raw and relative terms (ibid.).

Extensive risk is also increasing rapidly and presents a formidable challenge for the rural and marginalized-urban poor, and for the international goals and efforts, such as the MDGs, aimed at promoting their sustainable livelihoods and development. Imposed over the existing burdens of rapid urbanization, social inequality, and environmental degradation, global climate change will only intensify these hardships. The solution envisioned by the reviewers calls for the current focus on saving lives to be matched by a commitment to development efforts to improve them:

The current emphasis on saving lives needs to be complemented with a vision of protecting and strengthening livelihoods and human development. Addressing localized, recurrent exposure to risk is the most effective way of protecting the livelihoods of vulnerable communities and realistically meeting the MDGs. . . . Alongside efforts to strengthen preparedness and response, local disaster risk reduction strategies should also address underlying risk factors through measures such as livelihood diversification and protection, environmental management, safe building and risk sensitive planning. (UN, 2007: 74)

The prevailing emphasis that remains fixated on response-centered, hazard-focused approaches, rather than on addressing underlying vulnerabilities, must change (UN, 2007).

Conclusion

Natural disasters, by their very definition, overwhelm the capacity of the affected community. In poor countries, disasters commonly overwhelm the national government as well, often times erasing years of hard-won development gains. With lower coping-capacity thresholds, poor peoples and nations are the most vulnerable; "Disasters seek out the poor and ensure that they stay poor" (International Federation of Red Cross and Red Crescent Societies, 2002: 11). International support for countries struck by disasters, therefore, is well warranted.

Furthermore, the international system is best positioned to gather and facilitate the sharing of experiences from all over the world, thereby, contributing to the development of a global knowledge bank of primary challenges and best practices in disaster disk reduction. The ISDR fulfills this role with competence and compassion. What is needed is the political will to undertake strong commitments to the requirements it identifies, and the financial resources to fund their achievement.

The Hyogo Framework for Action 2005–2015 and the ISDR guidance for its implementation identify the tools for achieving its objective of substantial reductions in lives and economic assets lost. Its greatest contribution, however, is not instruction on early warning, telecommunication, and information systems; rather, it is the recognition that, so long as people remain powerless in slums, degraded ecosystems, and poverty, technology will never be enough. The test of the Framework, and the international consensus and commitment from which it came, will lie in the development of a culture of prevention that extends to human livelihoods and sustainable development.

References

Braine, T. (2006). Was 2005 the year of natural disasters? *Bulletin of the World Health Organization* 84 (1): 4–8.

Bureau for Crisis Prevention and Recovery, United Nations Development Programme (2004). Reducing disaster risk: A challenge for development. New York: United Nations Development Programme. Retrieved August 1, 2007, from www.undp.org/cpr/whats_new/rdr_english.pdf.

de Ville de Goyet, C., and Griekspoor, A. (2007). Natural disasters, the best friend of poverty. *Georgetown Journal on Poverty Law and Policy* 14: 61–94.

Global Platform for Disaster Risk Reduction (n.d.). About the global platform. Retrieved August 22, 2007, from www.preventionweb.net/globalplatform/gp-abouth.html.

International Federation of Red Cross and Red Crescent Societies (2002). *World disasters report*. London: Earthspan.

International Institute for Sustainable Development (2005). The world conference on disaster reduction: 18–22 January 2005. *Earth Negotiations Bulletin* 26 (4): 1–2.

National Oceanic and Atmospheric Administration (2006a). NOAA reviews record-setting 2005 Atlantic hurricane season, active hurricane era likely to continue. Retrieved August 4, 2007, from www.noaanews.noaa.gov/stories2005/s2540.htm.

National Oceanic and Atmospheric Administration (2006b). Noteworthy records of the 2005 Atlantic hurricane season. Retrieved August 4, 2007, from www.noaanews.noaa.gov/stories2005/s2540b.htm.

Schipper, L., and Pelling, M. (2006). Disaster risk, climate change and international development: Scope for, and challenges to, integration. *Disasters* 30, no. 1: 19–38.

United Kingdom Department for International Development (2004). *Disaster risk reduction: A development concern, a scoping study on links between disaster risk reduction, poverty and development.* Norwich, England: Overseas Development Group.

United Nations (1989). International Decade for Natural Disaster Reduction (A/RES/44/236). Retrieved August 1, 2007, from http://daccessdds.un.org/doc/resolution/gen/nro/549/95/img/nro54995.pdf?OpenElement.

United Nations (1991). International Decade for Natural Disaster Reduction (A/RES/46/149). Retrieved August 1, 2007, from http://daccessdds.un.org/doc/resolution/gen/nro/582/37/img/nro58237.pdf?OpenElement.

United Nations (1994a). International Decade for Natural Disaster Reduction (A/RES/49/22). Retrieved August 1, 2007, from http://daccessdds.un.org/doc/undoc/gen/n94/601/17/pdf/n9460117.pdf?OpenElement.

United Nations (1994b). Report of the World Conference on Natural Disaster Reduction (A/CONF.172/9). Retrieved August 1, 2007, from http://documents-dds-ny.un.org/doc/undoc/gen/n94/376/04/pdf/n9437604.pdf?OpenElement.

United Nations (1999a). Activities of the International Decade for Natural Disaster Reduction, Report of the Secretary-General (A/54/132–E/1999/80). Retrieved August 1, 2007, from www.un.org/ga/search/view _ doc.asp?symbol = A%2F54%2F132&Submit = Search &Lang = E.

United Nations (1999b). International decade for natural disaster deduction: Successor arrangements, report of the Secretary-General (A/54/497). Retrieved August 1, 2007, from www.un.org/ga/search/view_doc.asp?symbol = A%2F54%2F497&Lang = E.

United Nations (1999c). Recommendations on institutional arrangements for disaster reduction activities of the United Nations system after the conclusion of the International Decade for Natural Disaster Reduction (A/54/136–E/1999/89). Retrieved August 1, 2007, from http://documents-dds-ny.un.org/doc/undoc/gen/n99/181/37/pdf/n9918137.pdf?OpenElement.

United Nations (2000a). International decade for natural disaster reduction: Successor arrangements (A/RES/54/219). Retrieved August 1, 2007, from http://daccessdds.un.org/doc/undoc/gen/n00/271/75/pdf/n0027175.pdf?OpenElement.

United Nations (2000b). List of participants at General Assembly's Millennium Summit, 6–8 September (GA/9744). Retrieved August 1, 2007, from www.un.org/millennium/participants.htm.

United Nations (2002). The millennium development goals and the United Nations role. Retrieved August 1, 2007, from www.un.org/millenniumgoals/mdgs-factsheet1.pdf.

United Nations (2004a). International strategy for disaster reduction (A/RES/58/214). Retrieved August 9, 2007, from http://daccessdds.un.org/doc/undoc/gen/n03/507/36/pdf/n0350736.pdf?OpenElement.

United Nations (2004b). Review of the Yokohama Strategy and Plan of Action for a Safer World: Note by the Secretariat (A/CONF.206/L.1). Retrieved August 9, 2007, from www.unisdr.org/wcdr/intergover/official-doc/L-docs/Yokohama-Strategy-English.pdf.

United Nations (2005). Report of the World Conference on Disaster Reduction (A/CONF.206/6). Retrieved August 10, 2007, from http://documents-dds-ny.un.org/doc/undoc/gen/g05/610/29/pdf/G0561029.pdf?OpenElement.

United Nations (2007). *Disaster risk reduction: 2007 global review*. Retrieved August 22, 2007, from www.preventionweb.net/english/documents/global-review-2007/Global-Review-2007.pdf.

United Nations Inter-Agency Task Force on Disaster Reduction (2001). *Framework for action for the implementation of the international strategy for disaster reduction (ISDR)*. Geneva: United Nations Inter-Agency Task Force on Disaster Reduction.

United Nations International Strategy for Disaster Reduction (n.d.a). Disaster occurrence, trends-century. Retrieved August 1, 2007, from www.unisdr.org/disaster-statistics/occurrence-trends-century.htm.

United Nations International Strategy for Disaster Reduction (n.d.b). Disaster occurrence, type of disasters 1991–2005. Retrieved August 1, 2007, from www.unisdr.org/disaster-statistics/occurrence-type-disas.htm.

United Nations International Strategy for Disaster Reduction (n.d.c). Disasters impact, affected. Retrieved August 1, 2007, from www.unisdr.org/disaster-statistics/impact-affected.htm.

United Nations International Strategy for Disaster Reduction (n.d.d). Disasters impact, economic damage. Retrieved August 1, 2007, from www.unisdr.org/disaster-statistics/impact-economic.htm.

United Nations International Strategy for Disaster Reduction (n.d.e). Disasters impact, killed. Retrieved August 1, 2007, from www.unisdr.org/disaster-statistics/impact-killed.htm.

United Nations International Strategy for Disaster Reduction (n.d.f). Review of 8 MDGs' relevance for disaster risk reduction and vice-versa. Retrieved August 1, 2007, from www.unisdr.org/eng/mdgs-drr/review-8mdgs.htm.

United Nations International Strategy for Disaster Reduction (n.d.g). Top 50 Countries. Retrieved August 1, 2007, from www.unisdr.org/disaster-statistics/top50.htm.

United Nations International Strategy for Disaster Reduction (2004). *Living with risk: A global review of disaster reduction initiatives.* Geneva: United Nations. Retrieved August 1, 2007, from www.unisdr.org/eng/about_isdr/bd-lwr-2004-eng.htm.

United Nations International Strategy for Disaster Reduction (2006a). Disaster Statistics 1991–2005, Introduction. Retrieved August 1, 2007, from www.unisdr.org/disaster-statistics/introduction.htm.

United Nations International Strategy for Disaster Reduction (2006b). NGOs and disaster risk reduction: A preliminary review of initiatives and progress made, background paper for a consultative meeting on a "global network of NGOs for community resilience to disasters." Retrieved August 1, 2007, from www.unisdr.org/eng/partner-netw/ngos/meeting1-october-2006/NGOs_and_DRR_Background_Paper.pdf.

United Nations International Strategy for Disaster Reduction (2007). *Words into action: A guide for implementing the Hyogo Framework.* Geneva: United Nations. Retrieved August 22, 2007, from www.unisdr.org/eng/about_isdr/isdr-publications/02-words-into-action/words-into-action.pdf.

United Nations Millennium Declaration (2000). Retrieved August 1, 2007, from www.un.org/millennium/declaration/ares552e.pdf.

United Nations Road Map Towards the Implementation of the United Nations Millennium Declaration (2001). Retrieved August 1, 2007, from www.un.org/documents/ga/docs/56/a56326.pdf.

United Nations World Conference on Disaster Reduction Secretariat (2004). *Review of the Yokohama strategy and plan of action for a safer world.* Geneva: United Nations World Conference on Disaster Reduction Secretariat.

United States Geological Survey (2007). Most destructive known earthquakes on record in the world. Retrieved July 27, 2007, from http://earthquake.usgs.gov/regional/world/most_destructive.php.

World Bank (2005, January 17). Tsunami provides an opportunity to rethink disaster management. *Press Release.* Retrieved August 1, 2007, http://web.worldbank.org/wbsite/external/countries/africaext/extafrsumaftps/0,,contentMDK:20321756~menuPK:2050032~pagePK:51246584~piPK:51241019~theSitePK:2049987,00.html.

7 | Unanticipated Consequences of the Tsunami Response

ANNIE MAXWELL

On December 26, 2004, a 9.0 earthquake centered off the coast of northern Sumatra, Indonesia, shook the floor of the Indian Ocean. This massive underwater earthquake triggered a tsunami that killed more than 200,000 people, making it one of the deadliest natural disasters in modern history (Secretary-General of the United Nations, 2005). The tsunami reshaped the physical characteristics of five Southeast Asian countries and permanently altered the way the world looks at the potential impact of natural hazards (ibid.).

The tsunami affected a total of twelve countries and 5,000 miles of coastline, causing more than ten billion dollars in damage (Clinton, 2005). Major infrastructure was severely affected, with more than 3,000 miles of roads and 118,000 fishing boats damaged or destroyed (ibid.). Some 400,000 houses were also seriously damaged or destroyed, displacing entire communities, and the economic well-being of many was threatened as 1.4 million people lost their source of livelihood (ibid.).

The tsunami, traveling at approximately 500 mph, carried a brute force with the potential to cause irreparable damage (Clinton, 2005). This force was intensified due to the lack of disaster prevention and

Annie Maxwell is the Chief Operating Officer of Direct Relief International. From August 2005 to October 2006, she was seconded to the United Nation's Office of the Special Envoy for Tsunami Recovery, as Partnerships and Outreach Officer.

reduction strategies, as well as an increase of human settlements in coastal belts, vulnerability of poor communities, substandard housing and construction standards, lack of a regional early warning system, and environmental degradation.

Immediately following the tsunami, there were discussions about how many lives could have been saved if there had been a regional early warning system. The discussion was not unwarranted: the tsunami did not hit Sri Lanka until almost two hours after its origination, and yet over 35,300 people died there (Steering Committee, 2006). An early warning system presumably would have saved many of those lives.

This chapter will focus on the recovery efforts made in response to the 2004 Indian Ocean tsunami and examine three key reasons why environmental concerns were not appropriately translated into efforts on the ground despite broad policy support for environmental issues. Examples will be taken from the five major tsunami-affected countries, with a focus on the response in Indonesia and Sri Lanka.

Unprecedented Response

The 2004 Indian Ocean tsunami shook the world in more ways than through its physical force. The tsunami set records in terms of damage as well as levels of funding for reconstruction. Unlike most humanitarian crises, the response to the tsunami was incredibly well financed. The tsunami triggered an outpouring of support from governments, private companies, development banks, and private citizens. The United Nations launched a flash appeal for USD 1.1 billion for emergency relief activities, 85 percent of which was pledged within four weeks (Clinton, 2005). Pledges from both official and private sources totaled USD 13.6 billion (ibid.).

With the disparate group of actors, enormous amounts of funds, and broad range of activities, new mechanisms were essential in order to manage the effort. In each of the affected countries, the national government established a centralized coordinating entity. In Sri

Lanka, the government established the Presidential Task Force for Rebuilding the Nation (TAFREN), the Presidential Task Force for Logistics, Law, and Order (TAFLOL), and the Presidential Task Force for Rescue and Relief (TAFRER), all of which were replaced ten months later after presidential elections by a single Authority for Reconstruction and Development (RADA) (Bennett et al., 2006). The government of Indonesia approved the establishment of the Aceh and Nias Rehabilitation and Reconstruction Agency (BRR) in September 2005, taking over for the central government's Coordinating Minister for People's Welfare, which had been brought from the capital city of Jakarta in Java north to Aceh in Sumatra (Bennett et al., 2006).

The UN established a distinctive coordinating mechanism with the Secretary General's Special Envoy for Tsunami Recovery. In February 2005, then-Secretary General Kofi Annan, at the request of the General Assembly, announced the appointment of former U.S. President Bill Clinton to this position. The appointment of President Clinton generated significant public attention, since he is arguably the highest-profile Special Envoy ever mandated by the Secretary-General.

As Special Envoy, President Clinton focused on four themes that defined the key policy areas for the recovery. First, he sought to "support coordination efforts at the country and global levels, to ensure that all the actors governmental and non governmental, public and private, local and international retain[ed] the spirit of teamwork that [had] characterized the operations . . . and that resources [were] used to maximum effect" (OSE, 2006).

Second, he sought to "keep the world's attention on the tsunami operations in order to avoid the short-term attention-span that has characterized so many previous international efforts of this kind, as well as ensure that promises-made [did] not become promises-forgotten" (OSE, 2006). Under this theme, he became involved in several key recovery priorities, intervening "where the regular systemic response had not proven successful"[1] (ibid).

---

1. For more information, please see "Key Initiatives by the Special Envoy and His Office." This document is available on the website of the Office of the Special Envoy, www.tsunamispecialenvoy.org.

Third, the Special Envoy sought to "promote transparency and accountability measures" that ensured resources were used effectively and helped "retain the engagement of the millions of 'investors'" (OSE, 2006).

Finally, he sought to act as a spokesperson for "a new kind of recovery, one that not only restores what existed previously, but goes beyond, seizing the moral, political, managerial and financial opportunities the crisis has offered governments to set these communities on a better and safer development path"—to champion a recovery that "builds back better" (OSE, 2006).

Relief, Recovery, and the Environment

The tsunami response posed immense challenges for those in the affected countries and the members of the international community who responded. More than three years later, there are many lessons to be learned for future humanitarian response efforts. The Tsunami Evaluation Coalition (TEC), a collaborative of representatives from donor countries, UN agencies, the Red Cross, and nongovernmental organizations, published five excellent thematic joint evaluations and a *Synthesis Report* that explore many of these issues.[2] Among the most interesting and important is the failure of many actors to integrate environmental considerations successfully into the recovery, and the reasons these critical issues were ignored.

While some have argued that environmental issues were not incorporated because the environment is not valued as important or as a significant issue in the world of international development, the evidence suggests the contrary. Although environmental conservation and protection were formerly viewed in opposition to international development, that tension has lessened over the past thirty years (Galizzi, 2006). Indeed, the environment has come to be recognized as

---

2. This chapter draws from the Tsunami Evaluation Coalition's reports "Coordination of International Humanitarian Assistance in Tsunami-Affected Countries" and "Impact of the Tsunami Response on Local and National Capacities."

central to development, as is clearly evidenced by the modern sustainable development approach (Galizzi, 2006). The tsunami revealed in dramatic fashion the powerful role that the condition of the environment plays in determining the consequences of natural hazards, and the correlation was not lost on those struggling to respond and rebuild.

Nature as Natural Protectorate

During the tsunami, the environment proved that it could serve as a natural protectorate. Communities where mangroves, coral reefs, and natural dunes were cared for were spared the most devastating ravages of the tsunami. In Sri Lanka, researchers found that intact coral reefs played an important role. In one rather dramatic example in the town of Peraliya, where the coral reef had been destroyed, water came 1.5 km inland, smashing a passenger train and killing 1,700 people (Liu et al., 2005). The old, crushed rail car became one of the more tragic iconic images of the tsunami. Down the coast in Hikkaduwa, where the coral reef had been protected, the tsunami hit the shore minimally, coming in only 50 meters, and it killed no one (ibid.).

While there are dramatic examples of coral reefs playing a role in protecting coastal communities, mangroves are the more prominent and widespread example of the benefits of holistic environmental management. Satellite images from before and after the tsunami offer visual evidence of the protective power of these trees. Those communities with protected mangrove forests along the coastline fared far better (UNEP, 2005). In India, researchers found that areas in the Cuddalore district with coastal vegetation were less damaged than others that had been environmentally degraded (Danielsen et al., 2005). The tsunami demonstrated an obvious physical connection between environmental protection and disaster risk reduction.

Government and Agency Response:
Linking the Environment to Risk Reduction and Recovery

With the existing and ever-expanding emphasis on sustainable development and the tsunami's forceful highlighting of the environment's role in disaster risk reduction, it is not surprising that nearly all governments and relief and recovery agencies declared support for environmental protection as part of the tsunami recovery efforts. They embraced the principle of sustainable development to promote both disaster risk reduction and economic growth, cognizant that environment and recovery are not in conflict but inextricably linked. In fact, governments and relief agencies acknowledged that the tsunami had created an unparalleled opportunity to get it right in terms of community development and the environment.

One of the most explicit examples of this kind of endorsement can be found in the Tsunami Green Reconstruction Policy Guidelines.[3] Prepared by the World Wildlife Fund (WWF), this document outlines policy and implementation plans for how those involved in the reconstruction of post-tsunami Aceh can mitigate the environmental impact. The Acting Governor of the province of Nanggroe Aceh Darussalam wrote the preface for this explicitly pro-environmental document. The Governor gave a strong personal endorsement of the document and its principles, writing, "No matter how urgent the need is to use our land for various purposes, ecosystem health and nature conservation should be our main concern, especially in the rehabilitation and reconstruction of Aceh after the tsunami" (WWF, 2005: 2).

The major actors in the reconstruction process voiced strong support for integration of environmental issues on a policy level, but several significant hurdles still remained to incorporate them fully into the recovery process. As the reconstruction process moved forward in earnest, governments and agencies faced at least three significant obstacles to incorporating environmental concerns into reconstruction: urgency, coordination, and local capacity.

Urgency

Many parties in the recovery faced such urgent demands to rebuild destroyed communities that they had little time to plan for some of

---

3. As defined by WWF, "green reconstruction aims to improve the quality of life for communities and affected individuals while minimizing the negative impacts of

the longer-term implications of their work, particularly with regards to environmental impact. Even those tasked with assessing and advocating for environmental issues recognized the challenge. The Office for the Coordination of Humanitarian Affairs (OCHA) and UNEP stated in their environmental assessment for Aceh, "that in the immediate term, efforts must remain concentrated on acute impacts and relieving more urgent needs" (Joint UNEP/OCHA Environment Unit, 2005: 11).

The best example of the crisis overwhelming the need for incorporating long-term environmental concerns was the enormous pressure to rebuild homes in Indonesia and the corresponding increase in demand for timber. Aceh, the Indonesian province on the island of Sumatra closest to the epicenter of the quake, was hardest hit by the tsunami. More than 130,000 people lost their lives in Indonesia, and those left had little to call home (Aceh and Nias Rehabilitation and Reconstruction Agency [BRR], 2006). In many areas, entire villages were completely flattened. More than 120,000 houses needed to be completely rebuilt and another nearly 85,000 homes suffered major damage and required serious rehabilitation (BRR, 2006).

Communities were in desperate need to return to normalcy and begin their lives anew. The government and NGOs were tasked with helping to rebuild long-term housing and to provide short-term housing for the more than 500,000 who had been left homeless and were in need of transitional shelter (BRR, 2006). The need for houses—both temporary and permanent—generated an enormous demand for timber. Wood was needed for both the framing of temporary and permanent houses and for firing bricks (BRR, 2006). According to early estimates by the Food and Agriculture Organization of the United Nations (FAO), the amount of wood required for reconstruction, plus annual consumption of wood for other purposes in Aceh, far exceeded the total annual legal cut (Kuru, 2005).[4] With demand for timber far

reconstruction on the environment and maintaining the long-term biological diversity and productivity of natural systems" (WWF, 2005: 7).

4. "The annual consumption of wood products in Aceh is estimated to be 551,620 roundwood equivalent (RWE). The reconstruction programme is estimated to require [sic] and an additional 415,000 m3 of solid wood products (740,000 m3 RWE) will be

outpacing supply, agencies faced a very real dilemma of where and how to source their wood.

The island of Sumatra is home to one of the world's most endangered tropical rain forests. While the island of Sumatra used to be covered in a lush forest, it has been heavily deforested over the last hundred years. Illegal logging is widespread in Sumatra and elsewhere in Indonesia. It is estimated that 40–80 percent of all logs consumed in Indonesia have been harvested illegally (Kuru, 2005). Aceh contains one of the last intact forests on Sumatra, mainly because Aceh has been a conflict zone for more than thirty years, and acute security concerns over the last ten years have reduced logging (Kuru, 2005). Preserving Sumatra's remaining forests is important for both people and other residents. While increased deforestation would exacerbate mudslides and flooding, already a problem in Sumatra, further threatening human communities, the old-growth forest is also home to some of the world's most endangered species, such as the Sumatran orangutan and tiger. The choice between the competing needs for timber and conservation was grim.

Finding legally and sustainably managed wood was only the first of many issues for those agencies rebuilding. Timber importation remained difficult, and transport into the affected area was a logistical nightmare due to damaged infrastructure (BRR, 2006; Kuru, 2005). The BRR established a Timber Help Desk (THD) to address these concerns and assist those rebuilding with information related to timber (BRR, 2006). One of THD's most important outreach activities was the Timber Trade Show, which was organized to assist NGOs with procuring wood. Unfortunately, this event was not held until late June 2006, well over a year after the tsunami (BRR, 2006).[5] Time constraints, the expense of imported wood, procurement policies, donor reporting requirements, and other operational requirements

---

required for reconstruction. Unfortunately total annual legal cut is only 50,000 m3 RWE and is insufficient for either domestic demand or reconstruction" (Kuru, 2005: 4).

5. The WWF was also highly active in trying to engage NGOs on the issue of sustainably managed timber (WWF, 2005a).

only further hindered on-the-ground efforts to use sustainably-sourced timber. Confronted with the other, very real, pressures that agencies faced, including issues of land tenure and ownership, damaged land, the rising cost of construction materials, and lack of adequate infrastructure, the best intentions succumbed to the harsh reality (BRR, 2006).

There had been broad support on a policy level to use timber from legally and responsibly managed sources, and endorsements came from a diverse group of actors including the BRR, the Governor of Aceh, NGOs, and the UN.[6] But tension arose operationally between the conflicting priorities of the humanitarian crisis and the environment to the point where the NGOs were forced to make a choice. As India's former Prime Minister Indira Gandhi famously observed, "We do not wish to impoverish the environment any further and yet we cannot forget the grim poverty of large numbers of people. Are not poverty and need the greatest polluters?" (Gandhi, 1972). Ultimately, NGOs were compelled to respond to the urgent humanitarian crisis of thousands of homeless Acehnese and were not able to solely use sustainably managed timber.

Coordination

A second challenge was coordination (Bennett et al., 2006). In Aceh, the relief effort brought thousands of foreign nationals and NGOs to this isolated part of northern Sumatra. The Indonesian government processed more than 7,000 individual immigration requests in 2006 (BRR, 2006). This was a huge inflow of human traffic for any region, but particularly for an area that had previously been closed off to foreigners and had little exposure to international NGOs.

With the onslaught of people came a multitude of different types of actors, including local and international NGOs, public and private donors, and private volunteers, with no real coordination in plans

---

6. For specific examples, see the preface to WWF's *Tsunami Green Reconstruction Policy Guidelines* (2005), and BRR (2006).

concerning the larger economic and reconstruction picture. The challenge of coordinating these entities, in terms of their effects on the environment, was most acutely demonstrated in the management of livelihood programs.

In Sri Lanka, almost 150,000 people lost their livelihoods. The majority of these people, not surprisingly due to their proximity to the ocean, were in the fisheries sector. The damage to the industry was immense. In all, 75 percent of boats were destroyed in the tsunami (Steering Committee, 2006). In Sri Lanka, the lack of coordination and integration of wide-ranging projects was evident in the public sector. The central government and its newly formed coordinating body, TAFREN and then RADA, understandably strained to address environmental concerns since there were so many other urgent and competing activities to coordinate. Reflecting on the relief effort two years after the tsunami, the government of Sri Lanka acknowledged this struggle: "TAFREN as well as RADA were focused on reconstruction of infrastructure, while cross cutting areas such as environment and gender related matters were expected to be led by the relevant line ministry and the expectation of the line ministries were for RADA to take a lead role in these areas. The result is that these two areas have not been adequately addressed" (Steering Committee, 2006: 59).

The situation was similar for NGOs. Hundreds of organizations came to Sri Lanka with a large number trying to assist the local communities through support for the fishing industry. Due to lack of coordination of needs assessments, beneficiary lists, and programmatic objectives, the area was inundated with fishing boats and nets. The result was a plethora of boats but very little coordination of decision making concerning who was to receive them, the type of boats being distributed, and if this was the proper response from an environmental, economic, and social standpoint.

The sudden influx of boats, distributed in a haphazard manner from various sources, caused a noticeable disruption within the fishing industry. There was a noted oversupply of single-day boats that threatened exploitation of fishing resources in the shallower waters and an undersupply of the multiday boats that bring in one-third of the total catch. In addition, many workers, who had previously served on boats owned by other people, became boat owners, causing a shortage in labor (Steering Committee, 2006).

At the end of 2006, the government of Sri Lanka recognized the adverse effects that the lack of coordination was having on the environment: "Appropriate measure should be undertaken immediately in a coordinated manner so that long-term sustainable economic revival and the well-being of affected communities are not jeopardized by depleting natural resources, environmental conditions, and increasing vulnerability to extreme weather conditions" (Steering Committee, 2006: xiv). Unfortunately, the government's warning on the two-year anniversary may have come too late. Experts believe that local fishing stocks have already been severely depleted (World Fish Center, 2005).

Local Capacity

Throughout the aftermath of the tsunami, local institutions lacked the capacity to address long-term environmental issues. This was the third challenge, one closely tied to the unmet need for close coordination among the responders. Aceh, as well as northeastern Sri Lanka, had been a conflict zone for an extended period, and this crippling factor was reflected in challenges facing both regions in managing their environmental issues.

Indonesia has a complicated decentralized structure for managing environmental issues, and Aceh, in particular, has greater autonomy over use of its natural resources (UNEP, 2005). Furthermore, Aceh does not have a strong or well-developed environmental NGO community. Aceh was considered a Military Operational Area from 1989 to 1998, and martial law was reimposed in 2003. Consequently, there were almost no international NGOs in Aceh before the tsunami, and local NGOs were slow to develop in the harsh political and social climate (Scheper, Parakrama, and Patel, 2006).

Exacerbating the problem, many of the casualties of the tsunami included civil servants and staff from NGOs (UNEP, 2005). In Aceh, 3,000 civil servants and 5,200 staff from local authorities died, with another 2,200 reported missing (Scheper et al., 2006). More than

10,000 civil servants, or approximately 21 percent of the government work force, were adversely affected (UNEP, 2005). Leadership was also heavily impacted, with an estimated sixty senior leaders of civil society having lost their lives in the disaster (Scheper et al., 2006).

Aceh's Dinhas Kehutanan (Provisional Forestry Department) alone lost forty-nine key staff members. Lacking a substantial portion of its staff, the department was significantly handicapped as it labored to meet the rising demands of reconstruction (Kuru, 2005). An assessment by FAO concluded that the Forestry Department did "not have the resources to fully support a comprehensive forest management and wood supply programme" (Kuru, 2005: 29).

While there have not been any detailed assessments of the effect of the tsunami on specific subgroups within the government's civil service, UNEP has surmised that "the same level of devastation has been experienced by civil society as a whole, including those groups with specific interests and expertise on environmental issues" (UNEP, 2005: 31). Those working in the government's environmental agencies who survived were overwhelmed by an unprecedented workload and additional responsibilities imposed in order to manage the recovery effort (UNEP, 2005). BRR, in its two-year report, stated that one of its major concerns as a coordinating body was the lack of capacity in environmental management, impact assessment, and spatial planning (BRR, 2006).

UNEP, recognizing that most of the affected governments were suffering gaps in their environment ministries, offered surge capacity by supporting local assessments and prioritization of need, training of government staff, and provision of technical advice. UNEP also urged the international community to combine its assistance with capacity building for all environmental agencies, with a focus on integrated planning and strategic assessments (UNEP, 2005).

The joint OCHA/UNEP team suggested, in its "Rapid Environmental Assessment for Aceh, Indonesia," that, for the short term,

> In view of the scale of the disaster and the losses sustained by local authorities, additional environmental capacity to support local staff is a key priority. Specifically, donors and the international community may wish

to consider providing experts in waste management (including landfill design and construction) and other environmental experts to support authorities such as BAPEDAL [Indonesia's Environmental Impact Agency], until staff shortfalls can be addressed on a permanent basis. Additional expertise is also required to ensure both environmentally sound population relocation, and safe environmental conditions in existing IDP camps (Joint UNEP/OCHA Environment Unit, 2005: 10).

This lack of local capacity came sharply into play with management of the fishing industry. As noted earlier, fishing fleets were rebuilt almost entirely in terms of numbers, though the type and capability of boats had shifted. As a result, the fishing industries for the affected countries were practically reinvented. With the recreation of the industry, it was critical for countries to examine their fishing practices and to develop a trained, local workforce to manage its future. UNEP argued in its initial assessment that those involved in the reconstruction needed to pay particular attention to the improved management of coastal fisheries (2005). However, with urgent demands to build capacity in other areas, like training new schoolteachers—2,500 schoolteachers died in the tsunami—and coordinating the distribution of survivor benefits, establishing best management practices for fishing was not a priority.

Balancing Act

To implement a successful reconstruction project in the wake of a natural disaster that incorporates the environment, planners must know how to respond effectively to a community's most urgent needs under a demanding schedule, while coordinating among agencies and ensuring local capacity to manage the effort.

One example of a project that incorporates two of these three elements is the World Conservation Union's (IUCN) Mangroves for the Future (MFF). This is a large-scale, multiagency coastal ecosystem restoration project. It emphasizes replanting mangrove trees, but also incorporates programs to strengthen local and national capacity that include the environment as a priority. With a funding target of USD 62 million and a five-year plan, the project focuses on long-term conservation and sustainable management of coastal ecosystems in twelve Indian Ocean countries. IUCN made a conscientious effort throughout the design process to integrate the project with already existing local, national, and regional plans by bringing together representatives from governments, UN agencies, and development and environmental NGOs in the initial design phase (IUCN, 2006).

Focusing on coordination and capacity building will be the key to MFF's success. Without local commitment to the project, the biggest threat to the mangroves will be the community itself (Check, 2005). Natural mangrove forests have been cut down for decades to make room for shrimp farms and tourist resorts. According to FAO (2003), in the five main tsunami-affected countries, over a quarter of the region's mangrove cover—1.5 million hectares—was eliminated as a result of development between 1980 and 2000. The problem lies not only with regional developers and government projects; it also rests with local communities that may not be aware of the long-term effects of their behavior or choices. Communities must be given viable alternatives and resources to shift their economic base. In short, the problem needs to be attacked at all levels—policy making, education, administration, enforcement, and local economics; and it needs to be done in a sustainable way so that the projects will survive even after the NGOs and external funding have disappeared.

MFF, however, is a long-term development project that did not formally launch until December 2006, two years after the tsunami struck. As a consequence, IUCN avoided having to deal with the issues of urgency that arose in the immediate aftermath of the disaster. It is unclear whether the timing has hindered the effort. Because IUCN did not initiate this project immediately after the tsunami, the vast majority of funding for the tsunami had already been pledged for other activities. In addition, the many decisions that had to be made as part of the short-term response to the crisis have inevitably affected aspects of IUCN's long-term plans. By the time MFF was launched, a patchwork of newly developed programs, institutions, and activities was already in place and had to be accommodated.

Conclusion

In order to address all three challenges (urgency, coordination, and local capacity) and thereby increase the potential for addressing wide-ranging ecological issues, environmental groups must tightly integrate their activities with those of development and relief NGOs and affected governments from the very beginning. They must be aggressive in input on initial-needs assessments, seek strong representation at sectoral/cluster coordination meetings, and make the secondment of experts available to large relief organizations and the government for advisement in the early stages. This will allow for planners to capture not only the benefits but also the costs of the environmental perspective, so that the costs can be accounted for in the initial phase. Likewise, development and relief organizations have their own responsibility to respect the environmental perspective and design programs that fund this aspect from the very beginning. Most important, they must look to environmental actors for not only policy and technical advice, but for assistance with operational activities.

The environmental movement is not alone in its struggle to integrate fully its agenda into relief and recovery activities. Gender issues went through a similar struggle in the 1990s, when women's empowerment groups struggled to mainstream the gender perspective. The concept of mainstreaming the environment was discussed immediately after the tsunami, but it typically remained at the policy level and did not reach the implementation stage.[7]

Ultimately, successful implementation of an objective requires adequate funding and funding sources that value this set of objectives. Just as certain donors push for gender equity in project design, donors must begin to drive agencies to prove that the funded activities will have minimal environmental impact. These agreements must come in the initial phase between funders, governments, and development and

---

7. For policy examples, please see WWF's *Green Reconstruction Guidelines* (2005b). UNEP also stated in its initial assessment, "Reconstruction and restoration begins now. Mainstreaming environmental concerns is a prerequisite for sustainable reconstruction" (2005).

relief NGOs, not just environmental organizations. As the government of Sri Lanka argued in its two-year report, "Even at this stage it is not too late for RADA to give due consideration to the subject of environment and incorporate environmental safeguards into the contracts signed with donors" (Steering Committee, 2006: 53).

While these three challenges—urgency, coordination, and local capacity—hindered the response to the tsunami, they are not insurmountable. Nor should these challenges overshadow the immense good that has come about in the recovery process. The tsunami brought together private citizens, NGOs, governments, and companies as never before. In remembering the tragedy, one must also remember the lessons learned regarding cooperation and generosity, and the efforts to establish a more peaceful future.

References

Aceh and Nias Rehabilitation and Reconstruction Agency, Asian Development Bank, and Office of the United Nations Recovery Coordinator for Aceh and Nias (2006). *Aceh and Nias two years after the tsunami: 2006 progress report.* Retrieved August 1, 2007, from www.e-aceh-nias.org/upload/Adv%202yr%20Report%20lo-res.pdf.

Bennett, J., Bertrand, W., Harkin, C., Samarasinghe, S., and Wickramatillake, H. (2006). *Coordination of international humanitarian assistance in tsunami-affected countries.* London: Tsunami Evaluation Coalition.

Check, E. (2005). Natural disasters: roots of recovery. *Nature* 438: 910–911.

Clinton, W. J. (2005). Tsunami recovery: Taking stock after 12 months. Retrieved July 27, 2007 from www.tsunamispecialenvoy.org/pdf/OSE_Final_Report.pdf.

Danielsen, F., Sorensen, M., Olwig, M., Selvan, V., Parish, F., Burgess, N., et al. (2005). The Asian tsunami: A protective role for coastal vegetation. *Science* 310: 643.

Fernando, H. J. S., McCulley, J. L., Mendis, S. G., and Perera, K. (2005). Coral poaching worsens tsunami destruction in Sri Lanka. *Eos* 86: 301–304.

Food and Agriculture Organization of the United Nations (2003). *State of the world's forests.* Rome: Food and Agriculture Organization of the United Nations.

Galizzi, P. (2006). From Stockholm to New York, via Rio and Johannesburg: has the environment lost its way on the global agenda? Fordham International Law Journal 29: 952–1008.

Gandhi, I. (1972). Address by the Prime Minister of India. In M. K. Tolba (Ed.) (1988), *Evolving environmental perceptions: From Stockholm to Nairobi,* 97. Boston: Butterworths.

Joint UNEP/OCHA Environment Unit (2005). *Indian Ocean tsunami disaster of December 2004, UNDAC rapid environmental assessment of Aceh, Indonesia.* Geneva: Joint UNEP/OCHA Environment Unit. Retrieved July 27, 2007, from www.benfieldhrc.org/disaster_studies/rea/indonesia_REA_tsunami_aceh.pdf.

Kuru, G. (2005). FAO assessment of timber demand and supply for post-tsunami reconstruction in Indonesia. Retrieved July 27, 2007, from www.humanitarianinfo.org/sumatra/reference/assessments/doc/shelter/FAOAssessmentOfTimberDemandSupply260405.pdf.

Liu, P., Lynett, P., Fernando, H., Jaffe, B., Fritz, H., Higman, B., Morton, R., et al. (2005). Observations by the international tsunami survey team in Sri Lanka. Science 308: 1595.

Office of the Special Envoy for Tsunami Recovery (OSE), United Nations (2006). About the office of the special envoy for tsunami recovery. Retrieved August 1, 2007, from www.tsunamispecialenvoy.org/about/default.asp.

Scheper, B., Parakrama, A., and Patel, S. (2006). *Impact of the tsunami response on local and national capacities.* London: Tsunami Evaluation Coalition.

Secretary-General of the United Nations (2005). *Strengthening emergency relief, rehabilitation, reconstruction, recovery and prevention in the aftermath of the Indian Ocean tsunami disaster* (Report of the Secretary-General to the Economic and Social Council and the General Assembly, A/60/86–E/2005/77). Retrieved July 27, 2007, from www.reliefweb.int/rw/RWB.NSF/db900SID/KHII-6EA3P8?OpenDocument.

Steering Committee (2006). Sri Lanka, post tsunami recovery and reconstruction (Joint Report of the Government of Sri Lank and Development Partners). Retrieved July 27, 2007, from http://site

resources.worldbank.org/INTTSUNAMI/Resources/srilankareport-deco5.pdf.

United Nations Environmental Programme (2005). After the tsunami: Rapid environmental assessment. Retrieved August 18, 2007, from www.unep.org/tsunami/reports/TSUNAMI_report_complete.pdf.

World Conservation Union (IUCN) (2006). Mangroves for the future, a partnership for coastal conservation. Retrieved August 1, 2007, from http://iucn.org/places/asia/assets_tsunami/Mangroves%20for%20the%20future_facts.doc.

World Fish Center (2005). *Rebuilding boats may not equal rebuilding livelihoods* (Consortium to Restore Shattered Livelihoods in Tsunami-Devastated Nations Policy Brief No.1). Retrieved July 27, 2007, from www.worldfishcenter.org/pdf/CONSRNbrief1.pdf.

World Wildlife Fund (2005a). Timber for Aceh: Implementation design. Retrieved July 27, 2007, from wwf.org.au/publications/WWFTimberForAceh.

World Wildlife Fund (2005b). Tsunami green reconstruction policy guidelines. Retrieved July 27, 2007 from www.worldwildlife.org/news/attachments/aceh_reconstruction.pdf.

8 | Lessons from Hurricane Katrina

SACHA THOMPSON

In October 2004, Joel K. Bourne published an article in *National Geographic* depicting the devastation caused by a major hurricane tearing through New Orleans. In his article, hundreds of thousands of New Orleanians, many unable to evacuate before the storm, are drowned or trapped on rooftops as storm surges drive walls of water over the city's levees. The floodwaters turn the city into a cesspool of contamination, toxic waste, decaying flesh, and disease. It is declared the worst natural disaster in the history of the United States (Bourne, 2004). As fantastic as this story seemed in 2004, the article is not entirely a work of fiction. Bourne interviewed several local engineers, fishermen, business owners, and scientists, all of whom agreed that Louisiana's severely eroded wetlands—which protect the low-lying city of New Orleans from the severity of hurricanes—made Bourne's disaster more of an inevitability than a possibility (Bourne, 2004). At 6:10 a.m. on August 29, 2005, Hurricane Katrina, a high-intensity Category 3 storm, made landfall in Louisiana (Knabb, Rhome, and Brown, 2005) and wreaked havoc on the city of New Orleans as if it used Bourne's article for a blueprint.[1]

Sacha Thompson is a Centennial Fellow with the Sustainable Development Legal Initiative of the Leitner Center for International Law and Justice at Fordham Law School.

1. Other authors have also penned eerily prophetic articles depicting the demise of New Orleans in a hypothetical hurricane. See Berger (2001) and Fischetti (2001). See also Blumenthal (2005).

The destruction of Katrina was written all over the levee walls. New Orleans is a culturally rich and vital city carved out of the wetlands of the Mississippi Delta. Since its founding, it has been struggling to tame its surrounding environment—namely, to prevent the wetlands from swallowing the city whole. The massive feats of engineering that keep New Orleans dry and prosperous are truly a marvel. Ironically, these human-made marvels also aided Hurricane Katrina to cause as much destruction as it did. The destruction of the wetlands, the growth and exploitation of the oil industry, a deeply rooted legacy of racism, and ineffective governance jointly contributed to making Katrina the worst natural disaster in U.S. history.

This chapter will explore the role that each of these factors played in the disaster. It will also offer suggestions drawn from the lessons of other disasters that may aid a rebuilt New Orleans in mitigating the devastation of the next, inevitable hurricane.

New Orleans: A Charmed City

New Orleans is often referred to as "the city that care forgot." It is the birthplace of jazz and the home of the world-famous Mardi Gras celebration and of Cajun culture and cuisine. There is no other place on earth quite like it. Apparent from its history, it is also the city where Lady Luck lives. From the very first day that it was settled, the city has stood on literally less than solid ground. Yet, it miraculously withstood the 1927 flood of the Mississippi River, one of the worst disasters in the nation, prior to Hurricane Katrina. So why did its luck suddenly run out that fateful August morning?

The city of New Orleans was settled in 1718 on a stretch of land that was bordered by Lake Pontchartrain to the north and coastal wetlands to the east, south, and west. The first problem that the settlers encountered was flooding; as O'Brien (2002) notes, there is on average a major flood every three years. By 1812, those settlers had built levees all the way up to Baton Rouge and Pointe Coupée, 130 and 165 miles upstream respectively (Tibbetts, 2006: A40). Next, the settlers drained the surrounding swamps in order to curb the mosquito-borne yellow fever outbreaks that had killed 40,000 people over an eighty-eight year period (Tibbetts, 2006). The drained swamplands created room for the city to expand and develop; they also created the unforeseen issue of subsidence. The drained swamps and wetlands are called lowlands because they lie mere inches above sea level. When the water is drained out of the soil, it compresses and sinks. Hence, the city of New Orleans was built on sinking land that continues to sink farther and farther below sea level. Almost 80 percent of New Orleans now lies below sea level—in some places, more than eight feet below (Bourne, 2004). The city of New Orleans continued to grow and development commercially; it also built more and more levees.

In 1927, the Mississippi River breached its levees in 145 different places and flooded 27,000 square miles of land. In some places the water was thirty feet deep. The flood cost USD 400 million in damage and killed hundreds of people in seven states, including Louisiana. The city of New Orleans was miraculously spared from most of the devastation of the flood when several levees farther north were destroyed with dynamite (Barry, 2002). "In 1928, Congress authorized major levee improvements, and the United States Army Corps of Engineers (USACE) began shoring up the flood control system, including levees, along the entire lower Mississippi and in New Orleans" (Tibbetts, 2006: A40).

In the 1950s, local engineers noticed that the Atchafalaya River was capturing more and more of the Mississippi River. They feared that the Mississippi would soon change course and take over the Atchafalaya River all together. The Atchafalaya lies roughly sixty miles northwest of New Orleans. If the Mississippi River were to change course and take over the Atchafalaya, the city would once again stand in the path of certain destruction. To prevent this almost certain demise, the USACE built the Old River Control Structure, a system of floodgates that could be opened and closed to maintain the 70–30 ratio of water distribution between the two rivers (Kemp, 2000).

While human ingenuity has been able to protect New Orleans from the river, it has done nothing to alter the fact that New Orleans continues to sink ever farther below sea level as the wetlands beneath it compress and subside. An average rainfall causes flooding throughout the city. To address this problem, the city's Sewer and Water Board, in conjunction with the USACE, dug a maze of canals to collect rainwater (Fischetti, 2001). This solution, however, presented its own problems. The collected water has to go somewhere, and the only feasible solution was that it go into Lake Pontchartrain, which lies at an elevation one foot above New Orleans (ibid.). The Corps and the Sewage and Water Board therefore built pumping stations at the head of every canal to push the rainwater uphill into the lake (ibid.).

Unfortunately, the canals collect not only rainwater but also groundwater from the surrounding wet, marshy soil. This poses yet another problem. If the canals are already full of groundwater, they cannot efficiently collect enough rainwater to prevent flooding. Thus, the pumps are run at regular intervals in order to keep the canals free of groundwater. The continuous pumping of the canals draws more and more water from the ground, which causes the land to dry out and subside; hence, the city sinks even deeper, and the lower the city sinks the more it floods (Fischetti, 2001). The solution has now become the problem. Meanwhile, the water that is continuously being removed is pumped into a nearby lake that sits at a higher elevation than the New Orleans, a precarious situation for a city prone to visits from hurricanes.

Pumping groundwater is only part of the equation that placed the city in such a dangerous position. As part of the nation's most vital shipping ports, New Orleans is home to both the Mississippi River Gulf Outlet (MRGO) and the Inner Harbor Navigational Canal, also called the Industrial Canal. Built in the 1960s, during the oil-industry boom, the MRGO was intended to provide a shortcut from the Gulf of Mexico directly into New Orleans Inner Harbor for the deep-sea vessels that would have difficulty with the navigation channel locks on the Mississippi River (Driesen et al., 2005). The MRGO bisects the Industrial Canal, which runs north to south across the city of New Orleans, connecting the Mississippi River to Lake Pontchartrain.

Commerce was not all the MRGO brought to the city. Wakes from the few vessels that use this channel are large enough to erode the nearby shoreline, disrupting local businesses (Bourne, 2004). The MRGO also brings large amounts of saltwater past the barrier islands and marshes and directly inland to freshwater habitats. The saltwater intrusion kills the freshwater vegetation, creating more unprotected bodies of open water in close proximity to the already soggy and sinking city (Caffey and LeBlanc, 2002).

Still more dangerous than its introduction of corrosive saltwater, the MRGO also acts as a water highway for storm surges. In 1965, Hurricane Betsy created a tidal surge that traveled up the MRGO and flooded Orleans Parish. The tidal surge killed eighty people and caused USD 2 billion in damages (Caffey and LeBlanc, 2002). When the MRGO was built, it was only 650 feet wide (Bourne, 2004). Over the years, soil erosion has widened the channel to its present width of more than 2,000 feet (Caffey and LeBlanc, 2002). In 2005, during Hurricane Katrina, a surge originating in the Gulf of Mexico entered Lake Borgne and traveled directly up the MRGO to the Industrial Canal juncture in the center of New Orleans. Almost all of the levees and floodwalls along the MRGO canal were destroyed (Driesen et al., 2005).

Destruction of the Wetlands

The Deltaic Process

The Louisiana coastal wetlands were created by the deltaic process. When the Mississippi River enters a smaller body of water, the bottleneck at the entrance to the smaller body of water forces the river's velocity to slow. The silt and sediment being transported by the faster current of the Mississippi is then deposited at the mouth of the smaller body. Over time, sediment is deposited and new land is created between the distributary rivers. Known as wetlands or deltas, these expanses of new soil will continue to expand seaward, fanlike, as long as the Mississippi River continues to deposit sediment in the area periodically.

Eventually, the new land created by the deltaic processes will cause the Mississippi River to change course. The river will break through a weak point in its bank or shift the majority of its flow into another distributary river as it seeks a shorter, steeper, and faster path to its final destination in the Gulf of Mexico (USACE, 2004).[2] When the Mississippi River changes course, it no longer feeds freshwater and sediment to the delta wetlands of its previous course. The new land on the interior of the delta will compact and sink through the subsidence process and, ultimately, be submerged and become the marshes and bayous of the coastal wetlands. The outer, gulf-facing edge of the delta evolves into barrier islands (USACE, 2004).

The Mississippi River shifts course, as part of the deltaic process, roughly every thousand years. It has happened at least six times in the last seven thousand years, creating six delta planes. Over the past five thousand years, this process has expanded the south Louisiana coastline roughly thirty-five miles into the Gulf of Mexico. It is also responsible for creating the rich range of biodiversity in southern Louisiana. The wetlands are home to forested swamps; freshwater, brackish water, and saltwater marshes; and natural levees, barrier islands, and beach ridges (Driesen et al., 2005). This range of habitats makes the wetlands a vibrant home to a phenomenal range of wildlife species, including nearly three-quarters of all migratory waterfowl species. The wetlands "make the Florida Everglades look like a petting zoo" (Bourne, 2004).

The wetlands also serve other functions that are vital to human existence. They filter out large quantities of nitrogen, phosphorus, and other pollutants from bodies of waters that enter the area (Driesen et al., 2005) and, most important, protect the inland area from storm surges caused by hurricanes.

> Imagine blasting water through a garden hose at full force onto a cement driveway. The water splashes and surges, fanning out in many directions. Now imagine spraying water from the same hose onto a thick, dense lawn. The difference between the cement and the lawn is the difference

---

2. The Mississippi River also deposits large amounts of sediment whenever it overflows its banks and floods surrounding areas. Springtime floods are a common occurrence along the Mississippi River. Floods often occur when the main river finds a weak point in a natural levee or the riverbank and the force of the main river causes it to burst through creating a sediment-rich crevasse (USACE, 2004).

between a storm path composed of open water and denuded coast and one composed of lush forests and marsh. Louisiana's coastlands act as vast sponges, absorbing billions of gallons of rainfall and shielding people and property from storms . . . *every two miles of wetlands south of New Orleans reduces tropical storm surges there by half a foot.* (Driesen et al., 2005: 10; emphasis added)

Hurricanes are a common occurrence in Louisiana. "Since 1893, approximately 135 tropical storms and hurricanes have struck or indirectly impacted Louisiana's coastline. On average, since 1871, a tropical storm or hurricane affects Louisiana every 1.2 years" (USACE, 2004: MR 2–2).

Taming the River

The deltaic process has made the southern Louisiana wetlands an incredibly prosperous region. It is home to one of the nation's most vital port complexes, which include the Port of New Orleans. "Through South Louisiana's ports the bulk commodities of U.S. agriculture—corn, wheat, and soybeans—are sent around the world, and the bulk commodities needed for American industry—steel and concrete, for instance—come into the country" (Tibbetts, 2006: A41). The Louisiana coastline also produces one-quarter of the nation's natural gas and one-fifth of its oil (Tibbetts, 2006: A41). The wetlands support a USD 300 million per year commercial fishing industry (Bourne, 2004). Tourism has also become a vital part of the Louisiana economy. Americans traveling to Louisiana spent approximately USD 8.1 billion in 2001 (USACE, 2004: MR 2–37). In 2000, recreational fisherman spent USD 1.2 billion in the state of Louisiana, and in 2001, big-game hunters spent USD 446 million and wildlife watchers spent USD 168 million in the state (USACE, 2004).

Residents of the South Louisiana area were happy to reap the benefits of the area's resources, but they were not willing to endure the constant and destructive flooding necessary to sustain them. "Today, South Louisiana is one of most intensively engineered places in the nation. Vast quantities of water are diverted or rerouted through a lacework of navigation corridors held in place by 2,000 miles of earthen, rock, and concrete levees" (Tibbetts, 2006: A41). Instead of depositing its sediment into the wetlands, the Mississippi River now carries the sediment into the Gulf of Mexico, to the edge of the continental shelf, where it falls over an underwater cliff, never to be recovered (Fischetti, 2001). Blocking the flow of the river to the floodplains and wetlands also blocks the arrival of the freshwater and sediment that sustains these areas; without the deltaic renewal process, the wetlands will begin to subside and give way to open saltwater (Tibbetts, 2006). This will cause loss of not only the rich biodiversity and natural resources but also the crucial natural protection from inevitable hurricanes and storm surges.

As early as 1950, geologists began to document dramatic loss of land in the Louisiana coastal plains—as much as 80 percent of the nation's total loss of coastal wetlands, amounting to 1,900 square miles from 1932 to 2000 (Tibbetts, 2006: A40). Louisiana is now losing approximately 6,600 acres of wetland per year (Driesen et al., 2005), or one acre every twenty-four minutes (Fischetti, 2001).

The Oil Industry and the Wetlands

The levees that prevent the Mississippi river from depositing new sediment in the wetlands are not solely to blame for their degradation. The Louisiana oil industry is a major factor in their destruction, attacking on more than one front.

Water, Water Everywhere

The oil industry claims to have provided the state with "low taxes and high-paying jobs," but the wetlands have paid the price while the state reaps the benefits (Bourne, 2004). A vast network of navigation channels, canals, pipelines, and production facilities was created to support this lucrative industry. "Today, an estimated 9,300 miles (14,973 kilometers) of oil and gas pipelines crisscross the coastal wetlands of

Louisiana. In addition, there are approximately 50,000 oil and gas production facilities located in the Louisiana coastal area" (USACE, 2004: MR 2–4). The navigation channels are especially threatening not only to the wetlands, but also to the developed city surrounding them. These channels often act as highways that allow stormwaters to bypass the protective marshes and funnel directly inland (Driesen et al., 2005).

The canals act as inroads connecting the bodies of freshwater to the saltwater in the Gulf of Mexico (USACE, 2004). The barrier islands and brackish water marshes created during the deltaic process normally buffer the inland bodies of freshwater from the corrosive saltwater in the Gulf. The introduction of saltwater into the freshwater habitats kills the freshwater plants and drives away the animals that depend on freshwater habitats. It also accelerates the second phase of the deltaic process: degradation and subsidence. In essence, the canals are turning the delta wetlands into dangerous bodies of open water (USACE, 2004). An industry-funded study of wetland losses found that oil and gas industry is responsible for one-third of the delta's land loss (Fischetti, 2001).

Death from Above, Death from Below

It has recently come to light that the oil industry's impact on the fading wetlands may be larger and more direct than previously thought. A recent study shows that the removal of "millions of barrels of oil, trillions of cubic feet of natural gas and tens of millions of barrels of saline formation water lying with the petroleum deposits caused a drop in the subsurface pressure" (Bourne, 2004). As one geologist describes it, "when you stick a straw in a soda and suck on it, everything goes down" (Bourne, 2004). The process, known as regional depressurization, induces subsidence throughout the region. Without the periodic river floods to reintroduce freshwater and sediment into the area, the wetlands will continue to subside and allow more saltwater into the area (Bourne, 2004).

Attempts to Restore the Wetlands Before Katrina

Local scientists, environmental groups and businessmen began campaigning to promote awareness about the destruction of the wetlands in the late 1980s (America's Wetlands, n.d.). In response, Congress passed the Coastal Wetlands Planning and Restoration Act of 1990. Named after Senator John Breaux of Louisiana, who presented it, the act created a USD 40 million fund for projects designed to save the wetlands (Driesen et al., 2005). It soon became clear that it would take much more money to halt the destruction.

In 1998, state and federal agencies, including the USACE, joined to create a comprehensive plan to combat wetland losses (Louisiana Coastal Wetlands Conservation and Restoration Task Force, 1998). The plan outlined a fifty-year plan to reintroduce or "mimic" the natural deltaic process that creates and sustain the wetlands (Driesen et al., 2005). The Coast 2050 plan came with a price tag of USD 14 billion (ibid.). It was never funded (ibid.).

Louisiana Governor Kathleen Blanco asked for her state to receive a small portion of the USD 5 billion that the federal government receives from the oil and gas leases off the coast of Louisiana and was again denied (Driesen et al., 2005). In 2005, President Bush's energy bill provided only USD 540 million for coastal restoration over four years (ibid.). Hurricane Katrina is estimated to have cost USD 150 billion of damage, making it the costliest Atlantic hurricane in history (Burton and Hicks, 2005; Knabb, Rhome, and Brown, 2005).

Hurricane Katrina

On August 29, 2005, following the landing of Hurricane Katrina, the MRGO Canal, the Industrial Canal, the 17th Street Canal, and the London Avenue Canal breached their respective levees and flooded 80 percent of the city of New Orleans. More than 1,500 residents of the state of Louisiana died directly or indirectly as a result of the hurricane, and approximately 26,000 people were forced to seek shelter in the Louisiana Superdome. Lake Pontchartrain rose above its bank, causing major flooding in the parishes of Slidell and Mandeville. Hurricane Katrina made it glaringly obvious that a system of levees and drainage canals is woefully inadequate protection from a major hurricane.

Damages to Oil and Gas Industry

Katrina, and subsequent Hurricane Rita, destroyed 113 of the Gulf's 4,000 oil and gas platforms and damaged 52 others. The hurricane also did severe damage to the Shell Oil Company's Mars Platform, one of the Gulf's biggest producers. "The storm's 175-mph winds and 75-foot waves broke the steel clamps that attached the 1,500-ton rig structure to the platform and knocked a 200-foot derrick into the water. The surge caused the rig to rise up and slam into the platform, causing heavy damage" (Porretto, 2007). Two years later, the industry had resumed most pre-Katrina production (ibid.). Most of the major producers in the Gulf region had begun to invest in improved methods of securing and minimizing possible hurricane damage to their rigs, platforms and pipelines. This is a particularly wise move, since recent reports show that global warming increases the intensity of hurricanes (Hecht, 2005).

Several days after Katrina, the International Energy Agency (IEA) stated that Katrina's damage to the Gulf oil industry could cause a worldwide energy crisis (Heinberg, 2005). To avert the crisis, twenty-six nations, including the Unites States, agreed to release 60 million barrels of oil, gasoline, and other petroleum products from their emergency reserves over the next thirty days (Heinberg, 2005).

Impact on the Environment: The Wetlands

Hurricane Katrina devastated the barrier islands. Dauphin Island suffered massive beach erosion when the hurricane traveled inland to the Mississippi Sound. The Chandeleur Islands, which were heavily damaged by a hurricane the year before, also suffered massive erosion. As much as a fifth of the local marshland area was permanently overrun. According to a 2006 survey of wetland changes, almost 140,000 acres of wetlands were converted to open water throughout the Louisiana coastal area (Zinn, 2007). The Breton Sound, southeast of New Orleans, sustained the most damage. Half of the land in the Breton National Wildlife Refuge was swept away during the hurricane. The Caernarvons Diversion restoration project also suffered damages from the hurricanes, losing roughly 25,000 acres of wetland. Damage to the present restoration efforts will only prolong the overall restoration period (Zinn, 2007).

Wetland Restoration After Katrina

After Hurricanes Katrina and Rita, advocates of Louisiana wetland restorations began to call for a bolder and more substantial plan than the one Congress had already authorized. Advocates argued that "hurricanes caused nationally significant disruptions, especially to energy supplies and bulk transportation, and therefore the restoration would have significant national benefits" (Zinn, 2007: 3). These advocates are pushing for Congress to reconsider funding the USD 14 billion Coast 2050 Plan, which outlines a strategy for protecting or restoring 450,000 acres of wetland throughout the Louisiana coastal area, but with a concentration on the coastal area south and southwest of New Orleans (Zinn, 2007).

New Orleans

It took forty-three days to pump all of the residual floodwater out of New Orleans and into Lake Pontchartrain. During those six weeks, the water had turned into a toxic soup made of raw sewage; oil from five major spills totaling 160,000 barrels; another five to seven million gallons of oil spilled from storage tanks and industrial plants and facilities; a vast array of toxic chemicals, pesticides, and deadly mold; and the bloated, decaying bodies of humans and animals (Driesen et al., 2005). Pumping this contaminated water into Lake Pontchartrain is expected to harm many of the fish and living organisms in the lake (Driesen et al., 2005).

Environmental Injustices

Louisiana's deeply rooted legacy of racism effects both the African American population and the environment. It also played a central part in the immense suffering caused by Hurricane Katrina. Many of the toxins that fouled the floodwaters came from dump sites within the city of New Orleans. Many of these sites were located in poor, African American communities. Most were either poorly cleaned or poorly contained. The Environmental Protection Agency (EPA) keeps a National Priority List (NPL) of known sites where hazardous substances have been released or are likely to be released (EPA, n.d.). The list shows five sites in the city of New Orleans and ten others in the state of Louisiana, all of which lay in Katrina-affected areas (Esworthy et al., 2005). The Agriculture Street Landfill is a prime example.

> The site that was the hardest hit by Katrina is the Agriculture Street Landfill, sometimes referred to as the "black Love Canal." The 95-acre site, located three miles south of Lake Pontchartrain in a community that is 60–80 percent African American, is an old municipal landfill where ordinary garbage was mixed together with liquid hazardous waste to a depth of between two and 32.5 feet. In 1969, the City of New Orleans built a low income housing project on top of the site, as well as the Moton Elementary School. In 1993–94, after community leaders demanded that EPA conduct a full investigation of the site, the Agency decided that [the] contamination at the site warranted an emergency cleanup and placement on the NPL. In a health assessment prepared for the site by the Agency for Toxic Substances and Disease Registry (ATSDR), a unit of the Centers for Disease Control, experts concluded that the undeveloped portions of the site posed a "public health hazard" and that if the land was ever used for residential housing, exposure to lead, arsenic, and polycyclic aromatic hydrocarbons (PAHs) in the soil could pose an "unacceptable health risk." All of those toxic materials are now floating through the streets of New Orleans. EPA's choice of a remedy for the site has significantly exacerbated this damage. Instead of excavating the site, treating contaminated soil in situ, or even installing a liner that would prevent the landfill's contents from washing away,

EPA decided that its final remedy would be limited excavation of less than two-thirds of the site and the placement of two feet of "clean fill" on top of the buried waste. (Driesen et al., 2005: 5)

In an interesting twist, the residents of the Agricultural Street Landfill asked the EPA to relocate them away from the landfill. The EPA refused. The cost of relocation was $12 million, while the partial treatment the landfill received cost $20 million.

The legacy of racism in New Orleans has led to a long, sordid history of ignoring the disproportionate impact that certain laws, policies, and industries have on its poor, the majority of whom are African American (Sze, 2006). Katrina shined a light on this injustice for the entire world to see. In the days following the hurricane, most Americans sat in front of their televisions stunned at the images of tens of thousands of people, mainly African American, seeking shelter in the Louisiana Superdome or stranded on rooftops. The immediate question that came to mind was, "Why didn't they evacuate the city before the storm arrived?" The answer, as economist Stephen Raphael discovered, is very simple: because they don't have cars (Becklund, 2005).

"It is no coincidence that the storm's most dramatic effects were felt in a city where black reliance on public transit was four times higher than that of whites, and where public plans for evacuation were tragically deficient" (Pastor et al., 2006). In New Orleans, "the water flows away from the money. Those with resources who control where the drainage goes have always chosen to live on the high ground. So the people in the low areas are hardest hit" (Driesen et al., 2005). In a city where the people with least access to a vehicle are also the same people who are most likely to be affected by a flood, how does an evacuation plan center on the ability to get into a car and drive away? The tragedy in the New Orleans evacuation plan is not that so many people were left behind; it is that city officials were aware of this fact in advance and did nothing about it.

Unfortunately, the lack of consideration for poor residents in the evacuation plan is not an isolated incident. Many of the policies that disparately burden poor, and often African American, communities are allegedly justified by their cost-effectiveness. "In the case of New

Orleans, rather than reduce the risks to the public by regulating activities that destroyed wetlands and other natural storm protections or funding adequate flood control measures, the government opted to rely on evacuation warnings leaving people to avoid the risks themselves" (Driesen et al., 2005: 37). This "risk avoidance" approach to policy making assumes that everyone is equally capable of avoiding the risks of an impending hurricane. The poor do not necessarily have expendable assets or resources available to rent cars and pay for motel rooms. These policies highlight the main argument of environmental justice advocates. "People of color and the poor disproportionately comprise the communities that are overburdened by pollution, underserved by public projects and amenities, and underserved by the government decision makers" (Driesen et al., 2005: 38).

An Incomplete Analysis

The haphazard approach to risk management reflected in the New Orleans evacuation plan is also reflected in its approach to flood management. The USACE based its decision to build many of the levees that surrounded New Orleans on a Cost Benefit Analysis (CBA). At the time, USACE flood-control projects were measured by their ability to produce net economic development (NED). This meant that the levee projects put in place were meant to "recover land" from the flood plains that could be developed for residential or commercial use (Verchick, 2007). Levee projects based on NED analysis gave a false sense of security that encouraged development, causing more communities and businesses to situate themselves in high risk localities (ibid.). This false sense of security was based on the dangerous assumption that the decision to build flood-protection levees and recover land took into account risk to human life and environmental degradation; unfortunately, that was not necessarily the case. At the time of their implementation, most of the levees built under this analysis were in rural areas far away from the large populations in the city. The lack of proximity to populations also explains why many of the levees were built to meet a lesser safety standard than the structures built in or near cities (ibid.). In southern Louisiana, the levees that once protected rural lands now protect a metropolitan area with a population of over 2 million people (ibid.).

The lack of consideration for human lives in the NED cost benefit analysis is also due, in part, to its secondary requirement that the state show its willingness to pay. The willingness-to-pay requirement was meant to measure how highly the public valued the protection and security of the levee system, assuming that if the public really wanted the levees in place they would be willing to pay for it. The USACE measured this willingness by requiring that the state housing the levees pay for 30 percent of the cost of construction. This analysis is deeply flawed because it equates willingness to pay with ability to pay.

Louisiana is one of the poorest states in the union. Furthermore, the parties that control the purse strings are not always "willing" to pay to protect the people who would be most affected. The low-lying areas of New Orleans, the areas with the worst drainage, are populated by poor African Americans (Driesen et al., 2005). The legacy of racism and years of inadequate public policy have disenfranchised the majority of poor African Americans from the political system (Driesen et al., 2005). This model could never accurately reflect their willingness or their ability to pay for protection.

Since Katrina, Congress has directed the USACE to revise its storm-protection systems. The model now being designed is not based on cost-benefit analysis and is designed to withstand a wide array of storm possibilities, including a Category 5 storm (Verchick, 2007). This new approach to storm protection is definitely a step in the right direction, but may not be enough. The new model should also accurately measure risk, fairness, and the inclusiveness of the decision-making process; create efficient and effective evacuation plans; and factor environmental degradation into the planning of new levees. It is possible that Hurricane Katrina will not be the worst hurricane to visit the Gulf Coast in the next hundred years. The reality of global warming and climate-change issues means that New Orleans, a low-lying, subtropical city at the mouth of a river delta, is very likely to be visited by stronger and more frequent storms in the future. The combination of further degradation to the wetlands combined with stronger storms makes for bigger problems than Katrina.

Suggestions for a Drier Future

Perhaps the biggest lesson Louisiana learned from Hurricane Katrina is that structural protection from floods is not enough. It is not alone in this realization. Worldwide, many governments, organizations, and communities are finding that structural protection elements do not provide as much protection as envisioned, but they do cause more damage to the environment than previously thought (Global Water Partnership [GWP], n.d.). Governments and communities have also noticed that past flood-protection practices tend to shift flood risks instead of mitigating them, leaving certain sectors of the population more vulnerable than others (ibid.).

In 2002, at the World Summit on Sustainable Development, the international community was called upon to create "integrated water resource management (IWRM) and efficiency plans" to better manage water resources and the risks and hazards that these resources cause (IWRM Questionnaire, n.d.). IWRM is defined as "a process which promotes the coordinated development and management of water, land and related resources in order to maximize the resultant economic and social welfare in an equitable manner without compromising the sustainability of vital ecosystems" (Jonch-Clausen, 2004: 14). It is not intended to be a static, replicable model. It is a "political process and involves conflicts of interest that must be mediated" (Jonch-Clausen, 2004: 15). The government and constituents of each locality must, collectively, create their own IWRM system.

In 2005, the United Nations declared 2005–2015 The International Decade of Action, Water for Life (GWP, n.d.). The UN's Water for Life program espoused the WSSD's call for integrated management systems, but it also included an integrated flood-management plan as part of that system. There are five key elements to the UN integrated flood management component:

- Adopting a best mix of strategies, both structural and nonstructural;

- Managing the water cycle as a whole while considering all floods, including both extremes;

- Integrating land and water management, as both have impacts on flood magnitudes and flood risks;

- Adopting integrated hazard-management approaches, taking into consideration the risks due to all related hazards such as landslides, mudflows, avalanches, storm surges, and tsunamis and creating synergies;

- Ensuring a participatory approach to develop a sense of ownership and reduce vulnerability. (GWP, n.d.).

IWRM systems are more widespread internationally than they are in the United States; however, several states are moving toward integrated systems (American Rivers, 2006; Kahan et al., 2006). Several organizations have conducted case studies of the worst floods throughout the United States and created an IWRM framework for the Gulf Coast based on the lessons learned from these disasters:

- *Modernize the policies and regulations that govern water-resource projects.* The past policies, as mentioned, have been destructive to both the environment and human life. The new policies and regulations must include protection and/or mitigation for nearby communities that lie in high-risk areas, higher levels of accountability, and an external system of review.

- *Restore, protect, and sustain natural flood protection, i.e. the coastal wetlands.* Sustainably managed wetlands provide a myriad of benefits to local communities other than flood protection, including clean water, sporting and recreation activities, and commercial fisheries.

- *Move away from structural reliance on levees and damns for flood protection.* The levees and canals are key contributors to the destruction of the wetlands as well as the destructive force of hurricanes and storm surges. They severely affect the health of rivers and surrounding wildlife. More important, they cause harm to human life by creating a false sense of security that draws people onto the floodplain.

- *Move certain high-risk communities farther inland away from the floodplains.* Several localities have opted to relocate high-risk communities as part of a plan to "give back some of the land," meaning that they would restore the surrounding floodplains to there former status as protective wetlands. In several locations, the state implemented buyout programs to encourage homeowners in the floodplains to move farther inland. The relocation/buyout programs also resulted in heavily discounted flood insurance in these regions. (American Rivers, 2006)

Each of these steps could greatly reduce the devastation caused by the next hurricane. Unlike the previous NED cost-benefit analysis, protection of the environment and human life are a central tenet of the IWRM flood-protection plan. Also, the government returns to its role as the protector of the public good, whereas, under the risk-avoidance approach, there was very little government protection or enforcement.

The aforementioned risk-management approach would lessen the destructive impact of hurricanes and floods, particularly on the most vulnerable sections of the population. However, there is still one element missing: an effective evacuation plan. Hurricanes in Louisiana are inevitable. The steps in this framework will reduce their severity, but without an effective evacuation policy, vulnerable sectors of the population are still proverbial sitting ducks.

Cuba: The Little Island That Could

IWRM systems are intended to meet the specific needs and uses of different communities, and therefore should not be wholly duplicated without reference to local conditions. Nonetheless, many valuable lessons can be learned by studying successful systems. Cuba offers one of the best examples of a comprehensive disaster preparation and response system.

Cuba is located at the mouth of the Gulf of Mexico. Geographically, it lies in the path of most Atlantic hurricanes as well as the hurricanes that cross over Central America. Between 1996 and 2002, Cuba was hit by six major hurricanes; however, there were only sixteen deaths reported. In 1998, Hurricane Georges caused 380 deaths in the Dominican Republic and 209 deaths in Haiti; only six Cubans died. In 2001, Hurricane Michelle, a Category 4 storm, touched down in Cuba; only five lives were lost. The United Nations and the International Federation of the Red Cross have repeatedly pointed to the Cuban model of risk reduction as a guideline for other countries (Thompson and Gaviria, 2004).

One of the goals of Cuba's socialist government is to reduce or remove socioeconomic disparities in the population. A positive side effect of this goal is that the most vulnerable sectors of the population receive more protection than they might under other forms of governance. Outside of its socialist agenda, the national government also took several key steps to protect the well-being of the population. The government invests in infrastructure in all sectors of the population, thereby creating resources that all citizens may utilize in the face of a natural disaster. In addition, Cuba's universal education and healthcare systems provide a wealth of well-trained and well-educated individuals who can assist with disaster preparation, response, and recovery throughout the island, while a planning board manages land use, as well as economic and environmental changes, in high-risk areas. Furthermore, Cuba's disaster preparation, response, and recovery plans are all written in the legal code, which creates a blueprint for action: all parties are clearly aware of their powers and responsibilities during a disaster. The Cuban government has also created a dedicated system and backup system for disseminating clear information and instructions (Thompson and Gaviria, 2004).

On a local level, the Cuban government also implemented several programs that promote its "culture of safety." Disaster preparation is taught in schools, workplaces, and institutions, so that every member of a community knows where to go and what to do in the face of a disaster. Local leaders and politicians also double as civil defense leaders (CDLs), responsible by law for coordinating local disaster response. In a disaster, CDLs mobilize all available resources in order to ensure that all of their constituents reach safe shelters, creating "both a centralized decision-making process, which is key for emergency situations, alongside a decentralized implementation process, providing the agility and adaptation equally necessary for effective emergency preparedness and response" (Thompson and Gaviria, 2004: 30).

Another result of the government's investment in social capital is the high level of social organization represented in every community. The heads of neighborhood organizations conduct community risk mapping to keep track of the most vulnerable members of their communities, and are responsible for seeing these members to safety during disasters (Thompson and Gaviria, 2004). The national government calls upon these local heads to assist in the annual updating of the national emergency plan, which helps to correct mistakes and implement new changes (ibid.).

Just before the start of every hurricane season, the entire nation participates in a two-day natural-disaster simulation (Thompson and Gaviria, 2004). The first day is a drill to refresh memories and implement any changes, while the second is spent in preparation, clearing dangerous trees and inspecting vital infrastructure (ibid.). When preparing for an actual disaster, all shelters are opened seventy-two hours in advance; stocked with food and water; and staffed by a director, a deputy director, a doctor, a nurse, police, and a representative of the Red Cross (ibid.).

The combination of these steps creates severable intangible assets that risk reduction experts consider vital to all disaster preparation:

- social cohesion and solidarity (self-help and citizen-based social protection at the neighborhood level);
- trust between authorities and civil society;

- political commitment to risk reduction;

- good coordination, information-sharing, and cooperation among institutions involved in risk reduction;

- attention to the most vulnerable populations;

- attention to lifeline structures (concrete procedures to save lives, evacuation plans, etc.);

- investment in human development;

- an effective risk communication system and institutionalized historical memory of disasters, laws, regulations, and directives to support all of the above;

- investments in economic development that explicitly take potential consequences for risk reduction or increase into account;

- investment in social capital;

- investment in institutional capital (e.g., capable, accountable, and transparent government institutions for mitigating disasters). (Thompson and Gaviria, 2004)

The common thread throughout every component of the Cuban disaster preparation plan is the reliance on social capital. All sections of the population are included in the decision-making process, informed of the evacuation plan, and involved in the evacuation process. The most vulnerable members of the society are accounted for and protected at both the local and the national level.

Of course, Louisiana need not take up socialism in order to implement the Cuban model of risk reduction. Clearly, it can adapt and substitute components of this plan to create its own disaster preparation and response methodology. However, "there is no comprehensive substitute for reducing poverty and promoting social and economic equity as the fundamental long-term strategies to reduce vulnerability to hazards" (Thompson and Gaviria, 2004: 53). This fact has already been painfully reflected in the disastrous evacuation policy enacted during Hurricane Katrina. Given the deep racial and economic divides that permeated every aspect of life in Louisiana, building social capital may be Louisiana's greatest obstacle in implementing a comprehensive IWRM plan. Nonetheless, it must be done. It is counterintuitive to implement changes that protect the environment in order to lessen the severity of natural disasters, if nothing is done to protect the people who will still be disparately affected.

Conclusion

Efforts to rebuild New Orleans are well under way, but it is still unclear whether city and state officials will use this opportunity to build a city that values and protects its natural environment and all of its residents or one that repeats its past transgressions. Early indicators seemed to suggest that New Orleans was headed toward the path of the latter. Less than a month after Katrina's arrival, the *Wall Street Journal* ran an article wherein wealthy white residents of New Orleans expressed their wish to build a New Orleans that was completely different demographically, geographically, and politically (Cooper, 2005). Before Katrina, 67 percent of New Orleanians were black (U.S. Census Bureau, 2000), and the city was a Democratic stronghold (Cooper, 2005).

This new vision of a completely different New Orleans may also have negative implications for the environment. Conservative politicians argued that Katrina's damage to the oil and gas industry may have sparked an energy crisis that would lead to higher fuel prices. In order to prevent such a crisis, these politicians have been pushing new legislation aimed at waiving existing environmental bans on drilling in protected areas (Abendschein, 2005). Denouncers of the new legislation claim that lifting bans on drilling will not ease fuel prices; instead, it will only cause further degradation of fragile environments and harm to public health (Abendschein, 2005).

Many oppose the "completely different" vision of New Orleans. Some organizations have put forth the argument that any plan to rebuild the city must incorporate plans for "fair distribution of environmental and other burdens" and a system that addresses the source of the problem instead of the victims (Driesen et al., 2005: 42; see also Brookings Institution, 2005; Berube and Katz, 2005). This argument is in direct response to city officials' risk-avoidance practices of the past. Instead of unfairly expecting citizens to be responsible for avoiding risks, officials should hold environmental polluters responsible for causing risks that endanger public health.

The "new" New Orleans sits at a fork in the road. It can heed the lessons of its past and create an IWRM system that restores and protects the wetlands, curbs industry abuse, implements inclusive decision-making processes built on social capital, and supports it all with stronger governance. On the other hand, politicians and officials may seize this opportunity to expand drilling areas, further degrade the wetlands, and widen class and race divides. The one thing that both these versions of New Orleans have in common is the next hurricane. Neither the many benefits of the wetlands nor the rich cultural diversity and history of New Orleans are replaceable. With luck, it will not take another Hurricane Katrina to drive that lesson home.

References

Abendschein, D. (2005). Capitalizing on catastrophe. *Los Angeles City Beat*. Retrieved August 8, 2007, from www.lacitybeat.com/article.php?id=2730&IssueNum=123.

American Rivers (2006). Unnatural disasters, natural solutions, lessons from the flooding of New Orleans. Retrieved September 8, 2007, from http://amr.convio.net/site/DocServer/Katrina_Publication-take2.pdf?docID=4 481.

America's Wetlands (n.d.). Retrieved August 8, 2007, from http://www.americaswetland.com.

Barry, J. (2002, October). The 1927 Mississippi river flood and its impact on U.S. society and flood management strategy. Paper presented at annual meeting of the Geological Society of America,

Denver. Retrieved August 8, 2007, from http://gsa.confex.com/gsa/2002AM/finalprogram/abstract_44272.htm.

Becklund, L. (2005). Listening to Katrina. *California Alumni Magazine*. Retrieved August 8, 2007, from http://alumni.berkeley.edu/Alumni/Cal_Monthly/November_2005/COVER_STORY-_Listening_to_Katrina.asp.

Berger, E. (2001). Keeping its head above water. *Houston Chronicle*. Retrieved August 31, 2007, from www.hurricane.lsu.edu/_in_the_news/houston.htm.

Berube, A., and Katz, B. (2005). *Katrina's window: Confronting concentrated poverty across America* (Brookings Institution Metropolitan Policy Program). Washington, D.C.: Brookings Institution. Retrieved August 8, 2007, from www3.brookings.edu/metro/pubs/20051012_concentratedpoverty.pdf.

Blumenthal, S. (2005). No one can say they didn't see it coming. *Salon*. Retrieved August 8, 2007, from http://dir.salon.com/story/opinion/blumenthal/2005/08/31/disaster_preparatio n/index.html.

Bourne, J. (2004). Gone with the water. *National Geographic Magazine*, October 2004.

Brookings Institution (2005). *New Orleans after the storm: Lessons from the past, a plan for the future* (Brookings Institution Metropolitan Policy Program, Special Analysis). Washington, D.C.: Brookings Institution. Retrieved August 8, 2007, from http://media.brookings.edu/mediaarchive/pubs/metro/pubs/20051012_NewOrle ans. pdf.

Burton, M., and Hicks, M. (2005). Hurricane Katrina: Preliminary estimates of commercial and public sector damages. Retrieved August 8, 2007, from www.marshall.edu/cber/research/katrina/Katrina Estimates.pdf.

Caffey, R. H., and LeBlanc, B. (2002). Closing the MRGO: Environmental and economic considerations (Interpretive Topic Series on Coastal Wetland Restoration in Louisiana, Coastal Wetland, Planning, Protection, and Restoration Act, National Sea Grant Library No. LSU-G-02–004). Retrieved August 8, 2007, from www.ccmrgo.org/documents/closing_the_mrgo.pdf.

Cooper, C. (2005, September 8). Old-line families escape worst of flood and plot the future. *Wall Street Journal*.

Driesen, D. M., Flourney, A., Foster, S., Gauna, E., Glicksman, R., Gonzalez, C. G., et al. (2005). *An unnatural disaster: The aftermath of Hurricane Katrina* (Center for Progressive Reform Publication No. 512). Washington, D.C.: Center for Progressive Reform. Retrieved August 8, 2007, from www.law.umn.edu/uploads/images/2233/ Unnatural_Disaster_512.pdf.

Esworthy, R., Schierow, L. J., Copeland, C., and Luther, L. (2005). Cleanup after Hurricane Katrina: Environmental considerations (Congressional Research Service Report for Congress). Retrieved August 8, 2007, from http://cnie.org/nle/crsreports/06may/ RL33115.pdf.

Fischetti, M. (2001). Drowning New Orleans. *Scientific American*, October. Retrieved August 8, 2007, from www.sciam.com/article.cfm.

Global Water Partnership and World Meteorological Organization (n.d.). Making integrated flood management part of the development agenda. Retrieved September 8, 2007, from www.apfm.info/ pdf/iwrm_floods.pdf.

Hecht, J. (2005). Global warming may pump up hurricane power. *New Scientist*. Retrieved August 8, 2007, from http://environment .newscientist.com/article/dn7769.

Heinberg, R. (2005). Katrina, New Orleans, and peak oil. *Global Public Media*. Retrieved August 8, 2007 from http://globalpublicmedia .com/richard_heinberg_katrina_new_orleans_a n d_peak_oil.

IWRM Questionnaire (n.d.) Retrieved September 8, 2007, from www .un.org/esa/sustdev/natlinfo/natlinfo_guidelines.htm.

Jonch-Clausen, T. (2004). *Integrated water resources management (IWRM) and water efficiency plans by 2005: Why, what and how?* (Global Water Partnership, TEC Background Paper No. 10). Stockholm: Global Water Partnership. Retrieved September 8, 2007, from www .gwpforum.org/gwp/library/TEC%2010.pdf.

Kahan, J., Wu, M., Hajiamiri, S., and Knopman, D. (2006). *From flood control to integrated water resource management: Lessons for the Gulf Coast from flooding in other places in the last sixty years* (Rand Corporation, Gulf States Policy Institute, Occasional Paper). Santa Monica:

RAND Corporation. Retrieved September 8, 2007, from www.rand.org/pubs/occasional_papers/2006/RAND_OP164.pdf.

Kemp, K. (2000). The Mississippi levee system and the Old River Control Structure. Retrieved August 8, 2007, from www.tulane.edu/~bfleury/envirobio/enviroweb/FloodControl.htm.

Knabb, R. D., Rhome, J. R., and Brown, D. P. (2005). Tropical cyclone report, Hurricane Katrina (National Hurricane Center). Retrieved August 8, 2007, from www.nhc.noaa.gov/pdf/TCR-AL122005_Katrina.pdf.

Louisiana Coastal Wetlands Conservation and Restoration Task Force and the Wetlands Conservation and Restoration Authority (1998). *Coast 2050: Toward a sustainable coastal Louisiana.* Baton Rouge: Louisiana Department of Natural Resources. Retrieved August 8, 2007, from www.lca.gov/net_prod_download/public/lca_net_pub_products/doc/2 050re port.pdf.

O'Brien, G. (2002). Making the Mississippi over again: The development of river control in Mississippi. *Mississippi History Now*, March 2002.

Pastor, M., Bullard, R., Boyce, J., Fothergill, A., Morello-Frosch, R., and Wright, B. (2006). Environment, Disaster, and Race After Katrina. *Race, Poverty and the Environment* 13 (1). Retrieved August 8, 2007, from http://urbanhabitat.org/files/Pastor.Bullard.etc.Env.Katrina.pdf.

Porretto, J. (2007). The petroleum industry adjusts, hopes to avoid Katrina-like damage. Retrieved August 8, 2007, from http://climate.weather.com/articles/petroleum052607.html.

Sze, J. (2006). Toxic soup redux: Why environmental racism and environmental justice matter after Katrina. Retrieved August 8, 2007, from http://understandingkatrina.ssrc.org/Sze/index.html#e1.

Thompson, M., and Gaviria, I. (2004). *Weathering the storm, lessons in risk reduction from Cuba.* Boston: Oxfam America. Retrieved September 8, 2007, from www.oxfamamerica.org/newsandpublications/publications/research_reports/art71 11.htm.

Tibbetts, J. (2006). A lesson in nature appreciation. *Environmental Health Perspectives* 114 (1): A40–A43. Retrieved August 8, 2007, from www.pubmedcentral.nih.gov/articlerender.fcgi?artid=1332684.

United States Army Corp of Engineers (2004). Louisiana coastal area (LCA), Louisiana. ecosystem restoration study. Retrieved from August 8, 2007, from http://data.lca.gov/Ivan6/main/main_report_all.pdf.

United States Environmental Protection Agency (n.d.). Summary of assessments at Superfund national priority list sites. Retrieved, August 8, 2007, from www.epa.gov/katrina/superfund.html#la.

Verchick, R. (2007). Risk, fairness, and the geography of disaster (International Law Forum of the Hebrew University of Jerusalem, Law Faculty Research Paper Series No. 01–07). Retrieved September 8, 2007, from http://ssrn.com/abstract=959247.

Zinn, J. (2007). Coastal Louisiana ecosystem restoration after Hurricanes Katrina and Rita (Congressional Research Service Report for Congress). Retrieved August 8, 2007, from www.ncseonline.org/NLE/CRSreports/07Jul/RS22276.pdf.

PART III KNOWLEDGE NECESSARY TO MEET POVERTY-ALLEVIATION GOALS

9 | Knowledge Necessary to Meet Poverty-Alleviation Goals

GILLIAN MARTIN MEHERS, ELISABETH CRUDGINGTON, AND KEITH WHEELER

What knowledge is necessary to meet poverty alleviation goals? What is the process by which this knowledge actually leads to real change? How is the knowledge marketplace changing, and how are institutions committed to poverty alleviation linking knowledge to action?

The knowledge needed to meet poverty-alleviation goals is found in the process of "knowledge to action." Learning is the link between knowledge and action, and between knowledge and change. We need to facilitate learning in order to put our knowledge to work for action to bring about the positive change that is crucial to alleviating poverty.

Things need to be done differently. We contend that knowledge about the role of the environment in meeting poverty alleviation goals will not, per se, lead to meeting these goals. We know that creating knowledge and making it generally available is not sufficient to bring about the changes necessary (although we still write a lot of books on the topic). To meet poverty alleviation goals, we need to look beyond the supply of facts, information, and skills and endeavor to understand this "knowledge to action" process better. In the process, we will generate new learning and knowledge.

Gillian Martin Mehers is the Head of Learning and Leadership at the World Conservation Union (IUCN). Elisabeth Crudgington is the IUCN Learning and Leadership Officer. Keith Wheeler is the Chair of the IUCN Commission on Education and Communication.

Traditionally, we think about knowledge from the supply side of knowledge creation and dissemination; it is helpful to broaden our thinking to include an understanding of the demand side. This brings the learner into the center of the discussion. This shift is warranted by changes in today's knowledge marketplace. Increasingly, the knowledge creation role is shared more broadly with the rise of user-generated content facilitated by the Internet. Another feature of this dynamic knowledge environment is that knowledge is increasingly "opt-in"; that is, the users decide what kind of knowledge they want, they go to many sources to find it, and then this aggregated information is managed by the user rather than the producer. Relevance, format, and context are of utmost importance in the new knowledge marketplace.

The knowledge needed to meet poverty-alleviation goals must be created with the learner/user in mind, or better yet, in collaboration with the learner/user. When knowledge is created in this way, the learning process itself becomes as important as the knowledge being shared. This process will reflect the diversity of learning preferences, contexts, time frames, and needs. Currently, this view of knowledge and learning is creating opportunities for organizations and individuals working in the field of environment and poverty alleviation.

Key questions about learning as the link between knowledge and change will be explored throughout this chapter, with reference to the environment and poverty alleviation work of the World Conservation Union (IUCN) in three areas:

- The Livelihoods and Landscapes Strategy (LLS), a newly launched strategy that has as its objective to improve poor people's livelihoods by effectively implementing national and local policies and programs that leverage real and meaningful change while advancing good environmental stewardship;

- The "Bitter Bamboo and Sweet Living" learning exercise, a part of IUCN's "Poverty Reduction and Conservation: Linking Sustainable Livelihoods and Ecosystem Management" Project, which aims to strengthen IUCN's ability to incorporate poverty reduction and livelihood considerations into conservation actions; and

- The Commission on Education and Communication (CEC), one of the six Commissions of the IUCN, and its new toolkit on Communication, Education, and Public Awareness.

Projects in these initiatives encompass many actors and many knowledge flows. In all three, the management of learning and knowledge has become a key feature of the work.

Poverty alleviation, like sustainable development more generally, can be viewed as a systems problem. Factors that create the conditions for poverty, and that exacerbate or alleviate them, interact dynamically through a complex interplay of cause and effect. Decisions and actions taken in different places (countries or continents), different arenas (policy or economic), and even in different time frames (thirty years ago or yesterday) can all be interacting to create conditions of poverty in a specific geographic location today. To what extent do our current knowledge-creation practices capture all of these interdisciplinary elements, make concrete recommendations, contextualize them, and then deliver them at the right time to the people who are actively working on change in that geographic location?

The traditional knowledge creators (whether local or international) are scientists, universities, research organizations, and to a certain extent, nongovernmental organizations (NGOs). Traditionally, these actors collect the data, analyze and publish it, make recommendations, test the recommendations in projects, and then perhaps publish them again as case studies and good practice (when the experience is a good one) in books and articles, or use them in courses and conferences.

What is the overall impact of this knowledge process? Sometimes knowledge is embedded in local practice and the change is permanent. Sometimes it is not, or we do not know. It has been only in the last ten years that outcome monitoring has been a feature of development activities. Increasingly, knowledge providers and creators do monitor for impacts and outcomes. IUCN started in its 2004–2008 program and is still learning how best to embed this practice. Yet, most of the time, the impacts of these interventions are seen elsewhere in place and time, making impact identification challenging, and making learning from these interventions weaker. The evaluation and monitoring field has flourished around the question of whether these kinds of development practices are making substantive change.

It can be challenging to learn from even one poverty alleviation intervention; this challenge is magnified when the desire is to learn from all poverty alleviation interventions worldwide. When our main pathway for knowledge dissemination is through books and articles, courses, and conferences, we encounter issues of both volume and time. For example, the U.S. Library of Congress online catalogue lists more than 4,200 items published on poverty since 2000, with roughly 600 new items added each year. That collection includes only copyrighted items from sixty countries—a fraction of what is produced globally in this field each year. These mechanisms take time to produce: university courses are developed over years, books and articles in refereed journals are scheduled in one- to two-year increments. Some journals, such as *Science*, have developed express publications; these new features of the knowledge-publication trade acknowledge the time delays that have constrained this knowledge-dissemination route in the past.

The growing volume of information globally is another aspect of the changing knowledge environment. With knowledge updating so rapidly, and with new options about how to receive it being created almost daily, information is increasingly a flow. Like electricity or water in your house, it is constantly available, you know where to get it, and you can turn it off or on whenever you like. You would never keep buckets and buckets of water around, all over your house, just in case you need it. Likewise, it is becoming more and more unrealistic to think that we should keep knowledge and information physically around us at all times. It is starting to make less and less sense to keep information around, carefully filed, just in case you need it—for example, keeping a library of reference books, a stack of every edition of a printed journal, or a copy of every newsletter published online. This is especially true when the information is constantly being updated on a website or portal, quickly changing, or becoming obsolete.

Perhaps new avenues for knowledge creation and learning are not only about information provision, but also building the skills to find it—to know where to go, whom to go to, and what to do with it when you finally get it. How might that change the way learning is approached in institutions?

What have we learned so far that has fostered the way in which we now think about the knowledge necessary to meet poverty alleviation goals? How are new programs working to link knowledge and learning for greater impact in meeting poverty alleviation goals? The following examples illustrate how IUCN is connecting knowledge to action and focusing on learning for more sustainable development.

Case Study: The IUCN Livelihoods and Landscapes Strategy

The Forest Conservation Programme of the World Conservation Union supports an integrated people-centered approach to conservation that ensures that forest resources are effectively restored, conserved and employed at the landscape-level to help secure sustainable and desirable livelihoods, particularly for the poor. In 2007, the Programme launched the Livelihoods and Landscapes Strategy to improve poor people's livelihoods in a set of pilot landscapes in three regions through leveraging real and meaningful change. "The strategy will represent a period of intense learning for the Union," explained Ibrahim Thiaw, former Acting Director General of IUCN, "establishing a systematic learning framework on which adaptive management decisions can be made. Livelihoods and Landscapes does not represent a magic formula, but is rather an innovative and exciting journey of practice and learning, during which long-held beliefs will be challenged and radical new insights emerge" (World Conservation Union, n.d.: 1).

The Livelihoods and Landscapes Strategy builds and draws on the many lessons learned in relevant existing field and policy work that the IUCN has and is currently undertaking. The value proposition of the strategy is not to create new stand-alone projects, but rather to connect past lessons and experience, and mobilize existing activities of IUCN and its members and partners, around the implementation of national priorities as they relate to rural poverty reduction and equitable and sustainable natural resource management. Central to this is the establishment of a process to share learning and knowledge among the partners at all levels to increase the effectiveness of the many pieces of the strategy.

The strategy emphasizes the delivery of locally-defined, landscape-level outcomes as well as program-level outcomes. Progress is assessed and success measured in terms of how well the program is facilitating learning and making real changes in the lives of Livelihoods and Landscapes beneficiaries. This is an innovation on past practice of limiting initiatives to the delivery of a series of predefined knowledge outputs, such as workshop studies and reports. The LLS supports the establishment of strong baselines around social, economic, and environmental indicators in order to improve lesson learning and outcome monitoring. An adaptive management approach will be employed in order to keep interventions responsive to changing circumstances and lessons learned.

The concept of leverage is a cornerstone of the strategy and has been an explicit founding principle. Among other ways, leverage will be demonstrated by upscaling results and lessons learned. It is therefore critically important that the lessons learned through LLS are captured and communicated, and LLS is investing in capacity building in order to translate field experience into meaningful policy change and to support effective on-the-ground implementation of progressive policies. By doing so, the various components will reinforce and add value to one another by effectively and consistently applying lessons learned both cross-thematically and cross-regionally—remembering that there is no single approach to capacity building and that this type of work must be done in a way that is appropriate to the context and target audience.

LLS is a learning initiative that builds knowledge about technical learning as well as process learning within its stakeholder team. It aims to increase its effectiveness and impact in contributing to poverty reduction in real terms within its landscapes, as well as to changes that support poverty reduction in the broader policy environment.

In the next case study, IUCN strives to scale up knowledge and follow it through the learning processes of other institutions and individuals, working toward real change on the ground.

Case Study: Using Learning to Scale Up and Scale Sideways—The IUCN Bitter Bamboo Project

From 1995 to 2002, a pilot project of the IUCN and the National Agriculture and Forestry Research Institute (NAFRI) promoted sustainable harvesting regimes for bitter bamboo shoots and wild cardamom in Nam Pheng, a village in the northern mountainous province of Oudomxay, Lao People's Democratic Republic (Lao PDR). The goal of the project was to conserve forest biodiversity by promoting sustainable economic exploitation of nontimber forest products at community and provincial levels.

The project's sustainable-harvesting regime for bitter bamboo generated impressive results in poverty alleviation and livelihood development, and it continues to be a shining example in the Lao PDR. In addition to providing villagers with enduring incentives and adequate capacities to manage village forests, the achievements were remarkable for their equitable distribution among villagers, capacity to reach the poorest households, and the interest that they raised in nontimber forest products among development and conservation organizations.

While positive and robust changes in forest-based livelihoods in a few pilot villages is good news, the more interesting and significant story concerns the extent to which learning about the successes at pilot sites was shared in a way that these successes could be replicated locally by others (scaling sideways), and the degree to which the project's learning and results influenced the way rural development is pursued nationally, through improved policies and programs in the forest sector (scaling upward).

Programs and projects that adopt the pilot-village or demonstration-site approach to influencing both the policy and practice for rural development need to take into consideration how they will use their learning to take lessons to scale sideways and upward. This is important to consider at both the design and implementation stage, since we need to document and consider how most effectively to share both the outcome and the process that takes us there.

As a contribution to improving forest-based livelihoods development interventions through learning, a study to "identify factors that determine how lessons learned from forest-based livelihoods development interventions are adopted into national policy frameworks or locally replicated at non-project sites in Lao PDR" began in early 2006. The study identified how the positive achievements at several pilot sites of the nontimber forest products project (including Nam Pheng) had been expanded locally (horizontally) and scaled up nationally (vertically) in Lao PDR.

The study found that government and nongovernment forest-sector projects in the Oudomxay Province were facilitating local replication of nontimber forest-product developments, but were focusing more on spreading the technical aspects of production and less on the social aspects such as collaborative-management agreements and the development of village organizations.

Similarly, the technical aspects of nontimber forest-product development were found to be replicated by local people in nearby villages about twice as much as the social aspects of collaborative management and marketing that were demonstrated in the pilot villages.

Furthermore, a clear disconnect was identified between the reason why people are attracted to replication and the progress of replication achieved so far. The establishment of marketing groups for nontimber forest products was one of the most successful in terms of helping to reduce poverty in the pilot village; however, it is the intervention that has been replicated the least by other projects and by local people themselves.

It appears that local replication at other places mostly involves separate interventions rather than a replication of the whole package of the project's interventions. The significance of replicating the package has been lost in sideways scaling. As a result, the full impact of a well-planned nontimber forest-product intervention may not eventuate to the extent that was hoped.

A major finding of the study was that serious national recognition is linked to a project's perceived success at the local level and to some extent the visibility of some sideways spread of some interventions. The involvement of division and departmental directors in the NAFRI and IUCN project activities facilitated the flow and exchange of project outcomes during and after the life of the project. This created a lasting impact on national policy, since they were the key people involved in drafting subsequent sector policy and strategy papers (Morris et al., 2004).

Enhance Capacities to Transfer, Adapt, and Use Knowledge by "Networking Knowledge"

In the Livelihoods and Landscapes Strategy and the Bitter Bamboo Project, strategic learning processes helped to increase the impact of the project. In the second case study, the people involved were critical in appropriately contextualizing and embedding the learning in further processes beyond the pilot activity. How can "networking knowledge" help increase the impact of poverty alleviation efforts in the current dynamic, knowledge environment?

George Siemens, author of *Knowing Knowledge*, observes that it is increasingly challenging for individuals to master a body of knowledge; there is too much information and it is becoming too complex (G. Siemens, personal communication, November 30, 2006). Therefore, we need to network our knowledge and rely on our network to collect and filter knowledge for us. Networked learning is knowing where to go, to whom to go, and how to learn as you go. This becomes particularly tactical in an environment where information changes rapidly, comes from distributed sources, and is for the most part itself technologically mediated (e.g., made publicly available through increasing numbers of portals, websites, and online libraries). It also means that learners are becoming more deliberate in their knowledge searches. They identify their knowledge needs and go for the specific information needed, rather than trying to keep up with the ocean of information flowing their way every day.

Case Study: Commission on Education and Communication

The World Conservation Union is the world's largest conservation network. The Union brings together 83 states, 110 government agencies, more than 800 NGOs, and some 10,000 scientists and experts from 181 countries in a unique worldwide partnership. The Union's mission is to influence, encourage and assist societies throughout the world to conserve the integrity and diversity of nature and to ensure that any use of natural resources is equitable and ecologically sustainable. It has 1,100 staff members located in forty countries, with headquarters in Gland, Switzerland. And it has six Commissions that unite volunteer experts from a range of disciplines. They assess the state of the world's natural resources and provide the Union with sound knowledge and policy advice on conservation issues.

The Commission on Education and Communication (CEC) is one of the six IUCN commissions, with an expert global network of 500 professionals who work in the fields of learning and strategic communication. The goal of this commission is to facilitate learning among diverse constituents in the sustainability community, to build capacity in knowledge and processes, and to create a dynamic interactive communication platform for its members and partners. In essence, the goal of the CEC is to network its knowledge so that the Commission is a community that is able to transfer, adapt and use knowledge to increase the impact of both IUCN and the broader sustainability community.

A growing focus of the Commission is on how to support and moderate the flow of knowledge, rather than to freeze it into knowledge products where the conversations have been stopped. In the past, CEC has produced many publications, and has collected the knowledge of its members into books and papers. Recently, however, a CEC initiative produced a new kind of knowledge product that is more of a knowledge process: a toolkit for Communication, Education, and Public Awareness (CEPA) (Commission on Education and Communication, n.d.). The toolkit is being produced in partnership with the

United Nations Convention on Biological Diversity. More than a hundred CEC members and partners participated in its development. A survey of the target user group identified a wide variety of learning needs and preferences, which is why the toolkit is being made available in multiple formats: printed on paper, published on CD-ROM, accessible online, and free to download as a PDF file. The website invites suggestions for the next edition, and makes a novel effort to keep the conversation flowing and to continue learning and exchange among contributors and users: a blog has been set up to capture people's thoughts as the CEPA Toolkit is implemented (ibid.).

The toolkit was developed to support the Convention on Biological Diversity's Article 15 on CEPA, for the benefit of the many national governments who have signed on to the convention. It provides a record of networked knowledge from the CEC members who contributed to its content; it also provides a forum to capture the ongoing exchange as users and subject matter experts keep the knowledge and learning flowing through the purpose-built blog. As new items and ideas come in from users and experts, the toolkit itself will be strengthened and become more and more useful. In this way, knowledge goes from being a publication product to a process product, and from something created by a few for many users, to something ultimately being created with them.

Conclusion: Knowledge and Learning

Today the discussion around the kind of knowledge needed for poverty alleviation and more sustainable development has expanded to include not only what kind of knowledge is needed, but also what kind of processes are needed to increase the impact of this knowledge and to keep the flow of knowledge constant, dynamic and contextually appropriate. We no longer necessarily need to freeze knowledge in products, print or otherwise, where the conversations and learning have momentarily stopped.

Today we have the means to keep these dynamic learning processes going. Technology has helped on one hand. Social software and Web

2.0 tools are increasingly coming into use in professional environments. Blogs are being used to continually report on project developments within distributed teams. Wikis are being developed to capture ongoing updates to team membership, background data, and central documentation. New portals and asynchronous dialogue websites are being captured for work-related as well as recreational networking.

The changing paradigm around learning has also helped. As the vocabulary of *training* slowly shifts into *learning*, where the emphasis moves from information and knowledge "push" to personal development and learner-centered "pull" for knowledge, we see the same dynamic as we are seeing with the Web-based learning tools. There is more emphasis on capturing and sharing what we already know, there is a democratization of knowledge creation and exchange, there is more attention to and discussion about what we are learning as we go about our work, and there are more opportunities for reflection and feedback of learning into the work we are doing. We understand that in today's knowledge economy, networked knowledge can help us have a more powerful impact on our goal of poverty alleviation and sustainable development worldwide. We are making our learning environments more process-oriented, more collaborative, and more targeted. We are no longer passive consumers of knowledge created for us.

Note on Informal Learning: Many of the ideas presented here are not referenced, as they come from innumerable conversations and informal learning opportunities that have shaped our thinking and behavior. In the absence of substantive formal references, we would therefore like to give credit to those individuals who have made significant contributions to the ideas captured within, as well as say thank you to the many others who have inspired us to learn something new every day.

References

Commission on Education and Communication (n.d.). Toolkit CEPA. Retrieved July 27, 2007, from www.cepatoolkit.org/html/topic_

EB4F6A65-6A05-419D-A5B2-C7EFA0C8734F_B6F868 C6-C970-41DD-BEC3-377E1EF7916D_1.htm.

Convention on Biological Diversity, Article 15: Access to Genetic Resources (1992). Retrieved on July 27, 2007, from www.cbd.int/convention/articles.shtml?a=cbd-15.

Morris, J., Hicks, E., Ingles, A., and Ketphanh, S. (2004). *Linking poverty reduction with forest conservation: Case studies from Lao PDR*. Bangkok: International Union for Conservation of Nature. Retrieved July 27, 2007, from http://app.iucn.org/dbtw-wpd/edocs/2004-090.pdf.

World Conservation Union (n.d.). Livelihoods and landscapes, a leverage programme to catalyse the sustainable use and conservation of forest biodiversity and ecosystem services for the benefit of the rural poor: Executive summary. Retrieved July 27, 2007, from http://livelihoodsandlandscapes.pbwiki.com/f/Executive%20Summary%20LLS%20October.pdf.

10 | Conservation Information

JEAN-LOUIS ECOCHARD

Lack of access to conservation information is clearly not an element of poverty in the way that food or water shortage is, but conservation information is increasingly essential for local communities to govern the natural resources they depend on. Poor people have the right to access and use the best information available to support their natural resource management decision making. Exercising that right, however, is a great challenge.

Successful conservation depends a great deal on the logical synthesis of data, information, expertise, and technology. Information tools such as land-use maps and species databases can dramatically facilitate community decision making and local management of resources, but the information they depend on is often fragmented, expensive, or hard to access (Stolton and Dudley, 2004). It is common for parties discussing an environmental problem to be unable to reach agreement because they anchor their positions on fundamentally different information, sometimes inaccessible to the other. For example, resource-extraction companies closely guard the proprietary information they have regarding mining sites. Likewise, many indigenous communities are reticent to share parts of their traditional knowledge with outsiders. The challenge for managers of natural resources, policymakers, and community leaders in developing countries alike who want to

Jean-Louis Ecochard is Chief Information Officer of The Nature Conservancy.

make science-based (evidence-based) decisions is to collate, synthesize, classify, and compile that fragmented information.

Broadly defined, conservation information includes all data necessary for the protection of biodiversity. Spreading over multiple sciences and practices, it contains datasets as varied as scribbles in a field scientist's notebook, colorful satellite images, global species databases, and "limited-edition," manually collated workshop proceedings. With so many sources, it is no wonder that conservation information is fragmented and difficult to access.

In this chapter, we will illustrate our point by presenting data on megadiverse countries with more than a quarter of their population living below the poverty line (US CIA World Factbook, 2006). These countries are the Democratic Republic of the Congo (DRC), Madagascar, Bolivia, Ecuador, Colombia, Peru, South Africa, Venezuela, Philippines, Mexico, Papua New Guinea, Indonesia, and India.

Fragmentation

Dealing with many of the most remote places on Earth, conservation information is naturally dispersed around the globe.

The theses of North American and European students who have researched biodiversity in the Southern Hemisphere all too often languish in the archives of universities in the Northern Hemisphere, thousands of miles away from those who could benefit from reading them most. Local species surveys often rot on paper, inaccessible in the damp libraries of field stations, and the most up-to-date data often remain protected (inaccessible) by scientists for future publication. Research notes are even less accessible.

This fragmentation is brought to light by global integration efforts such as the one conducted by the Global Biodiversity Information Facility (GBIF), an international nonprofit organization providing free and universal access to data regarding the world's biodiversity. The GBIF database holds more than 130 million individual specimen records from 204 providers representing a wide range of countries and organizations (GBIF, n.d.). It is one of the few leading integration efforts of conservation information.

Analysis of GBIF data reveals that for poor, megadiverse countries, biodiversity information is extremely fragmented over multiple collections and housed in many institutions; with almost none in the country of origin (GBIF, n.d.). For example, the quarter of a million species records on Madagascar are held in 146 collections of 73 institutions, none in Madagascar (Table 10.1).

With such a wide dispersion of records, poor countries have an almost impossible challenge to obtain and analyze local biodiversity information on their own. GBIF thus plays an important role in data

Table 10.1. Fragmentation of biodiversity data for poor megadiverse countries

COUNTRY	TOTAL NUMBER OF RECORDS AVAILABLE VIA GBIF	NUMBER OF INSTITUTIONAL RECORD PROVIDERS	NUMBER OF RECORD COLLECTIONS
Democratic Republic of the Congo	101,168	61	109
Madagascar	258,392	73	146
Bolivia	221,822	76	167
Ecuador	574,367	86	211
Colombia	261,618	88	194
Peru	403,819	90	215
South Africa	2,387,043	84	193
Venezuela	151,686	89	184
Philippines	137,952	84	183
Mexico	1,075,952	98	238
Papua New Guinea	350,090	75	142
Indonesia	233,066	84	197
India	203,805	87	213

Data adapted from GBIF Secretariat.

repatriation to countries of origin, and additional integration efforts are needed to facilitate the task for other data sets.

The Cost of Conservation Information

When conservation information is published it is often copyrighted, which makes its unauthorized duplication or literal translation illegal inasmuch as most countries (163 as of 2007) have signed some form of international copyright treaty.

Copyrighted scientific journals are sold for considerable profit by large publishing houses, notably Reed Elsevier, Blackwell, Harvard University Press, and Australia's CSIRO, and are often beyond the reach of low-paid conservationists in poor nations. Reed Elsevier, which under the Elsevier brand publishes a thousand journals in the category of agricultural and biological sciences, reported 2005 profits of 29 percent, comparable with past results (Table 10.2). Similarly,

Table 10.2. Elsevier Profits, 1999–2006

YEAR	REED ELSEVIER PROFIT MARGIN ON SCIENTIFIC PUBLISHING (%)
1999	35
2000	36
2001	37
2002	32
2003	34
2004	32
2005	29
2006	29

Adapted from Elsevier (1999, 4; 2000, 5; 2001, 5; 2002, 5; 2003, 5; 2004, 5; 2005, 4; 2006, 9).

Blackwell Publishing, another leading publisher of environmental journals, enjoyed 2005 profits of 27 percent (Rushton and Pickering, 2005). As a point of comparison, Coca Cola's 2005 net profit margin was 20.97 percent and Microsoft's 27.96 percent (Google, 2007a, 2007b).

These profits have been made at the expense of readers who, in the last two decades, have seen a dramatic increase in the price of scientific journals. For example, the price of zoology journals increased eightfold between 1984 and 2005, while during the same period general interest periodicals only doubled in price (Dingley, 2005).

Bergstrom and Bergstrom also point out the substantial price difference of academic journals owned by commercial publishers and those owned by nonprofits: "These price differences have grown rapidly over the past 15 years. In economics, for example, the average inflation-adjusted price per page charged by commercial publishers has increased by 300 percent since 1985, whereas that of nonprofit economics journals has increased by 'only' 50 percent. Studies of journal production costs indicate that the price differences over time and among journal types do not reflect differences in production and distribution costs" (2004). One can question if, for the sake of profit, commercial publishers are abusing the tradition of scientists to publish their research in scholarly journals without payment—for the sake of inquiry and knowledge.

The primary distribution channels for periodicals are libraries and academic institutions. Many field conservationists work far from a library and are not affiliated with institutions, and thus cannot benefit from institutional pricing of journals. An individual subscription to the leading publication Nature costs £203 (USD 402) in Congo, Madagascar, South Africa, Venezuela, Philippines, Papua New Guinea, Indonesia, and India, more than twice the price charged in the United States (Nature, 2007). Given that a wildlife biologist in the Congo is paid half the salary of a U.S. wildlife biologist, these costs are prohibitively expensive. As a result, copyrighted publications are often illegally duplicated or translated, and offending conservationists risk legal action.

Copyrighting conservation information as a mean to secure profits can have a devastating effect on the dissemination of scientific information about new species discoveries. Of the 472 new ant taxa described in the year 2003 and entered into the Ohio State University Insect Collection database in 2005, only 34 were openly available online without copyright. The rest, 83 percent, were available from publishers—for a fee (Agosti, 2005).

Finally, the public funds most scientific research. It is disconcerting when that same public is required to pay many times over the printing cost to use it, only because it is copyrighted. Likewise, it is ethically questionable to prohibit access by poor communities by charging high costs while making significant profits. Fundamentally, scientific publications should be free for the poor.

Conservation Information Is Hard to Access

For the last decade, the exponential growth of digital data and access to information via the Internet have made the use of technology in conservation no longer optional. For example, unparalleled efficiencies in content searches have made using a printed glossary to retrieve scientific literature obsolete. Information such as short-term, medium-term, and long-term weather forecasts, critical to the subsistence of most of the poor who live in rural areas and are dependent on agriculture, is increasingly distributed electronically.

There is also evidence that increased access to information leads to improved decision making. In rural areas near Pondicherry in southern India, a network installed by the M.S. Swaminathan Research Foundation broadcasts to villagers accurate predictions of wave heights in the sea thirty-six to forty-eight hours in advance. Based on this information, fisherfolk can safely decide to put to sea on a given day. Since this service was started, there has not been a single death at sea (Arunachalam, 2004). Indeed, the right information (sea conditions) put into the hands of motivated individuals (seafarers who do not want to die) results in effective decision making. A similar hope exists that conservation information will make a difference in the hands of motivated individuals.

Yet information is not that easily accessible across the globe. The world penetration of the Internet (16.6 percent) greatly varies over geographies, from 69.4 percent in North America to 3.5 percent in Africa (Internet World Stats, 2007). The Internet in megadiverse countries is equally unevenly distributed, ranging from 70.2 percent in Australia (one of the wealthiest) to less than 0.2 percent in the Democratic Republic of the Congo (the poorest) (Table 10.3). Not only is there an inequality of access, but the Internet can also exacerbate differences. For instance, in the case where a centralized government or a corporation, connected to Internet databases, has access to more information than the rural stakeholders who may be negatively affected by infrastructure decisions.

With such a disparity of access, stakeholders in poor countries often experience a lack of empowerment in policy debates, since there is an imbalance of information in negotiations. With restricted access to conservation information, the poor are thus challenged to begin organizing themselves for collective action to influence the decisions affecting their environment and potentially their lives. This fact is particularly onerous when one considers the importance of the environment in supporting and maintaining the livelihoods of much of the world's poor. Similarly, conservationists in developing countries without ubiquitous Internet access are at a competitive disadvantage for grants and foreign aid as compared to their better-connected colleagues in the north.

More than 80 percent of people in the world have never even heard a dial tone (Black, 1999), let alone accessed the Internet, and the gap between the information haves and have-nots is not closing fast. More efforts to bridge the digital divide are needed to give poor people access to the information they need, as UN Secretary-General Kofi Annan reflected in 1999, "People lack many things: jobs, shelter, food, health care and drinkable water. Today, being cut off from basic telecommunications services is a hardship almost as acute as these other deprivations, and may indeed reduce the chances of finding remedies to them" (Annan, 1999).

Table 10.3. Poverty and Internet penetration in megadiverse countries

MEGADIVERSE COUNTRY	% PEOPLE BELOW POVERTY LINE	% INTERNET PENETRATION OF POPULATION
Democratic Republic of the Congo	80	0.2
Madagascar	71	0.5
Bolivia	70	5.1
Ecuador	65	5.9
Colombia	55	12.9
Peru	54	15.8
South Africa	50	10.3
Venezuela	47	11.8
Philippines	40	9
Mexico	40	19
Papua New Guinea	27	2.8
Indonesia	27	8
India	25	3.5
Brazil	22	13.9
Australia	13	70.2
China	10	10
Malaysia	83	8.9

Adapted from CIA World Factbook, 2006; Internet World Stats, 2007.

Movements of Hope

There is an increased awareness that conservation decisions should be based on the best available data and information, and that this information should be easily and freely available to all stakeholders. This recognition coincides within the general context of liberalization of scientific information.

In recent years, numerous efforts have started to make scientific information free and open, particularly when funded by public sources. The Budapest Open Access Initiative, Committee on Data for Science and Technology (CODATA), Global Information Commons for Science Initiative, Organisation for Economic Co-operation and Development (OECD), GBIF, and others have called for free and open access to scientific literature (Budapest Open Access Initiative, 2001; Committee on Data for Science and Technology, 2006; GBIF, 2007; OECD, 2005). The 1998 United Nations Economic Commission for Europe (UNECE) Convention on Access to Information, Public Participation in Decision-making and Access to Justice in Environmental Matters, usually known as the Aarhus Convention, has been key to open environmental information.

As a demonstration of public interest, the European Union's Petition for guaranteed public access to publicly funded research results had been signed by more than ten thousand individuals mere weeks after its publication (Petition, 2007). The OECD even reports that governments would boost innovation and get a better return on their investment in publicly funded research if they made research findings more widely available (OECD, 2005). The democratization of scientific information is happening even in the largest institutions. In May 2005, the National Institute of Health (NIH), the largest funder of medical research in the world, adopted policies to provide free online access to full-text, peer-reviewed journal articles arising from taxpayer-funded research (NIH, 2005).

In addition, open-access journals such as the *Journal of Insect Sciences* and the *Public Library of Science* are emerging. Free software such as Linux rivals dominant commercial products, and free online encyclopedias like Wikipedia, produced by social networks of volunteers, surpass printed volumes in terms of quantity (but not quality) of information available. Free and open scientific information is becoming reality.

The need for the broad availability of conservation information has been captured in numerous agreements. Principle 10 of the Rio Declaration proposes that "environmental issues are best handled with participation of all concerned citizens, at the relevant level, and that at the national level each individual shall have appropriate access to information concerning the environment that is held by public authorities" (1992). Article 17 of the Convention on Biological Diversity (CBD) advises that "parties shall facilitate the exchange of information, from all publicly available sources, relevant to the conservation and sustainable use of biological diversity, taking into account the special needs of developing countries" (1992). Article 8(j) of the Convention on Biological Diversity says that "each party shall, subject to its national legislation, respect, preserve and maintain knowledge, innovations and practices of indigenous and local communities embodying traditional lifestyles relevant for the conservation and sustainable use of biological diversity" (1992). Chapter 40 of U.N. Agenda 21, emphasizes that, "in sustainable development, everyone is a user and provider of information considered in the broad sense and that the need for information arises at all levels, from that of senior decision-makers at the national and international levels to the grass-roots and individual levels" (1992).

Making conservation information freely and openly available creates the highest potential for people to translate these assets into local actions. This is particularly true for poor communities that do not have the funds to experiment with trial and error (learn by doing) conservation and must instead leverage existing knowledge.

The Conservation Commons of the World Conservation Union (IUCN) is leading an effort to make conservation information as free and open as possible. Launched in 2004 (Moritz, 2004), it is a collaborative effort by more than eighty organizations to promote conscious, effective, and equitable sharing of knowledge resources to advance conservation (Conservation Commons, 2007b). At its simplest, it encourages organizations and individuals alike to ensure open access to data, information, expertise, and knowledge related to the conservation of biodiversity.

The principles of the Conservation Commons promote free and open access; encourage mutual benefits through use and contribution, and support the recognition of rights and responsibilities—such as the respect of attribution and integrity of the work (Conservation Commons, 2004). These principles have been recognized by governments at the World Conservation Congress and the 8th Convention of the Parties to the Convention on Biological Diversity, and they have been tabled to the Global Ministerial Forum at the 24th UNEP Governing Council (Conservation Commons, 2004, 2006, 2007a).

The Conservation Commons has also endorsed numerous projects to make more conservation information available to the public, such as ConserveOnline, the Conservation Geoportal, antbase.org, and the World Database on Protected Areas. It is also addressing the complex aspects of access to traditional knowledge for conservation purposes.

While the Conservation Commons is making rapid progress, the challenge remains to transform engrained institutional cultures and practices that effectively act to withhold these assets from the conservation community. In the end, it is changes in behavior by individuals, conservation organizations, governments, and the public that will ensure that conservation information is freely accessible—creating space for effective conservation.

Easier Access

Open and easy access to conservation information remains a challenge that numerous collaborative efforts are addressing—particularly for documents, maps, and species.

Documents

ConserveOnline is a meeting place for the conservation community, open to anyone who wants to find or share information relevant to conservation science and practice. The website offers a public library, discussion groups and the ability to create small, easy-to-use websites where teams can collaborate on the conservation problems they face, solicit feedback, find experts, and announce events. It is intended to help improve the practice of conservation across organizations and national boundaries and is used by many people from developing countries. As many organizations in poor megadiverse countries cannot afford to build a website, they find in ConserveOnline both a practical tool to disseminate their conservation results and a forum to reach peers (Table 10.4) (ConserveOnline, n.d.).

Table 10.4. ConserveOnline users from megadiverse countries

COUNTRY	NUMBER OF UNIQUE CONSERVEONLINE USERS JULY 2006–JANUARY 2007
Democratic Republic of the Congo	4
Madagascar	68
Bolivia	150
Ecuador	488
Colombia	2,085
Peru	3,689
South Africa	883
Venezuela	192
Philippines	145
Mexico	8,897
Papua New Guinea	56
Indonesia	485
India	1,312
Brazil	2,382
Australia	2,983
China	9,447
Malaysia	301

The Protected Areas Learning Network (PalNet) of the IUCN started in 2003 as a component of the United Nations Environment Programme's Ecosystems, Protected Areas, and People (EPP) project (n.d.). It is now a website designed to engage protected-area managers globally and promote knowledge exchange to foster experimentation with adaptive management techniques. It offers access to a directory of experts, protected area projects lists and protected area documents. Most notably, PalNet has scanned and made available online many management plans of protected areas that had never been available to the public before. That effort was in response to the request from park managers in developing countries to have access to already published documents from which they derive inspiration.

Greenfacts brings complex scientific consensus reports on health and the environment to the reach of nonspecialists (Greenfacts, 2007). It provides easy access to faithful summaries of scientific reports, in lay terms and in several languages. One benefit for poor communities, many of which comprehend complex documents in English, is its summarization of the "Ecosystems and Human Well-being: Biodiversity Synthesis" of the Millennium Ecosystem Assessment (MA), a leading scientific report produced in 2005 (Greenfacts, 2006). It openly conveyed this complex material in an easily accessible format to nonspecialists and the general public.

Maps

The Conservation Geoportal, launched in 2006, is a collaborative effort by and for the conservation community to facilitate the discovery and publishing of geographic information systems (GIS) data and maps in order to support conservation decision making and education (Conservation Geoportal, n.d.). It functions as a data catalog, intended to provide a comprehensive listing of GIS data sets and map services relevant to biodiversity conservation. It does not actually store maps and data, but rather the descriptions and links to those resources. This leaves resource owners in control of their data. As a free tool, it hopes to minimize the proliferation of geospatial data catalogs and reduce duplication of effort in building and maintaining metadata catalogs and map viewers (ibid.).

The Global Land Cover Facility (GLCF) is a NASA-funded project at the University of Maryland, which openly distributes remotely sensed satellite data and products that explain land cover from the local to global scales (GLCF, n.d.a; n.d.c). GLCF currently hosts more than thirteen terabytes of data, including 28,558 Landsat images, 235 MODIS composites, and 803 ASTER scenes (n.d.b). Satellite imagery taken at different times is essential to understand local and regional environmental changes. Poor communities (with a link to the Internet) can easily download images of their area from GLCF.

Species

As mentioned earlier, GBIF, launched in 2001, is the leading integrator of biodiversity specimen data. GBIF participants include governments and organizations that hold primary scientific data on biodiversity, and seek to make it electronically available to policymakers and decision makers, research scientists, and the public (GBIF, n.d.). Its current prototype portal allows users to search, by scientific name, more than 118 million records.

GBIF has commissioned a yet unpublished special study through the University of Kansas to generate time series data from the millions of records stored in its database with the purpose of assessing biodiversity status (are we losing it, and if so at what rate?). The study will be of direct benefit to poor and developing countries that have to report progress on the 2010 Biodiversity Target of the Convention on Biodiversity. The target report requires several internationally agreed time series indicators for which poor and developing countries might not have the data or find it impossible to compile. Developing countries will be able to freely and openly use GBIF data to assess use and conservation of biodiversity, as well as its affect and influence on poverty.

Ants are ecologically one of the most important groups of animals worldwide. As previously mentioned, however, information on ant species is heavily copyrighted. Antbase.org is a collaborative effort between scientists from around the world, aiming at providing the best possible access to the wealth of information on ants (Antbase.org, 2005). It is now providing, for the first time, access to all the ant species of the world freely and openly.

Lower Costs

There is an increasing recognition that scientific information should be freely available to developing countries. Providing free access to conservation publications will greatly benefit conservationists in developing countries where cost represents a significant barrier to access to the best available knowledge and information for the conservation of biodiversity.

The Society for Conservation Biology (SCB) was the first to provide free access to conservation publications for conservationists in developing countries. In collaboration with Blackwell and Elsevier it has made its online access to *Conservation Biology*, *Conservation in Practice*, and *Biological Conservation* free to members in developing countries. Elsevier has also added *Ecological Indicators*, *Ecological Complexity*, and *Ecological Informatics* to this list of free publications. This marks a hopeful change in the practices of the two leading publishers of biodiversity journals. "The destruction of biodiversity worldwide is so rapid that there is no time to waste. Information must get out to conservationists who otherwise would not have access. SCB is leading the way in making scientific information available to conservation professionals and students in developing countries," said SCB Executive Director Alan Thornhill (SCB, 2006: 1). To illustrate the need, hundreds of conservationists registered within the first few weeks.

The Online Access to Research in the Environment (OARE) is an international public-private consortium coordinated by the United Nations Environment Programme (UNEP), Yale University, and leading science and technology publishers. Matching SCB's program offered to individuals, OARE enables institutions in more than one hundred developing countries to gain free access to large collections of environmental science literature. It is the sister program of HINARI (medical information), and AGORA (agriculture information). The complementary efforts to democratize scientific information are closely tied to the United Nations' Millennium Development Goals.

Even publishers, increasingly under pressure from the public to freely open scientific publications, are liberalizing authors' rights. For example, Elsevier now allows authors to post a preprint version of their journal articles on their own Internet web sites (Elsevier, 2007). Although all of these initiatives are positive developments, it is important to work with commercial publishers to create effective business models that ensure broader and more permanent access to conservation and research results in developing countries.

While we see positive efforts to freely and openly publish or integrate conservation information such as the Conservation Commons and GBIF, these efforts are not enough. Open access also depends on increasing accessibility, such as reducing the digital divide (Wikipedia, 2006) or more simply finding ways to deliver the right information on paper or CD-ROM to remote villages.

There is hope that one day it will be easy for everyone to access digital information. The One Laptop Per Child project, which aims to deliver a USD 100 laptop to children of the world, is greatly enabling low cost computing for the masses (One Laptop Per Child, n.d.). Likewise, to lower connectivity barriers, the People First Network permits remote locations to have Internet and email access using a computer, shortwave radio, and solar power (PFN, 2006).

It is essential to remember that when people struggle to secure basic daily needs such as food or water, information is perceived as a luxury, particularly to those who lack education. Indeed, education is often a prerequisite to understanding the value information provides. As Andres Ferrer, Director of the Dominican Republic program of the Nature Conservancy, remarks, "There is a lady I know who is poor; she lives with probably USD 10 a day plus her husband's income. They have three sons that made it to higher education and became professionals. At their household they have Internet, which may be costly by their standards, but the fact that they got an education resulted in prioritizing access to information" (A. Ferrer, personal communication, February 2, 2006).

More so, scientific publications often require tertiary education levels matching that of the authors. While literacy rates of adults in poor megadiverse countries average 82 percent (United Nations Educational, Scientific and Cultural Organization [UNESCO] Institute for Statistics, 2006), the tertiary-level gross enrollment ratio[1] (GER) for

---

1. Tertiary-level gross enrollment ratio is the sum of all tertiary-level students enrolled at the start of the school year, expressed as a percentage of the midyear population in the 5-year age group after the official secondary-school leaving age.

the same countries averages only 23 percent, in contrast to 70 percent in North America and Western Europe, where a majority of the scientific literature is authored (Table 10.5).

Conclusion

Conservation information alone is not a cure for poverty, of course, but it can greatly assist people to find sustainable solutions to the challenges they are facing. Today, this information is extremely fragmented, expensive, and hard to access, which makes it almost unusable by most developing country communities, even though they are

Table 10.5. Education statistics for poor megadiverse countries

COUNTRY	ADULTS (AGED 15 AND OVER) LITERACY 2000–2004 (%)	TERTIARY ENROLLMENT 2004 (GROSS ENROLLMENT RATIO %)
Democratic Republic of the Congo	67	—
Madagascar	71	3
Bolivia	87	41
Ecuador	91	—
Colombia	93	27
Peru	88	32
South Africa	82	15
Venezuela	93	39
Philippines	93	29
Mexico	91	22
Papua New Guinea	57	—
Indonesia	90	16
India	61	11

Adapted from UNESCO Institute for Statistics, 2006.

the ones who need it most to improve the effectiveness of their conservation and development efforts. Perhaps more than any other variable, effective conservation efforts depend on the actions of poor rural communities throughout the developing world. These actions are more effective when they are informed by science and offer choices to communities.

At the very least, conservation information should be free and openly available to poor countries without restriction of adaptation to local needs, such as translation into local languages, and without dependence on foreign publishers. Users can then democratically choose the type of information they wish to access, produce new information or package it into services, as an example, for access by remote communities. Free of restrictions, conservation information has the potential to knit conservationists together globally, similar to the manner in which free information fueled the explosive growth of the Web.

Making conservation information freely and openly available to the developing world will depend on the conscious effort of authors, organizations, donors, and the public alike to enable and demand that their data, tools, and publications be freely and openly available to those individuals and institutions who cannot afford to pay for them. Anything less will significantly compromise the success of conservation efforts, and, along with it, the success of sustainable development efforts predicated on a healthy environment providing a steady flow of goods and services. As Michael Tiemann, Vice President of Open Source Affairs at Red Hat, suggests, "We will not save biodiversity with a data restricting license. We will only save it by putting information about it to use" (Tiemann, 2004).

References

Agenda 21, Art. 40. (1992). Retrieved January 19, 2007, from www.un.org/esa/sustdev/documents/agenda21.
Agosti, D. (2005, January 26). Is copyright undermining biodiversity research and conservation? Preprint of lecture presented at

UNESCO International Conference on Biodiversity: Science and Governance, Paris. Retrieved January 23, 2007, from http://users.ecs.soton.ac.uk/harnad/Hypermail/Amsci/att-4330/agosti_paris_biodiv.pdf.

Annan, K. (1999). Remarks at the ITU Telecom opening ceremony. Retrieved June 28, 2007, from www.itu.int/telecom-wt99/press_service/information_for_the_press/press_kit/speeches/annan_ceremony.html.

Antbase.org (2005). Welcome to antbase.org. Retrieved January 23, 2007 from http://antbase.org.

Arunachalam, S. (2004). What can ICTs do? Perspectives from the developing world. Paper presented at the 2004 Aachen Colloquium on Click—A Split World, November 2004. Retrieved January 23, 2007, from http://dlist.sir.arizona.edu/1256/01/AachenSA.pdf.

Bergstrom, C., and Bergstrom, T. (2004). The costs and benefits of library site licenses to academic journals. *Proceedings of the National Academy of Sciences of the United States of America* 101 (3): 897–902.

Black, J. (1999). Losing ground bit by bit. BBC News Special Report. Retrieved January 23, 2007, from http://news.bbc.co.uk/1/hi/special_report/1999/10/99/information_rich_information_poor/472621.stm.

Budapest Open Access Initiative (2001). Retrieved January 23, 2007, from www.soros.org/openaccess.

Committee on Data for Science and Technology (2006). Creating the information commons for e-science. Retrieved January 23, 2007, from www.codataweb.org/UNESCOmtg/index.html.

Conservation Commons (2004). Principles of knowledge sharing of the Conservation Commons. Retrieved June 28, 2007, from www.conservationcommons.org/media/document/docu-1fouyx.pdf.

Conservation Commons (2006). Conservation Commons at the CBD. Retrieved June 28, 2007, from www.iucn.org/en/news/archive/2006/03/28_commons.htm.

Conservation Commons (2007a). Joint statement—24th UNEP Governing Council, 2007. Retrieved June 28, 2007, from www.conservationcommons.org/section.php?section = work&sous-section = endorsement&langue = en.

Conservation Commons (2007b). Organizations that have formally endorsed the principles. Retrieved June 28, 2007, from www.conservationcommons.org/section.php?section=member&langue=en.

Conservation Geoportal (n.d.). What is it? Retrieved June 28, 2007, from www.conservationmaps.org/about/index.jsp?cmd=whatitis.

ConserveOnline (n.d.). What is ConserveOnline? Retrieved June 28, 2007 from http://conserveonline.org/faq/what_is_col.

Convention on Biological Diversity, Art. 17. Exchange of Information (1992). Retrieved January 23, 2007, from www.cbd.int/convention/convention.shtml.

Dingley, B. (2005). *U.S. Periodical Prices—2005. U.S. Periodicals Price Index.* Retrieved January 23, 2007, from www.ala.org/ala/alctscontent/alctspubsbucket/alctsresources/general/periodicalsindex/05USPPI.pdf.

Elsevier (1999). Preliminary results 1999 strategy. Retrieved January 23, 2007, from www.reed-elsevier.com/media/powerpoint/g/3/ResPre99.ppt.

Elsevier (2000). Preliminary results 2000. Retrieved January 23, 2007, from www.reed-elsevier.com/media/pdf/t/i/Website%20-%202000%20preliminary%20results%20presentation.pdf.

Elsevier (2001). Interim results 2001. Retrieved January 23, 2007, from www.reed-elsevier.com/media/powerpoint/RE_InterimResult_June2001.ppt.

Elsevier (2002). Interim results 2002. Retrieved January 23, 2007, from www.reed-elsevier.com/media/pdf/q/g/Website%20-%20interim%20results%202002.pdf.

Elsevier (2003). Preliminary results 2003. Retrieved January 23, 2007, from www.reed-elsevier.com/media/powerpoint/m/1/FINALpresentationforwebsite.pdf.

Elsevier (2004). Interim results 2004. Retrieved January 23, 2007, from www.reed-elsevier.com/media/powerpoint/q/6/Interim_Presentation_2004.ppt.

Elsevier (2005). Interim results 2005. Retrieved January 23, 2007, from www.reedelsevier.com/media/pdf/8/e/Interim%20results%202005%20analysts%20presentation.pdf.

Elsevier (2006). Interim results presentation 2006. Retrieved January 23, 2007, from www.reed-elsevier.com/media/pdf/s/a/2006_Interim_results_presentation_final_for_website/pdf.

Elsevier (2007). Copyright. Retrieved January 23, 2007, from, www.elsevier.com/wps/Find/authorsview.authors/copyright.

Global Biodiversity Information Facility (n.d.). About GBIF. Retrieved June 28, 2007, from www.secretariat.gbif.net/portal/index.jsp.

Global Land Cover Facility [GLCF] (n.d.a). About GLCF. Retrieved June 28, 2007, from http://glcf.umiacs.umd.edu/aboutUs.

GLCF (n.d.b). Data and products. Retrieved June 28, 2007, from http://glcf.umiacs.umd.edu/data.

GLCF (n.d.c). Sponsors of the GLCF. Retrieved June 28, 2007, from http://glcf.umiacs.umd.edu/aboutUs/sponsors.shtml.

Google Finance (2007a). Net profit margin, Microsoft Corporation. Retrieved June 28, 2007, from http://finance.google.com/finance?q=MSFT.

Google Finance (2007b). Net profit margin, The Coca-Cola Company. Retrieved June 28, 2007, from http://finance.google.com/finance?q=KO.

Greenfacts (2006). Scientific facts on biodiversity and human well-being. Retrieved June 28, 2007, from www.greenfacts.org/en/biodiversity/biodiversity-greenfacts.pdf.

Greenfacts (2007). About Greenfacts. Retrieved June 28, 2007, from http://about.greenfacts.org/index.htm.

Internet World Stats (2007). Internet usage statistics. Retrieved January 11, 2007, from www.internetworldstats.com.

Moritz, T. D. (2004). Conservation partnerships in the commons? Sharing data and information, experience and knowledge, as the essence of partnerships. *Museum International* 56 (4): 24–31.

National Institute of Health (2005). Policy on enhancing public access to archived publications resulting from NIH-funded research. Retrieved January 23, 2007, from http://grants.nih.gov/grants/guide/notice-files/NOT-OD-05-022.html.

Nature (2007). Subscribe. Retrieved January 23, 2007, from https://secure.nature.com/subscribe/nature.

One Laptop Per Child (n.d.). Progress. Retrieved June 28, 2007, from www.laptop.org/en/vision/progress.

Organisation for Economic Co-operation and Development (2005). Digital broadband content: Scientific publishing (Report of the Working Party on the Information Economy, Committee for Information, Computer and Communications Policy, Directorate for Science, Technology and Industry). Retrieved January 23, 2007, from www.oecd.org/dataoecd/42/12/35393145.pdf.

People First Network (2006). The Solomon Islands Remote E-mail Network +. Retrieved June 28, 2007, from www.peoplefirst.net.sb/downloads/pfnet_brochure.pdf.

Petition for guaranteed public access to publicly funded research results (2007). Retrieved June 11, 2007, from www.ec-petition.eu.

Protected Areas Learning Network (n.d.). About PALNet. Retrieved June 28, 2007, from www.parksnet.org/index.php?globalnav=aboutandsectionnav=intro.

Rio Declaration on Environment and Development, Principle 10 (1992). Retrieved January 19, 2007, from www.unep.org/Documents.Multilingual/Default.asp?DocumentID=78&&ArticleID=1163.

Rushton, K., and Pickering, R. (2005, June 22). Blackwell publishing journals boost profits. *Information World Review*. Retrieved January 23, 2007, from www.iwr.co.uk/information-world-review/news/2138587/blackwell-publishing-journals.

Society for Conservation Biology (2006, August 24). Society for conservation biology to provide free access to publications for developing country members. *Society for Conservation Biology Press Release*. Retrieved January 23, 2007, from www.conbio.org/media/benefits/NewBenefitsDevelopingCountrySCBmembers.pdf.

Stolton, S., and Dudley, N. (2004). Sharing information with confidence. Retrieved January 23, 2007, from http://conserveonline.org/workspaces/commons/PDF%20Background%20Documents/Sharing_Information_with_Confidence.pdf.

Tiemann, M. (2004, June 1). Remarks at IUCN Biodiversity Commons Workshop.

United Nations Educational, Scientific and Cultural Organization, Institute for Statistics (2006). *Global education digest 2006. Comparing education statistics across the world.* Montreal: UNESCO Institute for Statistics. Retrieved June 8, 2007, from www.uis.unesco.org/TEMPLATE/pdf/ged/2006/GED2006.pdf.

United States Central Intelligence Agency (2006). World Factbook. Retrieved January 19, 2007, from https://www.cia.gov/cia/publications/factbook/index.html.

Wikipedia (2006). Digital divide. Retrieved January 23, 2007, from http://en.wikipedia.org/wiki/Digital_divide.

World Intellectual Property Organization (2007). Contracting parties to the Berne Convention. Retrieved June 28, 2007, from www.wipo.int/portal/index.html.en.

11 | Building Enterprises to Reach Low-Income Markets

YASMINA ZAIDMAN, HELEN NG, AND ADRIEN COUTON

A small-scale farmer in Maharashtra, India, had planted a quarter-acre of chilies, a valuable cash crop, hoping that rainfall and the limited water resources from his well would allow him to grow this crop for his family's annual income. Unfortunately, neither the rainfall nor the well delivered what he had hoped. Falling water tables meant that even with a pump, he could not access sufficient water from his well to keep the crop alive. The scarce water resources he could access fell to waste through "flood irrigation," a process that involves pouring water into the furrows between crops. Before deciding to pull up the doomed plants, he heard about an affordable new product called Krishak Bandhu, which means "the farmer's friend" in Hindi. For USD 60, he purchased this low-cost irrigation technology that could potentially revive his chili plants. Using this technology, not only did he keep his chili crop alive, but he also expanded the amount of land he was irrigating to half an acre by using his existing water resources more efficiently. His investment paid off, and that year he doubled his income from USD 150 to USD 300, instead of facing a total loss. Within one more season, this farmer will have crossed India's official poverty line.

Yasmina Zaidman is the Director of Knowledge and Communications at Acumen Fund. Helen Ng is the Portfolio Manager for Housing, and Adrien Couton is the Water Portfolio Manager, both at Acumen Fund.

This is an example of an approach to addressing poverty and environmental challenges using market mechanisms and design innovations. The notion that provision of charitable or government aid and expertise from the developed world can solve the problems of the global poor is no longer taken for granted, as fifty years of top-down development have failed to close the gap between the rich and the extremely poor (Danaher, 1994). India, for example, is now considered by many to be an "emerging economy" with a huge and growing middle class, but 35 percent of the population still live below the national poverty line, which is around USD 1 per day, and 80 percent live on less than USD 2 per day (World Bank, 2005). Closing these gaps is likely to require not only philanthropic and governmental intervention, but also the development of markets that can reach underserved segments with affordable goods and services.

The emergence of social entrepreneurship as a recognized and respected profession has emphasized the importance of leadership and innovation, but has not always addressed the need for financial sustainability. On the other hand, a growing interest in business models that serve the Base of the Pyramid (BoP) or those who earn less than USD 4 per day indicates increased interest by the private sector in business models that are viable and reach the poor (Emmons, 2007). However, efforts to achieve both social and financial objectives have often failed, in some cases because traditional products or business models are only slightly modified and not redesigned to address the needs of completely distinct markets. As an example, the introduction of sachets to sell products like shampoo and detergent have certainly penetrated BoP markets, but it is not clear how these products are impacting the lives of the poor (Karnani, 2006).

This chapter focuses on social entrepreneurs that seek to balance these social and financial objectives. To ensure the delivery of critical products and services to the poor at the required scale and on a sustainable basis, these entrepreneurs are reducing their reliance on grants and subsidies and instead seeking out financial viability and access to capital markets.

In addition, as market forces alone are insufficient to address critical development challenges and deliver value to the BoP, entrepreneurs are forced to develop innovative business models that build on social capital. Their innovations are not exclusively of a technological nature, but also include innovations within the realm of product and service delivery models. According to Acumen's experience, some of the challenges involved in reaching markets that have been previously overlooked include: a lack of distribution channels, a lack of communication channels, and lack of basic infrastructure that could support the introduction of new products or services. Infrastructure such as roads, running water, electricity, phone lines, and other communications tools cannot be taken for granted. The entrepreneurs that can successfully address these challenges will look not only at social and financial parameters, but also at innovative design solutions that can function within a BoP context. By highlighting these success stories, we seek to demonstrate the potential for these models to succeed on social and financial terms, and reduce the perception of risk that is often associated with innovation.

One such success involves small-scale farmers that are severely affected by water scarcity but unable to afford water-saving technologies, and hence use water inefficiently. This case closely examines how access to technology can help alleviate the social and economic strains of water scarcity. Another case examines the affordable housing sector, where families that have been left out of the formal housing market are looking to build a community that is sustainable financially and environmentally. The case also presents the potential conflict between supporting development and preserving the environment.

Each entrepreneur that takes on the challenges involved in this sector has taken on a great degree of risk and has, explicitly or implicitly, volunteered to become a role model for entrepreneurs that will follow. For this reason, we also highlight the importance of leadership in the form of moral imagination, integrity, courage, and humility. Though leadership may be more innate than taught, the presence of real-world examples and role models clearly pave the way for a new kind of leader to emerge. This is an opportunity to showcase examples of a new kind of leadership, which combines pragmatic skills with a commitment to bring about dramatic change, despite significant risks. By focusing on examples of enterprises that achieve this delicate balance, we begin to see models and insights that can be applied more broadly.

Case Study 1: International Development Enterprises in India

The following case study on a drip-irrigation enterprise shows how design innovation, combined with a sustainable business model can help small-scale farmers manage scarce water resources, increasing their incomes while reducing their environmental impact.[1]

Water Management Challenges in India

Over the last fifty years, global grain production has tripled, the result of a "Green Revolution" relying on improved seeds, fertilizers, and irrigation (Brown et al., 1999, as cited in Postel et al., 2001). The revolution was strikingly successful, transforming India from a country with a food deficit to one with a food surplus. Today, Indian agriculture accounts for over 25 percent of the nation's GDP and 15 percent of its exports (Ministry of Agriculture, 2004). It is the mainstay of the economy; 65 percent of India's population and 75 percent of the poor, who earn less than a dollar a day, are directly dependent on agriculture (Ministry of Agriculture, 2004). For the sector's success story to continue, India will need to tackle looming environmental and social challenges.

On the environmental front, water availability has emerged as the most crucial issue for the future of the agricultural sector. India accounts for 16 percent of the world's human population, but it has only 4 percent of the water resources (SSKI, 2004). The share of water available for agriculture is bound to fall as the industrial sector grows. Growing water needs for domestic and industrial uses will reduce the quantity of water available for farming at a time when increasing

---

1. Unless otherwise cited, the information provided in the case is sourced from the management team at IDEI and Acumen Fund's internal documents.

quantities of food will need to support the rapid growth of the population.

Challenges are equally daunting on the social front. The Green Revolution completely bypassed the majority of the country's small-scale farmers and their families. Over 75 percent of India's farmers are small-scale farmers, cultivating plots of less than two hectares in size (Polak, 2006, as cited in Kulkarni et al., 2006). Yet, in order to achieve quick impact, the agricultural innovations of the Green Revolution were typically designed for large, mechanized farms. Most of the commercially available microirrigation systems, for example, are optimized for fields of four hectares or larger, requiring expensive emitters to do the dripping and highly qualified staff to operate and maintain the system. As a result, these systems are too expensive and impractical to operate on small plots, and irrelevant to the majority of poor farmers. As water resources become increasingly scarce, small-scale farmers are unable to afford the water saving technologies that would enable them to maximize their productivity, and would enable India to tackle its burning issue of water scarcity.

Ironically, an irrigation technology long viewed as appropriate only for wealthier farmers now appears to hold great promise for small-scale, poor farmers: drip irrigation. The organization that has brightened the prospects for India's small-scale farmers is a nonprofit enterprise, established in India in 2001, International Development Enterprises India (IDEI). IDEI recognized that the lack of agricultural innovations targeting low-income farmers with small plots of land was both a serious unmet social and environmental need and a significant untapped market. As a result, IDEI developed a range of drip irrigation systems affordable to small-scale farmers. The first system was priced at approximately USD 4. Through continued innovation in its technologies, IDEI has grown the prospects for the expansion of drip irrigation. IDEI has already sold more than 150,000 drip systems in seven Indian states, directly impacting the lives of 750,000 people. The accelerating spread of affordable drip-irrigation technologies to small-scale farmers can form the backbone of a second green revolution—this one socially and environmentally sustainable.

The Social and Economic Benefits of Drip Irrigation

Drip irrigation enables farmers to produce more crops with less water. The system consists of a network of pipes combined with suitable water emitting mechanisms based on the simple idea of irrigating only the root zone of the crop rather than the entire land surface as is the case with surface-irrigation methods. As a result, farmers achieve higher crop yields and better product quality for less water. Figures vary by crop, but the impact is generally dramatic: for example, IDEI systems increase the yield of banana plantations by 50 percent with 45 percent less water use, of sugarcane by 33 percent with 55 percent less water use, and of chilies by 45 percent with 60 percent less water use.

By unlocking access to drip technologies, IDEI generates dramatic economic benefits for small-scale farmers. For them, access to drip systems has three consequences. First, they can immediately begin harvesting three crops each year instead of one. In monsoon climates with a long dry season and in semiarid and arid regions where IDEI operates, poor farmers were traditionally bound to cultivate their land only during the rainy season; drip systems now make agriculture possible during the rest of the year and also irrigate larger tracts of land. Second, it allows them to grow higher-value crops, such as sugarcane, which is a twelve-month crop that requires continuous irrigation. Third, it brings higher yields, as irrigated plots commonly yield twice as much as that of rainfed plots.

Between a farmer with no irrigation who is able to grow only one low-yield crop (e.g., wheat) in his acre field and a farmer who is taking advantage of a water source to grow three high-yield, higher-value crops (e.g., wheat, chilies, tomatoes), the commercial value of the production varies by a factor of 20, an amount enough to build a bridge out of poverty. IDEI's involvement in drip irrigation also illustrates how an entrepreneurial approach can deliver social and economic value to low-income communities, while reducing the consumption of scarce resources.

By raising small-farm productivity, drip irrigation has the power to revitalize rural communities. It creates jobs for people both with and without land, since more productive farms need more hands to harvest, process and market the crops grown, and to supply farm inputs such as seeds and fertilizers. Access to drip irrigation broadens farmers' crop choices and enables them to grow higher value vegetables and fruits for the marketplace. Access to productivity-enhancing irrigation tools contributes to the rise of more secure and stable rural communities, which helps stem the tide of migration to already overcrowded cities and slums. It can also reduce seasonal migrations that see millions of Indian farming families leave their village for several months in search of wage labor. It also creates the possibility of an education for their children and improves the condition of women. The first decision made by a farmer with the extra income generated by the irrigation system is often to withdraw his wife, daughters, and daughters-in-law from working as hired-wage hands on others' farms.

The use of drip irrigation systems also has significant environmental benefits that go far beyond the typical water savings of 30 to 50 percent. Water-soluble fertilizers can be applied through the microirrigation system, resulting in targeted fertilizer application and reduced harm to the environment. Precision irrigation also saves energy: studies have revealed that 85 percent of energy requirements in a village are used for irrigation. The use of the drip-irrigation system reduces farmers' electricity consumption by 50 percent.

Design Principles to Serve Farmers Earning Less Than USD 1 Per Day

Designing irrigation systems tailored to the needs of small-scale farmers requires an entirely new approach. An appropriate design needs to consider small farmers' circumstances from the outset. Taking an existing design that has served large-scale farmers and modifying it to meet the needs of small-scale farmers would result in a product that is effective for neither group. IDEI places the small-scale farmer at the core of its design process and follows three principles in the design of agricultural technologies tailored to their needs: miniaturizing, aggressively pursuing affordability, and making expandable systems.

IDEI's first objective is to design for small-scale farms. Not only do small-scale farmers in India typically have less than two hectares of land, but most of these farms are also divided into five or six separate plots. For IDEI, the quarter-acre plot is therefore the measuring stick against which new technology for small farmers should be evaluated.

Second, systems must be affordable to farmers earning no more than one dollar a day by ensuring that the payback period for a drip system is no more than one crop season. Products are tailored to the farmer's willingness to pay and designed to generate rapid payback; farmers must recoup their investment in the systems within a year. IDEI discovered that the most effective way to optimize affordability can often be found in tracing back the history of the design process leading to the current form of the technology. For example, early drip systems used holes or microtubes instead of sophisticated emitters. Ultimately, these simple designs were abandoned because they did not fit the needs of modern medium and large-scale farming in developed countries. However, the designs are well-suited for drip-irrigation of small plots where elevation differences within the plot are minimal, so that pressure losses are small. Moreover, sufficient low-cost labor is available in India to allow for periodic inspection and maintenance of the less sophisticated emitters designed by IDEI. Where high-cost designs assume that it is cheaper to minimize labor, a low-cost design is often more labor-intensive.

IDEI's third golden rule is that systems must be expandable. Farmers must be able to purchase a small affordable system and gradually expand it as their income grows. IDEI achieves this in two ways: first, by making the systems modular and customizable based on the size of the plot to be irrigated; and second, by allowing farmers to choose which plastic tubing they buy based on the desired durability of the system and their ability to pay. An extremely poor farmer can thus buy a small system designed to last two seasons and invest in a larger, more durable system as his or her income increases.

Design, of course, does not end with the creation of a good product. Another major aspect of IDEI's product development work is an active marketing effort needed to convince the highly risk-averse small-scale farmers living at or below the poverty line to adopt new technologies. To make sure that agricultural innovations reach small-scale farmers, IDEI works simultaneously on two fronts, educating poor people about the benefits of new irrigation technologies and working with local manufacturers to build a strong production base. This approach establishes a network of dealers and builds the expertise of local technicians who are able to install, maintain, and repair the products.

Leading Innovation to Build Markets for Small-Scale Farmers

Amitabha Sadangi has brought his own leadership vision to his work at IDEI. Amitabha is a social entrepreneur who sees the value that entrepreneurial models can have in achieving sustainable and scalable results for the poor, and is committed to a market-based approach to addressing the critical needs of small-scale farmers. Amitabha understands the need for systems that support the delivery not only of needed technologies to small-scale farmers, but also of postharvest support to bring products to market and help farmers earn more by processing their crops into value-added products. Thus, under Amitabha's leadership, IDEI created an exemplary treadle pump supply chain that has provided more than 350,000 treadle pumps in India.

In its focus on bringing products to market, over the past several years IDEI has invested in the development of a supply chain for the manufacturing of the drip systems, and broad distribution networks to make them available to farmers across states with water scarcity challenges. IDEI continues to refine its product offerings and its approach to marketing, distribution, and supply-chain management.

IDEI's emphasis on the need for continual innovation includes a culture of accountability to customers, no matter how poor, that extends all the way to senior management. As a result, IDEI has created a continuously expanding portfolio of products designed for low-income farmers. The design process is largely informed by feedback from farmers, and IDEI's field staff is required to interact with dozens of farmers every year to get feedback on specific products, local challenges, and service issues.

Design innovations are complemented by an approach to supply chain management that builds on the incentives of local manufacturers and dealers, and allows IDEI to remain lean and focused on design, distribution and marketing. In addition, commitment to fiscal independence has led IDEI to rely less on grants and more on capital markets. With support from Acumen Fund, IDEI extended its focus on fiscal sustainability and created a quickly expanding for-profit entity, GEWP, to accelerate the distribution of IDEI's water management technologies for small-scale farmers. As this business model evolves and grows, IDEI's impact extends well beyond the individual customers it serves. IDEI's mission to address the needs of India's poorest through innovative design and market-driven strategies makes it a powerful example of a new kind of leadership that transcends the dynamic tension between social and financial objectives.

Case Study 2: Jamii Bora

This case study shares the history of an organization that brought financial products and a spirit of entrepreneurship to communities profoundly marginalized by poverty in the slums of Nairobi, Kenya. Now expanding its scope to affordable and sustainable housing, Jamii Bora exhibits the potential to balance social, financial, and environmental goals.

About Jamii Bora

Jamii Bora—which means "good families" in Swahili—is one of Kenya's fastest growing microfinance institutions (MFIs). Its mission is to assist members to get out of poverty and build a better life for their families through self-empowerment. To Jamii Bora members, this translates into the ability to borrow money to develop enterprises that can sustain families and communities. Headquartered in Nairobi, Jamii Bora serves both the urban and rural poor throughout the country, with concentrated operations in the slums of Nairobi: Kibera (the largest slum in sub-Saharan Africa), Mathare, and Soweto.[2]

Founded in 1999 by Ingrid Munro, Jamii Bora has expanded rapidly over the past seven years, having grown from fifty informal participants to more than 130,000 members in seven years. Jamii Bora opened its first three branches in 2000, and now has 61 branches located primarily in urban areas with a distribution system that reaches 140 communities in the outlying and rural areas. Jamii Bora has a staff of more than two hundred, about a third of them finance specialists. As of June 30, 2006, the Trust had USD 9.0 million in assets, of which USD 5.0 million are net loans under management. By December 2005, Jamii Bora had provided more than 144,700 loans, totaling more than USD 17.0 million. Its portfolio at risk is 1.27 percent and its loss recovery rate nearly 100 percent.

Jamii Bora is now planning to build Kaputiei, a completely new town with two thousand homes that will be inhabited by its members, many of whom are from the largest and poorest slums in Nairobi. In this innovative low-income housing development scheme, Jamii Bora plays the role of both developer and mortgage provider. This is a new role for Jamii Bora, which has traditionally focused on microfinance. Other offerings include education, health (life and health insurance), counseling, and a program for orphans as well as street children.

A New Kind of Leader: Ingrid Munro

One of the forces behind Jamii Bora's rapid rise is its founder, Ingrid Munro. Through a combination of spiritual will and practical action, she has helped Jamii Bora grow to more than 130,000 members in less than seven years. Affectionately dubbed "Mama Ingrid" by her inner circle, Ingrid is a charismatic and inspirational leader and a personal mentor to those in need. However, underneath the "Mama Ingrid" embrace is a hands-on, detail-oriented pragmatist with a track record of stoic discipline, resourcefulness, and practical action. From

---

2. All information in this case study has been sourced from Jamii Bora management and Acumen Fund's internal documents.

actively engaging in Jamii Bora's daily operations and overseeing every step in the construction of Kaputiei to maintaining a rigorous and strict membership policy based on loan repayments, Ingrid ensures that Jamii Bora is in order.

Yet beyond the persona, it is perhaps the philosophy that has been manifested in Jamii Bora that provides the foundation of the organization. Simple, and yet universally compelling, her message is that nobody is too poor to take themselves out of poverty. From the very beginning, Ingrid counseled her impoverished friends to have confidence in their own abilities and not be dependent on others. With a simple micro loan, the members of these communities—many of whom had traditionally resorted to begging, prostitution and stealing to support themselves—were able to carve out new livelihoods by creating their own businesses. As they became successful and gained greater confidence, they in turn became leaders and mentors within the slums of Nairobi. They inspired new members to join Jamii Bora and follow suit.

Building a Community Through Entrepreneurship

One by one, Jamii Bora has helped to change minds deep within the slums of Nairobi and beyond—effectively transforming a mindset of desperation into one of entrepreneurship through its "can do" attitude. Aside from its positive messaging, the basic tool that Jamii Bora uses to change minds is its micro loan products. Backed by a group guarantee, micro loans are issued to members based on the amount they have saved and are to be repaid within approximately one year. This plan requires learning how to save money as a first step. Learning to save is a difficult exercise for those who make less than USD 1 per day, but Ingrid instituted a strict, yet simple, system of incentives where saving a certain amount of money would earn a member access to twice the amount in the form of a loan. The next step towards entrepreneurship is learning how to manage a loan and use it toward creating a business. If a member can demonstrate successful repayment, he or she can then qualify for a larger loan depending on the level of savings. However, members who do not pay back their loans can be penalized and ultimately requested to leave the organization. Not only does this system instill a new sense of discipline with money management, but it also provides members with a lifeline out of poverty. Over time, many of its members have completely transformed their lives and have become successful businesspeople, running activities ranging from beauty salons, clothing design, and grocery stores to construction and real estate.

Kaputiei Town Development

Taking this spirit of entrepreneurship to scale, Jamii Bora has decided to build an entirely new town for their own members through private financing. Jamii Bora is launching a major private-sector led housing program to move communities out of slums and into safe and affordable homes in Kaputiei Town, a 293-acre site located approximately sixty kilometers south of the center of Nairobi. A pioneering urban-planning project in Africa, the Kaputiei model is the first known instance in which a sizeable town has been planned in its entirety within Kenya. Approximately 60 percent of Nairobi's population—roughly 1.9 million people—lives in slums (Brown et al., 2003). Affordable housing solutions have been difficult to implement due to the high cost of land and construction, lack of finance products for the poor, and lack of a strong legal and policy framework.

Using the principles of holistic sustainable development, the Kaputiei project will consist of two thousand homes that will be affordable for the BoP and be built with "green" materials. The town center will be "eco-friendly" with commercial, cultural, and social facilities including shops, administrative offices, a health center, banking facilities, and training centers. Once completed, the Jamii Bora philosophy and example of low-cost self-help housing is expected to serve as an inspiration and a role model for the rest of Africa.

Kaputiei will be a town of entrepreneurs. By default, prospective households are proven businesspeople. All those who wish to secure a home in Kaputiei need to meet the following criteria:

- Be a member of Jamii Bora;

- Register for a home;

- Have at least 10 percent of the potential cost of the homes in savings; and

- Indicate a perfect repayment record of at least three business loans.

These criteria are intended to help make the town the foundation of an economically viable community. Even though the new town is located one hour's drive away from Nairobi's city center, most prospective homeowners have existing businesses and sufficient entrepreneurial savvy to build an internal market, expand to neighboring communities, and continue to serve their existing clients in Nairobi. In addition, as the new town will be built by members, the project will provide marketable skills that can help community members find work elsewhere in the surrounding community, such as construction and plumbing. To strengthen its capital base further, Jamii Bora is also considering the establishment of a milk plant in Kaputiei that will help the local community to preserve and sell its milk to the town and to the greater metropolitan area of Nairobi.

The development of this town has faced challenges due to local environmental NGOs that are holding up this development because they fear that it interrupts wildlife migration patterns in a neighboring wildlife corridor. They have brought a case against Jamii Bora that is currently underway. Since the migratory pathways are in fact several kilometers away from the site selected for the town, these groups could also represent communities that are resistant to the idea of a new community consisting entirely of extremely poor people that are former residents of some of the most dangerous slums in Kenya. The creation of Kaputiei town represents a pioneering effort to provide marginalized communities with formal housing and economic opportunity, but is also confronting existing tensions about sharing space and land to support economic development.

Design for Price and Efficiency

To serve the poor, it is important to tailor solutions to their respective needs and behaviors. Since market-based solutions for the BoP are rare, it is important that the voices of the poor be incorporated into the design and that there be constant innovation and testing. For example, when Jamii Bora designed its basic micro loan product, it found that when the size of the loan increased from two to three times the amount of a member's savings, the program failed as the debt burden placed too great a strain on the user.

With the Kaputiei project, Jamii Bora incorporated the perspectives of the poor at all levels of the town design, from conceptual planning to construction processes to pricing and mortgage financing. In all cases, Jamii Bora borrowed from and experimented with established practices while organically tailoring solutions to the needs of the poor.

Kaputiei Town: Social and Economic Sustainability

In planning Kaputiei town, Jamii Bora dedicated significant time to consulting potential Kaputiei residents on the design of housing, community and commercial resources, and desired infrastructure. On one hand, the design of the town plan reflected the community values of the member households, such as providing schools and playgrounds for children. Homes would have two bedrooms on fifty square meters of land. On the other hand, it was important to balance the economic sustainability of the town. Therefore, a certain amount of land was reserved for retail activities. Also, to help keep the maintenance costs of the town low, Jamii Bora is seeking to lower its overall energy usage by using "green" technology. For instance, it has incorporated a hybrid wetlands/wastewater system into the town plan to help recycle water in an energy-efficient manner.

At approximately USD 3,500, the purchase price of each home is designed to be affordable to its members. This price will account for the cost of constructing the home as well as parts of the infrastructure and a modest markup. Part of the reason Jamii Bora is able to offer this low price is because of its efforts to balance the overall retail and residential mix of the community and to use local materials and labor for housing construction.

With respect to labor, Jamii Bora is keeping construction costs low by using member labor rather than conventional contractors. To address the low-skill level of its members, Jamii Bora developed processes to mass-manufacture building parts involving simple low-technology techniques that a small team could successfully carry out. For example, Kaputiei has a temporary on-site factory for the manufacture of hollow cement blocks and concrete roof tiles. The processes are simple to follow with clear instructions. Machines are small and low-tech and can often be handled by one to two people. Supervisors oversee production to ensure a smooth workflow. The units are subjected to rigorous quality control tests to ensure that they meet the required standards. Laborers are paid per unit produced, which has created an incentive to maximize productivity. To date, building materials for over half of the homes have been produced on-site.

Jamii Bora's other innovation is the design of affordable mortgages for its member households. The key challenges Jamii Bora faced in designing a mortgage product were to create a loan repayment schedule that is reasonable for the borrower and to find long-term sources of self-financing, so that it can offer long-term mortgages.

Jamii Bora relied on feedback from its members to design affordable mortgages. To qualify for the mortgage, families must be members and have set aside a certain amount in escrow. To ensure that this will be an attractive product, Jamii Bora aims to structure the monthly mortgage payment to be comparable to the typical rent in slums by current members. Rent for a permanent single room in "planned estates" (slum housing upgraded by the government of Kenya in 1970s and 1980s), typically located in periurban areas, ranges from USD 30 to USD 60 per month; rent for a single room in urban slums such as Kibera are comparable.

The other challenge Jamii Bora faces with offering affordable mortgages is the need to offer long-term mortgages—on the order of fifteen years—which would require Jamii Bora to also obtain sources for long-term self-financing. Although this endeavor is still underway and the financing structure is beyond the scope of this chapter, the fact that Jamii Bora holds title to Kaputiei will enable it to secure some of the long-term debt necessary to back its affordable mortgage program.

The work of Jamii Bora to extend housing to extremely low-income communities is a breakthrough in a region where 71 percent of urban populations live in slums (Tibaijuka, 2005), and the total number of slum dwellers in Africa is expected to reach 332 million by 2015 (Auclair, 2005). Through its commitment to engaging slum dwellers as customers and potential homeowners, Jamii Bora is demonstrating a path forward. It is also critical to note that when communities have a say in the design of their communities, they make choices that incorporate sustainability as well as affordability.

Conclusion

The cases of IDEI and Jamii Bora demonstrate viable, scalable, and sustainable models for social change and show that these seemingly divergent goals can be met simultaneously. These examples demonstrate four distinct drivers that, when combined, can deliver scalable models for social change:

- Social impact, which takes into account the needs and preferences of the poor;

- Financial viability, which allows for growth and releases high-potential organizations from a reliance on grants and subsidies;

- Design innovation, which can leapfrog existing approaches and achieve greater scale, as well as resource efficiency; and

- Leadership that can balance those issues and break down traditional boundaries between social and financial imperatives.

The potential success stories that could come out of these models will require a tolerance for risk, as well as the development of skills that draw from a cross-section of disciplines. A better understanding of these four components of a sustainable and scalable social innovation on the part of social entrepreneurs and investors will help drive the emergence of more models like these. It will also ultimately lead to the inclusion of the poor in markets and systems that currently exclude them, thus closing the gap in access to basic goods and services.

References

Auclair, C. (2005). Charting a framework for sustainable urban centres in Africa. *United Nations Chronicle* 42 (2). Retrieved June 15, 2007, from www.un.org/Pubs/chronicle/2005/issue2/0205p26.html.

Brown, W., Tilock, K., Mule, N., and Anyango, E. (2003). The enabling environment for housing finance in Kenya. *Cities Alliance Shelter Finance for the Poor Series*, 4. Retrieved June 15, 2007, from www.citiesalliance.org/citiesalliancehomepage.nsf/Attachments kenya + civis/$File/Kenya + CIVIS + April03.pdf.

Danaher, K. (Ed.). (1994). *Fifty years is enough: The case against the World Bank and the International Monetary Fund.* New York: Global Exchange.

Emmons, G. (2007, April 4). The business of global poverty. *Harvard Business School Working Knowledge Forum.* Retrieved June 15, 2007, from http://hbswk.hbs.edu/item/5656.html.

Karnani, A. (2006). Fortune at the bottom of the pyramid: A mirage. *Ross School of Business Working Paper Series*, No. 1035. Retrieved June 15, 2007, from http://papers.ssrn.com/sol3/papers.cfm?abstract_id=914518.

Kulkarni, S. A., Reinders, F. B., and Ligetvari, F. (2006). Global scenario of sprinkler and micro irrigated areas (Contribution to the 7th International Micro Irrigation Congress, Kuala Lumpur). Retrieved June 15, 2007, from www.irncid.org/DWNLoad.aspx?FN=192411605438766968.pdf.

Ministry of Agriculture, Government of India (2004). Agricultural statistics at a glance 2004. Retrieved June 15, 2007, from: http://agricoop.nic.in/Agristatistics.htm.

Postel, S., Polak, P., Gonzales, F., and Keller, J. (2001). Drip irrigation for small farmers: A new initiative to alleviate hunger and poverty. *Water International* 26 (1): 3–13.

Tibaijuka, A. (2005). Cities without slums. *Our Planet* 16 (1): 12–13. Retrieved June 15, 2007, from www.ourplanet.com/imgversn/11/tibaijuka.html.

World Bank (2005). World development indicators. Retrieved June 15, 2007, from http://devdata.worldbank.org/wdi2005/Table2_5.htm.

12 | Including Natural Capital in Environmental Decision Making

HEATHER TALLIS, GRETCHEN C. DAILY, JOY GRANT, PETER KAREIVA, AND TAYLOR RICKETTS

For as long as there has been commerce, nature's goods have been bought and sold in markets. Although often undervalued, these and many other "ecosystem goods" can be ascribed economic value and at least partly accounted for based on market prices. The majority of the earth's natural capital and the services it provides are not explicitly valued in economic terms, however, and are commonly overlooked in decision making. A failure to account for natural capital creates problems not only for the environment but also for people. Overlooking the potential downstream effects of timber harvest, for example, can promote practices that cause high sediment loads in rivers (Chamberlin, Harr, and Everest, 1991; Lewis et al., 2001; Miller, Williamson, and Silen, 1974), with wide-reaching impacts. High sediment loads can reduce fish abundance (Beechie and Sibley, 1997; Kiffney, Richardson, and Bull, 2003; Murphy and Hall, 1981), shorten the lifetime of hydropower production facilities, and require expensive filtering

Heather Tallis is Lead Scientist for the Natural Capital Project. Gretchen C. Daily is a Senior Fellow at the Woods Institute for the Environment of Stanford University, cofounder of the Natural Capital Project, and the Project's Stanford Liaison. Joy Grant is the Director of Global Partnerships for The Nature Conservancy. Peter Kareiva is Chief Scientist for The Nature Conservancy, cofounder of the Natural Capital Project, and the Project's TNC Liaison. Taylor Ricketts is the Director of Conservation Science for the World Wildlife Fund, cofounder of the Natural Capital Project, and the Project's WWF Liaison.

before water is used for irrigation and drinking supply (Emerton and Bos, 2004).

Increasingly, policymakers are recognizing the merits of maximizing the so-called triple bottom line: social and environmental performance, in addition to financial performance. There is growing recognition that all three dimensions of performance are essential to human well-being. Sustainable forestry practices can not only provide higher timber profits over the long term but can also protect biodiversity[1] and provide cheaper hydropower, cleaner water supply, and greater aesthetic beauty, while mitigating the impacts of global climate change by preventing further release of CO_2 into the atmosphere. In fact, the Millennium Ecosystem Assessment (2005) concludes that nations will not be able to meet the United Nations Millennium Development Goals for moving the majority of the global population out of extreme poverty, hunger and health risk by 2015 without an environment that can sustain human activity (Millennium Ecosystem Assessment, 2005).

The same lines of reasoning that connect poverty and development to the environment also connect the private sector to the environment. When British Petroleum left the National Association of Manufacturers' Global Climate Coalition in 1997, it became the first energy company to recognize publicly the connection between fossil fuels and climate change, a step that gained it support from consumers and environmental groups alike (Lowe and Harris, 1998). Ten years later (January 2007), the United Kingdom's largest grocery chain, Tesco, took a similar step by vowing to become a leader in helping to create a low-carbon economy (Finch and Vidal, 2007). In addition to curbing their own emissions and waste production, the chain will label all of its 70,000 products with a carbon cost, alongside calorie counts and fat content (Finch and Vidal, 2007). Moving beyond a sole focus on climate change, the World Business Council for Sustainable Development has elevated ecosystems to the level of a focal area (in 2006),

---

1. Biodiversity is defined as "the variability among living organisms from all sources, including *inter alia*, terrestrial, marine and other aquatic ecosystems and the ecological complexes of which they are part; this includes diversity within species, between species and of ecosystems" (Convention on Biological Diversity, Article 2).

giving them the same attention as energy, development, and the role of business in society (Stigson, 2007). The Council's CEO, Bjorn Stigson, concluded that business cannot succeed within failing ecosystems (Stigson, 2007).

Making decisions that meet the triple bottom line, however, requires types of information that are not readily available. Moreover, environmentalists may have undercut their own efforts by proposing frameworks for policy or regulatory decisions that pit the environment against development. This chapter provides a conceptual model that illustrates the importance of ecological and social information in decision making. We then discuss the current major challenges to providing these types of information and present several cases in which the inclusion of such information has led to decisions that improved environmental, social, and financial conditions alike.

From Traditional Environmental Decision Making to Ecosystem Service–Based Cost-Benefit Analyses

Environmentalists have championed the precautionary principle and risk-analysis as a decision-making framework for critical resource management and environmental policy. For example, the Convention on Biological Diversity advocates the precautionary principle as one of its central tenets, arguing that any new technology or human impact should be considered potentially harmful to people and to biodiversity. Consequently, the principle states that caution should be exercised in the absence of proof of safety. Risk analysis seeks to quantify the risk to the environment of any course of action.

The problem with both risk analysis and the precautionary principle is that they are pessimistic approaches to decision making. Neither approach considers the possible economic benefits of actions. As a result, environmentalists that rely on these frameworks in discussing public policy are automatically at odds with business interests and economic development.

A more appropriate segue to policy development is cost-benefit analysis. The best cost-benefit analyses not only sum up the net costs and net benefits of a choice but also examine fairness and "distributional effects," or who benefits and who pays the costs. Until recently, environmentalists have shunned cost-benefit analyses as anti-environmental because all calculations are in terms of the dollar. This is a misguided reaction. As this chapter will demonstrate, real progress in decision making aimed at a sustainable environment must embrace cost-benefit analyses. The key, however, will be to develop a cost-benefit analysis that includes costs in terms of degraded ecosystem services and benefits in terms of protected ecosystem services. If ecosystem services can be made routine components of cost-benefit analyses, then policy debates regarding environmental decisions will not be couched in terms of the environment versus the economy. Instead, the debates will be framed as finding what is best in the long run for the economy, but with a full accounting of all pertinent sources of capital including ecosystem services.

One early effort to move in the direction of environmentally sound cost-benefit analyses assigned values to major habitat types, with the values essentially representing what would be lost if an archetypal (or average acre) habitat were converted to nothingness (Costanza et al., 1997). The obvious problems with this approach are that no habitat is ever converted into nothingness, and the economic value for any habitat type varies across the world depending on local ecological processes and the socioeconomic context. Thus, the relevant question concerns the marginal value of habitat loss or enhancement—that is, what costs and benefits would result from an incremental change, in a particular place? Nonetheless, Costanza and colleagues (1997) provided a wake-up call about the potentially huge value of ecosystem services.

In order to go beyond simple metaphors and dramatic statements about nature's value, one needs much more sophisticated analyses that meld ecological production functions with an appreciation for how people and institutions make decisions. An idealized path forward is illustrated in Figure 12.1, showing how decisions drive a cycle of actions and consequences that feed back to impact the economy, the environment, and society.

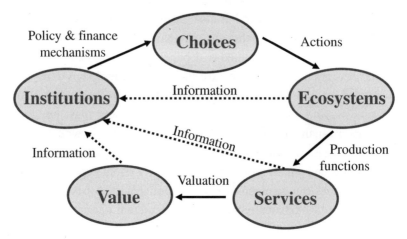

Figure 12.1. The decision cycle. Information and understanding needed to spur the development of institutions that align economic forces with social and environmental well-being.

Cost-benefit analyses of alternative policy options in the context of Figure 12.1 may be the best hope for aligning environmental health and conservation with human interests and economic development (Berkes and Folke, 1998; Levin, 1999; Daily et al., 2000; Heal, 2000; Committee on Assessing and Valuing the Services of Aquatic and Related Terrestrial Ecosystems, 2004). As the figure shows, the choices people make about resource use lead to actions that influence the condition of ecosystems. The level and type of service provision varies with ecosystem condition, and one can understand and represent this relationship through ecological production functions. Economic approaches can be used to assign societal values to services, including dollar values in some cases. Then, information about the dynamic status and values of natural capital (Figure 12.1, dotted lines) feeds into the political decision-making process that shapes new institutions, such as norms, regulations, and laws. These institutions provide positive or negative incentives for particular choices about resource use. The incentives close the loop in the decision cycle.

Employing this approach will be difficult, however, because in most places people lack critical elements of understanding in the following:

- ecological production functions for most natural capital and ecosystem services;

- values of natural capital and ecosystem services to multiple sectors in common terms (biophysical or economic) that allow the identification of trade-offs;

- information at the scale of policy decisions (usually local, regional or national);

- maps of natural capital and ecosystem services that locate providers and beneficiaries;

- a clear understanding of the institutional and financial mechanisms that will successfully change people's management decisions under different social, economic, and ecological conditions.

We discuss each of these information gaps, highlighting progress to date and persistent needs.

Information on Ecological Production Functions

Decision making weighs the differences between gains and losses on multiple fronts. Production functions showing the relationship between inputs, such as fertilizer and labor, and outputs, such as crop production, are extremely useful in visualizing these differences (Heal et al., 2001). Production functions are used extensively in agriculture, manufacturing, and other sectors of the economy today. The same types of relationships exist between natural inputs, such as extent of wetlands, and natural capital, or ecosystem services, such as water filtration (Committee on Assessing and Valuing the Services of Aquatic and Related Terrestrial Ecosystems, 2004).

In many cases, we do not know what the ecological production function looks like. In other words, we do not clearly understand how much natural capital of a particular type will be gained or lost as people change the condition of land or seascapes. We often know how ecological processes change in response to human actions, but we still need to make the link to human well-being (Tallis and Kareiva, 2006). There is extensive evidence that nitrogen and phosphorus concentrations in rivers can increase dramatically as a result of fertilizer applied to agricultural fields. But do higher nutrient concentrations increase the cost of water treatment? By how much? How much does nutrient-driven eutrophication cost the tourism or sport fishing industries? The answers to these largely economic questions require credible production functions. Without these production functions, it is impossible to predict how any action will affect ecosystem services. Unfortunately, we usually lack production functions and, hence, we cannot easily consider choices in a useful way.

Despite the nascent understanding of production functions and trade-offs, we may be able to jumpstart rapid progress by drawing from the extensive work done over decades in resource-use disciplines like fisheries, forestry, agroforestry, and agriculture. As may be expected, elegant work has already advanced the use of a production function approach for the few services directly linked to markets today. For instance, fisheries in freshwater and marine systems have been the target of many of the existing production function approaches to modeling ecosystem services (Carpenter, Ludwig, and Brock, 1999; Wu et al., 2003).

For many services outside existing markets, or attached to newly emerging markets, we simply need to make the link between existing ecological data and economic valuation methods. To demonstrate the types of advances that are immediately possible for these services, we created a sample ecological production function for soil carbon sequestration value in Nigerian agroforestry systems (Figure 12.2). We used published data on soil organic carbon stocks in different cultivation systems to calculate soil carbon sequestration rates over a thirteen-year period (Lal, 2005).

We ordered the cropping systems to represent a gradient from most impacted to most similar to native vegetation for the region, so that the x-axis of the production function is a relative measure of environmental condition, or environmental input. We then calculated

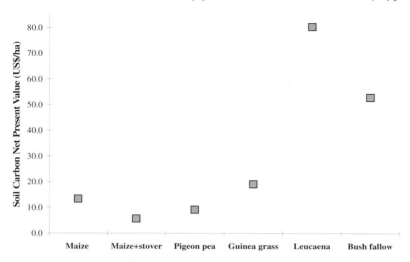

Figure 12.2. Ecological production function for soil carbon NPV in Nigerian agroforestry. We used thirteen years of soil organic carbon storage data to calculate sequestration rates, then made a crude approximation of the value this sequestration could bring on the European Union Emissions Trading Scheme market over the next ten years.

the net present value of a ten-year stream of payments that a farmer could expect from each cultivation system in a global market for soil-sequestered carbon. We used the 2006 median price for a ton of carbon on the European Union Emissions Trading Scheme market (USD 11.30/tCO_2) and assumed a 10 percent discount rate. The production function graph clearly shows the differences in soil carbon value among cultivation systems. Since the cropping systems are arranged in order of decreasing impact to natural systems, this graph also serves as a crude production function relating carbon sequestration value to ecosystem condition.

Information on Values to Multiple Sectors to Allow Assessment of Trade-offs

While the scientific community can make some rapid advances in the realm of single-service production functions, resource management decisions affect multiple ecosystem services at once, often in interactive ways. There is a need to understand how multiple services interact and have simple ways to communicate the potential trade-offs or synergies between services and sectors (Balvanera et al., 2001; Heal et al., 2001). We have this type of information in a few cases that demonstrate the essential role it could play in decision making.

The Kyoto Protocol was designed to encourage the management of land for carbon storage, but its tenets influence other forms of natural capital including water quality and soil fertility. The writers of the Kyoto Protocol fostered revolutionary institutional change that is now, unfortunately, one of the greatest threats to native grasslands in South America and elsewhere because they did not have a clear understanding of the conflicts between alternative services. The Kyoto Protocol currently awards carbon emission offsets for afforestation of open lands including native grasslands, which use far less water than trees planted for carbon sequestration. Planting trees in native grasslands does indeed enhance climate regulation services, but it also decreases soil productivity and water quality (Jackson et al., 2005).

Although the Kyoto Protocol may create unintended costs for agriculture and water sectors in some places, it is not all bad. Reforestation in areas where forests are the native vegetation actually provides multiple benefits, by increasing climate regulation and improving soil fertility and water quality (Jackson et al., 2005). Trade-offs are therefore context-dependent, and mechanisms like those within the Kyoto Protocol need to be adjusted accordingly. With this information in hand, the writers of the Protocol could change the stipulations to allow carbon credits only for reforestation of native forest lands, ensuring synergies and avoiding conflicts in service provision.

Assessing trade-offs will be easiest when multiple forms of capital are converted to a common currency (Daily et al., 2000). Often this common currency will be dollars, but biophysical measures will also be useful. The advantage of using money is that costs and benefits associated with changing natural capital can be compared directly to other inputs for mainstream sectors, such as agriculture and hydropower. Few studies have estimated economic values of multiple services to allow this type of integration, but those that have provided compelling examples. The invasion of tamarisk, a water-thirsty shrub, in the western United States has affected many sectors of the region's economy. An economic assessment of the costs of tamarisk invasion included up to USD 121 million in irrigation water, as much as USD 68 million in municipal water, up to USD 44 million in hydropower and USD 52 million in flood control (Zavaleta, 2000). The synergies among water-related services and native vegetation in this case make ecosystem restoration a win-win-win-win-win situation. Removing tamarisk would obviously benefit all four sectors and the environment. Having all four services assessed in common terms may also help decision-makers allocate responsibility for restoration, since it is easy to see which sector has the most to gain.

Information at the Scale of Policy Decisions

Since ecological and social conditions vary dramatically across levels of decision making from local to global, we need information about natural capital across scales. Gaps in information often exist because, while the right questions may have been asked, they were posed at the wrong scale for a decision at hand. The Millennium Ecosystem Assessment (MA) provides a clear example (2005). This recent global assessment of the status and trends of ecosystem services made a phenomenal contribution to the three international conventions it targeted: the Convention on Biological Diversity, the Ramsar Convention on Wetlands, and the Convention to Combat Desertification (Reid, 2006). The information in the MA reports was well aligned with the types of decisions the conventions address, and each is already moving to endorse and incorporate their findings (Reid, 2006); however, national, regional and local governments that have read the MA and become inspired by its findings find their hands tied in moving forward. General results of the MA, such as "60 percent of the ecosystem services assessed are being degraded" give little guidance to policy makers on how to act to curb such trends (2005, 1). Additional smaller-scale assessments, like the MA's subglobal assessments still underway, are essential for governments that generally have more power over ecosystem change than international conventions.

For a more concrete example, consider the case of climate change in California. The Intergovernmental Panel on Climate Change (IPCC) has been producing reports on likely future climate conditions for more than eighteen years. Most are familiar with the types of predictions they produce: 1–2° C increases in temperature, more severe droughts, and so on. These global predictions have done little to influence state policy in the United States. In 2004, this situation changed when a group of California scientists took climate prediction one step further. They downscaled the IPCC's global predictions, made new predictions for the state of California, and reported the outcomes in socially relevant terms. What did they find? Without aggressive emissions regulations over the next hundred years, California will experience heat waves that would kill 165–330 additional people a year in Los Angeles and cause annual losses of USD 266–836 million to the dairy industry, a shorter ski season, and bad-tasting wine from Napa Valley (Hayhoe et al., 2004). Massive media coverage followed the release of their report and the state legislature passed the Pavley Bill (State Assembly Bill 1493), the first emissions bill ever to restrict car emissions (Hakim, 2002).

Maps of Natural Capital Providers and Users

Many of the questions decision makers should be asking about natural capital have a spatial component. What part of the watershed contributes most of the nonpoint source pollutants to waterways? Are the people who benefit from tax breaks for endangered species protection living in low or high income areas? How will urban expansion affect water treatment costs? What kind of mechanism should one create to encourage planting trees on sloping lands to slow erosion and flooding in the region's three major cities? Where should one target management to increase water quality and decrease flood risk? Answering these questions requires knowledge of where natural capital is supplied across a land or seascape, and where it is consumed.

Ideally, one would have spatially explicit information on current and likely future status, value and sensitivity of all types of natural capital (Balvanera et al., 2000; Daily et al., 2000). We are far from this lofty goal, but exciting progress has been made on some fronts. Several efforts have now mapped the values of natural capital in some places, giving a sense of how natural capital varies across a landscape, and which types of capital coexist in space. Chan et al. (2006) mapped the current or potential provision of six services in the Central Coast ecoregion of California (e.g., Figure 12.3): crop pollination, carbon storage, forage production, water provision, flood control, and outdoor recreation. They also mapped biodiversity as a critical attribute of ecosystems, or natural capital.

The value of some services was spatially congruent, like those of carbon storage, water provision, and outdoor recreation. Other values

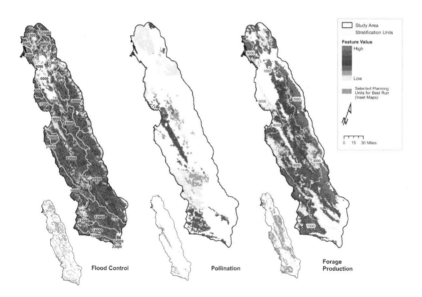

Figure 12.3. Selected maps of three different ecosystem services in the Central Coast ecoregion of California. Each map shows the current or potential level of service provided by each part of the landscape. The smaller gray maps show the parts of the landscape that were identified as priority areas for service provision in a planning exercise. Figure replicated from Chan et al. (2006) with permission.

conflicted, like those of biodiversity conservation and pollination, because high biodiversity values are found in large expanses of native habitat, while pollination can be provided by small patches of somewhat modified habitat in heavily managed landscapes (Chan et al., 2006). The calculation of service values did include consideration of the demand for services and the ability of providers to access markets, but services were not valued in economic terms. Some decision makers, like conservation planners, will find these maps very useful in prioritization processes, but these maps are still missing a link to other decision makers that rely mostly on cost-benefit approaches.

Shortcomings in the field of economics have made mapping natural-capital values difficult, but some attempts have been made. The most recent effort used a spatially explicit value transfer approach for total value of all services in three case studies (Troy and Wilson, 2006). The value transfer approach assigns total ecosystem service or natural capital value to a land cover category based on all previous valuations done on similar land-cover types under similar environmental, social, and economic conditions. This approach can be challenging since past work has placed emphasis on economic valuation of a small subset of all land cover types (Troy and Wilson, 2006). Value transfer is further limited by the requirement that the conditions in the case used as source data be compatible with the current study where the value is to be applied. It is often difficult to find valuation exercises done under similar conditions, adding to the lack of data available for this approach. Nevertheless, Troy and Wilson (2006) were able to create maps of total natural capital value for three landscapes in the United States ranging from ~2000 to 2,000,000 hectares.

Another recent study combined value transfer with production functions to estimate and map the value of five ecosystem services in the Mbaracayu Biosphere Reserve in Paraguay (Naidoo et al., 2006). For three of these services, production functions related input variables to the resulting economic value of the ecosystem service. For example, the value of sustainable bushmeat production was estimated as a function of habitat type and size, probability of species occurrence, average weight of hunted species, and market prices for meat. By applying these production functions in a spatial manner, and by expressing the value of each ecosystem service in terms of dollars, the study was able to produce maps of combined economic value for all five services (Figure 12.4). (To view this image in more detail, see the original cited work.)

Comparing this map with an equivalent map of the costs of conservation allowed a spatial cost-benefit analysis, showing where the benefits of conservation could outweigh the costs. This production function approach avoids some of the pitfalls of value transfer, such as the difficulty involved in matching existing estimates of value to the target landscape. A production function approach is also helpful to decision makers because it identifies the most important factors to manage for each service. As noted above, however, production functions are still difficult to construct, and Naidoo et al. (2006) were able to use this approach for only three of the five services considered.

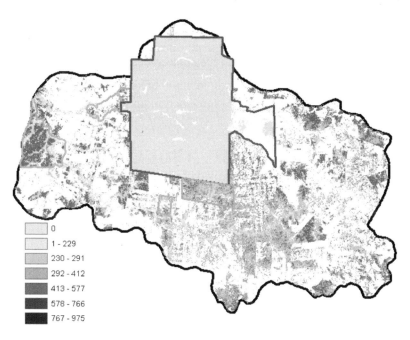

Figure 12.4. Economic values (in $, net present value per hectare) of five ecosystem services across the Mharacayu Biosphere Reserve, Paraguay. Ecosystem services included are sustainable bushmeat harvest, sustainable timber harvest, bioprospecting, existence value, and carbon storage. Figure replicated from Naidoo and Ricketts (2006) with permission.

While some progress has been made in mapping natural capital itself, it is equally important to have spatial information on the people who benefit from it. Without consideration of spatial social effects, decisions may have unintended distributional effects, or influence on social groups with different rights, income levels, education, and so on. Cost-benefit analyses for the sake of public policy must include a concern for fairness and equitability. In order to meet this requirement, analyses must be spatially explicit and track the flow of ecosystem services and money to different people. For example, community-based management is seen as a strategy with the potential to provide a win-win outcome for local communities that benefit from forest products while maintaining biodiversity and ecosystems (Antinori and Bray, 2005). However, poorer user groups near a forest in Tanzania received the least benefit from community-based management efforts. This inequity in benefit flow meant that the transaction costs of community-based forest management outweighed the benefits of forest use for these lower income groups (Meshack et al., 2006). Economic valuation approaches still remain predominantly nonspatial, leaving a critical gap in relevant information unfilled.

The Scope and Limitations of Financial and Institutional Mechanisms

Once decision makers have all the information described earlier, they still have choices to make about how to enact change. Various financial and institutional mechanisms have been employed to align economic forces with conservation of ecosystem services. The most widely used categories include regulations (ownership rights; use rights), government-driven markets (cap-and-trade; mitigation), and government investments (subsidies; other payments). The success of any one of these mechanisms in achieving its intended purpose is influenced by a milieu of environmental, social, and economic characteristics.

Some choices about which mechanisms to apply will be obvious. For instance, water use fees cannot be levied if individuals do not have water access rights. Many choices will be much more subtle. Would water quality improve more if improved management practices were subsidized, or if polluters were charged a fee? Would timber poaching decline more if additional funds were put into enforcement by government officials or communities living adjacent to forests? Could the hunting of bats be curbed through education about their role in pest control and pollination, or does hunting have to be outlawed? Would tourism revenues increase if property rights were changed to allocate ownership of tourism revenues and wildlife products to a different sector of society?

There are no single answers to these questions and we know little about what drives differences because most institutional research has focused on case studies or comparisons (Agrawal and Chatre, 2006). However, some attempts have been made to draw out the relative importance of different factors in the success of particular mechanisms for resource management. The allocation of forest management rights to communities is a popular mechanism for aligning conservation and poverty alleviation applied around the globe. Agrawal and Chatre (2006) have identified twenty-six demographic, institutional, sociopolitical, economic, and biophysical variables that significantly affect the success of community-based management in maintaining or enhancing forest condition in northern India. With these variables in hand, an Indian policymaker could examine the conditions in a particular region and identify the mechanism that would provide the best means to the desired end.

Similarly, Jack, Kousky, and Sims (2007) drew from the long history of incentive-based mechanisms in the United States to identify biophysical, social, and economic conditions that made incentives cost-effective, environmentally effective, efficient, and equitable. They found that spatial proximity between service providers and beneficiaries will be important for services that are not well mixed or homogenously distributed. For instance, carbon sequestration affects the global atmosphere, so proximity of the person receiving the incentive and the person receiving the service is not important. However, services such as water quality that vary dramatically across space will best be regulated with incentives when providers and users are close to each other. Incentives will also work well when the cost of providing a service varies substantially among providers, allowing those with the lowest costs to produce more of the service. Aligning this kind of information with characteristics of services and the landscapes of interest will lead to more successful design of institutions.

Putting Theory into Practice: Models of Success

The arguments for a cost-benefit approach are advancing from theory to evidence as the docket of real world examples backing the promise of economic valuation of nature's assets continues to build. The excitement and attention garnered by New York City's decision to invest in natural, rather than built, capital as a cost-saving measure for clean water provision has grown as the general concept has been applied to more creative and well-funded transactions around the globe. Over the past decade, an astonishing number and diversity of efforts to align economic forces with ecosystem service protection have emerged worldwide. Individually, most of these efforts are small and idiosyncratic. Yet, collectively, they represent a powerful shift in the focus of conservation organizations toward a more inclusive, integrated, and effective set of strategies (Daily and Ellison, 2002). Recent reviews of ecosystem service efforts by the world's two largest conservation NGOs, The Nature Conservancy and World Wildlife Fund, revealed dozens of major projects implementing an ecosystem services focus in some way (Yuan-Farrell and Kareiva, 2006a; Yuan-Farrell and Kareiva, 2006b). Taken together, these efforts span the globe and target multiple ecosystem services, including forest-generated services of carbon sequestration, water supply, flood control, biodiversity conservation, and enhancement of scenic beauty and associated recreation and tourism values.

Many ecosystem service efforts focus on a single service that stands out as sufficiently important, from economic and political perspectives, to overcome the activation energy required to protect it. Under the institutional umbrella created for the focal service, it is possible that other services may be at least partially protected (Balvanera et al., 2000).

To illustrate the range of efforts underway, in contrasting biophysical, economic, and institutional environments, we first examine two pioneering single-service examples, water purification in New York City and flood control in Napa, California, and the way they're being replicated and adapted creatively. We then review two larger-scale investments in natural capital: water flow regulation in China and tourism and hunting in Namibia. Finally, we present a pioneering case in Costa Rica that points to the future, in which a suite of ecosystem services is targeted simultaneously.

Water Supply from the Catskills

In 1997, a water pollution crisis led New York City to a bold experiment. Instead of relying on technology, in the form of a new water filtration plant, the city invested in natural capital (Chichilnisky and Heal, 1998). The decision saved several billion dollars and set a global precedent (ibid.). This investment is restoring the natural purification services of the Catskills-Delaware watershed, the heart of the water purification and delivery system supporting some 10 million water consumers (ibid.).

Less appreciated is the machinery of the watershed, pumping out as much as 6.8 billion liters of purified water daily and supplying 90 percent of the needs of New York City's residents (Chichilnisky and Heal, 1998). For decades, these urbanites have relied on the 5,000-square-kilometer watershed to purify their drinking water, a product of exceptional quality (ibid.). The forest provides this valuable service for free, cleansing the water as it sifts through roots and soil. The forest also metes out water gradually, stabilizing drinking supply and mitigating flooding; prevents soil erosion; shelters wildlife; stores carbon, helping to stabilize global climate; and graces the region with stunning beauty (ibid.).

In the late 1980s, the United States Congress and Environmental Protection Agency had started to react to perceived threats to surface-water supplies—rivers, lakes, and reservoirs. Two-thirds of the U.S.

population depends on such systems, rather than groundwater. Yet the great majority of them have been degraded by years of urban sprawl and runoff from second homes, farms, and golf courses. In 1991, the EPA ordered New York City to build a water filtration plant—unless the city could prove it could maintain water quality without it. Presented with the budget-breaking costs of that investment, up to USD 8 billion by some estimates plus at least USD 300 million in annual operating costs, city officials took a revolutionary approach and invested instead in restoring the natural asset, the watershed (Chichilnisky and Heal, 1998).

Since 1997, the city has invested nearly USD 2 billion in land-management changes and innovative tactics, such as purchasing land around reservoirs to preserve forests and wetlands that buffer against pollution, paying landowners to restore forest along streams, and offering technical aid and infrastructure to farmers and foresters. This summary does not capture either of the ongoing experiments relating to whether these natural capital approaches will work relative to, or in addition to, ever-changing technological alternatives (Committee to Review the New York City Watershed Management Strategy, 2000) and to the complex political negotiations inherent in land-use decisions. The bottom line is that, to many, the outcome is a triple win: urban people receive pure water at lower cost; rural people are rewarded for good land stewardship; and visitors and rural residents alike enjoy the spectacular landscape, saved from out-of-control urban development (Daily and Ellison, 2002). Cities worldwide are attempting to implement similar approaches (United Nations Food and Agriculture Organization, 2004).

Flood Mitigation in Napa Valley

Over the past decade, pioneering ecosystem approaches to flood control have been emerging as well. Throughout the world, floods are by far the most common natural disaster, invited by settlements that, throughout human history, sprang up in fertile floodplains. In the United States alone, floods cost dozens of lives and USD 4 billion in damages in the average year (Daily and Ellison, 2002). Worldwide, flooding caused some 40,000 deaths and USD 29 billion in economic losses in 1999–2000, as estimated by Munich Re (ibid.).

The revolutionary approach of Napa, California, is now touted as a model of success, though the revolution came only after enduring twenty-eight major floods and well over USD 500 million in damages, since recordkeeping began in 1862 (Brauman, 2006; Daily and Ellison, 2002). By the late 1990s, some residents proposed a new plan: a living river approach to flooding. Instead of investing in physical capital, by reinforcing the levees and concrete barriers that had served to control the river's surges in the past, they proposed using an ecosystem services focus to guide investments in a living river approach. This meant moving nine bridges and more than a hundred buildings and restoring 250 hectares of floodplain, instead of a deep, straight, concrete channel, yearly dredging, and tall floodwalls. Bridge replacement removed obstacles to high flows, bank terracing reconnected the river to its historic flood plain, and easements and acquisitions removed especially vulnerable structures from harm's way (Brauman, 2006).

Interestingly, residents approved of the ecosystem approach even though it was projected to cost more than USD 200 million more than the physical capital approach estimated at USD 150 million, and they had to pay for part of it. This choice was made in anticipation of the many benefits that would result from an investment in natural capital under the umbrella of flood control, that were not valued explicitly. These benefits include the restoration of fish, wildlife and scenic beauty, and all the recreation, tourism, and related commerce that residents hoped would follow. Indeed, as reported in popular magazines, the town was revitalized by this investment, with boating, hiking, fine dining, and other amenities unimagined when the town battled the river as an enemy (Cusumano, 2004). The City of Napa's Economic Development Office confirms that a major increase in private investment occurred after the flood plan was approved, amounting to USD 193 million in private construction from 1999 to 2005 (Brauman, 2006).

In total, Napa's plan will mitigate flooding over six of the fifty-five miles of the Napa River, and one mile of Napa Creek, a tributary in town. Having proven success at the local scale, the success of Napa's efforts now hinge critically on whether upstream management of the river improves (Jeffrey Mount, professor and director of the Center for Watershed Sciences at the University of California, Davis, personal communication, February 15, 2007). This still-fragile situation highlights the dependence of local efforts on support at larger scales.

Quito's Water Fund

Both the New York City and Napa cases have been replicated elsewhere. For instance, the capital city of Ecuador, Quito, has developed a water fund to safeguard the city's watershed. The fund, now valued at USD 4.9 million, is used to reforest and monitor the watershed and provide environmental education (Echavarria, 2002). Money flows into the fund from the major water users including the municipal water supplier, Quito's hydropower electricity firm, a major beer bottler called Andina and the 1.2 million residents of Quito (ibid.). Money has flowed back to communities and natural capital in the form of 500,000 trees planted a year; education of 6,000 children a year; the establishment of three dozen dialogues, meetings, or forums open to the public; the addition of jobs for new park guards to protect the Condor Bioreserve, a major protected area in the Quito watershed; capacity-building, such as water monitoring training, and conflict resolution, for municipalities and local organizations; and funding for the creation of hydrologic models, monitoring programs and social work (Pablo Lloret, Director of the Water Conservation Fund of Ecuador (FONAG), personal communication, February 15, 2007). The Quito model is being exported to two other cities in Ecuador as well as to Bogotá, the capital of Colombia (Echavarria, 2002).

Floods and Forests in China

China illustrates work at large scales perhaps better than any other part of the world. Prompted by massive flooding in 1998, at a cost of USD 20 billion in damages, the Chinese government enacted a sweeping land-use policy entitled the National Forest Conservation Program (NFCP). The policy is intended to regulate water flow and promote soil retention, primarily by conserving and restoring natural forests while increasing timber production in plantations (Zhang et al., 2000). A key component of the policy is a logging ban on 30 million hectares of natural forests in the upper reaches of the Yangtze River and upper and middle reaches of the Yellow River (Xu et al., 2006). The government has invested billions in the program since its inception, through a wide array of public policy instruments including training, resettlement, direct compensation of forest dwellers, and mandatory conversion of marginal farmlands to forest lands (Zhang et al., 2000).

An incentive for water flow regulation, in addition to flood control, is increasing hydropower production efficiency. Forests in the Yangtze River watersheds decrease flow in the wet season and enhance it in the dry season. As a result, researchers estimate that a single hydroelectric power plant in Gezhouba increases its annual electricity production by up to 40 million kWh per year, at a net benefit of ca. USD 610,000 each year (Guo, Xiao, and Li, 2000). This value may increase fivefold when the Three Gorges hydroelectric power plant is operational (Guo, Xiao, and Li, 2000). In total, there are actually about eight major, interlinked policies in China controlling land use and forest cover, with the aim of enhancing the supply of these vital watershed services (Xu et al., 2006).

Tourism and Hunting in Namibia's Conservancies

Going further in adapting ecosystem service approaches to the nuances of particular places, efforts in Namibia show how poverty alleviation, social justice and conservation can be aligned. The 1996 Nature Conservation Act set up the framework to move wildlife management from public to private hands by transferring ownership of huntable game and revenue from wildlife products and tourism. The first community to enter into the program to attain these rights was a group of San people, the most marginalized social group in the country. This San community in the northeast region of Namibia created the Nyae Nyae Conservancy. Over a four year period, income to this community increased eight fold. In the initial three years of the project, wildlife populations saw an incredible resurgence, with springbok abundance increasing 150 percent and oryx and zebra increasing 300 percent. Wildlife populations have now increased sixfold here (World Wildlife Fund [WWF] and Rossing Foundation, 2004).

There are now more than thirty conservancies in Namibia, covering 13 percent of the country's land and delivering monetary and social benefits to the nation's rural poor, while improving the environment (WWF and Rossing Foundation, 2004). Communities in the program earned USD 2.5 million in 2004 and contributed USD 9.6 million to Namibia's net national income in 2003 (ibid.). Nearly 550 full-time and 3,250 part-time jobs have been created since 1998 (ibid.). Women fill the majority of part-time posts and over half of the full-time jobs (ibid.). In 2004, the United Nations recognized the merits of this program by awarding one of the larger conservancies, the Torra Conservancy, the Equator Initiative Prize, a prize that recognizes projects that reduce poverty through the sustainable use of biodiversity. The largest free-roaming population of black rhino in the world now lives in northwest Namibia, and seasonal game migrations between Botswana and Namibia have resumed for the first time since the early 1970s (ibid.).

Costa Rica's Fuel Tax for Forests

For the provision of ecosystem services to be efficient, it is important that investments in particular land covers and uses be strategically oriented around the full suite of desired services and coordinated across landscapes. This is because the spatial configuration of ecosystems greatly influences the production of services (Goldman, Thompson, and Daily, forthcoming). Under existing incentive programs, this level of coordination is typically neither required nor encouraged. But in Costa Rica, the government does target a broad suite of important services, and the program is becoming more efficient with respect to landscape coordination as well.

In 1997, Costa Rica launched a nationwide scheme of payments for the provision of ecosystem services, known as Pago por Servicios Ambientales (PSA). The PSA targets carbon sequestration; water quality and quantity, for drinking and irrigation supply, and hydropower;

biodiversity conservation; and scenic beauty for ecotourism. Funds from a diversity of sources, including the private sector, World Bank, and a gasoline tax in Costa Rica, are pooled and distributed to voluntary participants at terms of ca. USD 50 per hectare per year (Pagiola, 2002). This program is seen as a model internationally, and is now being replicated in Mexico.

Conclusion

The practical experiments described here highlight the clear and powerful ethic—and increasingly, an economic rationale—for protecting people from unsafe drinking water, flooding, and climate change. How far can ecosystem-services approaches be taken? Stanford University recently joined forces with the world's two largest conservation groups, The Nature Conservancy and the World Wildlife Fund, to form an ecosystem-services-approach partnership, the Natural Capital Project. The project, which began in November 2006, is currently working worldwide to mainstream conservation through the development of new tools, approaches, and programs that align the environment and the economy. Networked with the Natural Capital Project is a complementary project funded by the United Kingdom's Leverhulme Trust and involving a group of British universities, specifically Cambridge, East Anglia, York, Cranfield, and Leeds, which is applying an ecosystem-service focus to the Eastern Arc biodiversity hotspot region of Tanzania. The overall aim is to replicate and scale up the many promising efforts underway that make conservation a compelling choice, on both moral and economic grounds.

Some have argued that if we vigorously pursue economic valuation of nature's assets we will be giving in to the dark side of economics with no ethics. While this concern is understandable, it ignores the current situation. The pertinent comparison is between a world that now routinely conducts cost-benefit analyses to make public decisions and assigns ecosystem services and natural capital no value, and a world that includes the economic benefits derived from undegraded nature into its cost-benefit analyses. Of course, nature will not always win when these calculations are made. But the types of information we advocate will allow us to identify cases where nature wins and economic development advances. Economic valuation of ecosystem services does not prohibit policy advocacy on the moral basis that protecting species and biodiversity is the right thing to do. By contrast, the moral argument will never allow us to identify the existence of win-win situations that do align the business and development worlds with the environment. The few examples of success that exist are inspiring, though including natural capital in environmental decision making is not a panacea and will not stand alone to guarantee a sustainable world. A continued failure to include natural capital in our environmental decisions, however, is a guarantee for an unstable and unsustainable planet.

References

Agrawal, A., and Chatre, A. (2006). Explaining success on the commons: Community forest governance in the Indian Himalaya. *World Development* 34: 149–166.

Antinori, C., and Bray, D. B. (2005). Community forest enterprises as entrepreneurial firms: Institutional and economic perspectives from Mexico. *World Development* 33 (9): 1529–1543.

Balvanera, P., Daily, G. C., Ehrlich, P. R., Ricketts, T. H., Bailey, S. A., Kark, S., et al. (2001). Conserving biodiversity and ecosystem services. *Science* 291: 2047.

Beechie, T. J., and Sibley, T. H. (1997). Relationships between channel characteristics, woody debris and fish habitat in Northwestern Washington streams. *Transactions of the American Fisheries Society* 126: 217–229.

Berkes, F., and Folke, C. (1998). *Linking social and ecological systems: Management practices and social mechanisms for building resilience.* Cambridge: Cambridge University Press.

Brauman, K. A. (2006, February 3). Napa River flood project put to the test. The Katoomba Group's Ecosystem Marketplace. Retrieved July 17, 2007, from http://ecosystemmarketplace.com/pages/

article.news.php?component_id=4125&component_version_id=5897&language_id=12.

Bray, D. B., Merino-Perez, L., Negreros-Castillo, P., Segura-Warnholtz, G., Torres-Rojo, M., and Vester, H. F. M. (2002). Mexico's community-managed forests as a global model for sustainable landscapes. *Conservation Biology* 17: 672–677.

Carpenter, S. R., Ludwig, D., and Brock, W. A. (1999). Management and eutrophication for lakes subject to potentially irreversible change. *Ecological Applications* 9: 751–771.

Chamberlin, T. W., Harr, R. D., and Everest, F. H. (1991). Timber harvesting, silviculture and watershed process. In W. R. Meehan (Ed.), *Influences of forest and rangeland management on salmonid fishes and their habitats*, 181–205. Bethesda, Md.: American Fisheries Society.

Chan, K. M. A., Shaw, M. R., Cameron, D. R., Underwood, E. C., and Daily, G. C. (2006). Conservation planning for ecosystem services. *PLoS Biology* 4: 2138–2152.

Chichilnisky, G., and Heal, G. (1998). Securitizing the biosphere. *Nature* 391: 629–630.

Committee on Assessing and Valuing the Services of Aquatic and Related Terrestrial Ecosystems, Water Science and Technology Board, Division on Earth and Life Studies, National Research Council of the National Academies (2004). *Valuing ecosystem services: Toward better environmental decision making*. Washington, D.C.: National Academies Press.

Committee to Review the New York City Watershed Management Strategy, Water Science and Technology Board, Commission on Geosciences, Environment, and Resources, National Research Council (2000). *Watershed management for potable water supply, assessing the New York City strategy*. Washington, D.C.: National Academies Press.

Costanza, R., d'Arge, R., de Groot, R., Farber, S., Grasso, M., Hannon, B., et al. (1997). The value of the world's ecosystem services and natural capital. *Nature* 387: 253–260.

Cusumano, C. (2004, September). Napa riverfront: Gourmets, culture hounds, and wine lovers gather at the river. *Via*, 31–32.

Daily, G. C., and Ellison, K. (2002) *The new economy of nature: The quest to make conservation profitable.* Washington, D.C.: Island Press.

Daily, G. C., Söderqvist, T., Aniyar, S., Arrow, K., Dasgupta, P., Ehrlich, P., et al. (2000). The value of nature and the nature of value. *Science* 289: 395–396.

Echavarria, M. (2002). Financing watershed conservation: The FONAG water fund in Quito, Ecuador. In S. Pagiola, J. Bishop, and N. Landell-Mills (Eds.), *Selling forest environmental services: Market-based mechanisms for conservation and development,* 91–102. London: Earthscan.

Emerton, L., and Bos, E. (2004). *Value: Counting ecosystems as water infrastructure.* Gland, Switzerland: World Conservation Union.

Finch, J., and Vidal, J. (2007, January 19). You've checked the price and calorie count, now here's the carbon cost. *The Guardian.* Retrieved July 17, 2007, from www.guardian.co.uk/supermarkets/story/0,,1994167,00.html#article_continue.

Goldman, R. L., Thompson, B. H., and Daily, G. C. (forthcoming). Institutional incentives for managing the landscape: Inducing cooperation for the production of ecosystem services. *Ecological Economics.*

Guo, Z., Xiao, X., and Li, D. (2000). An assessment of ecosystem services: Water flow regulation and hydroelectric power production. *Ecological Applications* 10 (3): 925–936.

Hakim, D. (2002, June 12). Curb on gas emissions is stalled in California. *New York Times.* Retrieved September 24, 2007, from http://query.nytimes.com/gst/fullpage.html?res=9C04E7D81F3DF931A25755C0A96 49 C8B63.

Hayhoe, K., Cayan, D., Field, C. B., Frumhoff, P. C., Maurer, E. P., Miller, N. L., et al. (2004). Emissions pathways, climate change and impacts on California. *Proceedings of the National Academy of Sciences* 101 (34): 12422–12427.

Heal, G. (2000). *Nature and the marketplace: Capturing the value of ecosystem services.* Washington, D.C.: Island Press.

Heal, G., Daily, G. C., Salzman, J., Ehrlich, P. R., Boggs, C., Hellmann, J., et al. (2001). Protecting natural capital through ecosystem service districts. *Stanford Environmental Law Journal* 20: 333–364.

Jack, B. K., Kousky, C., and Sims, K. E. (2007). *Lessons relearned: Can previous research on incentive-based mechanisms point the way for payments for ecosystem services?* (Center for International Development at Harvard University Graduate Student and Postdoctoral Fellow Working Paper, No. 15). Cambridge, Mass.: Center for International Development.

Jackson, R. B., Jobbagy, E. G., Avissar, R., Roy, S. B., Barrett, D. J., Cook, C. W., et al. (2005). Trading water for carbon with biological carbon sequestration. *Science* 310: 1944–1947.

Kiffney, P. M., Richardson, J. S., and Bull, J. P. (2003). Response of periphyton and insects to experimental manipulation of riparian buffer width along forest streams. *Journal of Applied Ecology* 40: 1060–1076.

Lal, R. (2005). Soil carbon sequestration in natural and managed tropical forest ecosystems. In F. Montagnini (Ed.), *Environmental services of agroforestry systems*, 1–30. New York: Food Production Press.

Levin, S. A. (1999). *Fragile dominion: Complexity and the commons*. New York: Perseus Books.

Lewis, J., Mori, S. R., Keppeler, E. T., and Ziemer, R. R. (2001). Impacts of logging on storm peak flows, flow volumes, and suspended sediment loads in Caspar Creek, California. In M. S. Wigmosta and S. J. Burges (Eds.), *Land use and watersheds: Human influence on hydrology and geomorphology in urban and forest areas*, 85–125 (Water Science Application Series, vol. 2). Washington, D.C.: American Geophysical Union.

Lowe, E. A., and Harris, R. J. (1998). Taking climate change seriously: British Petroleum's business strategy. *Corporate Environmental Strategy* 5: 22–31.

Meshack, C. K, Ahdikari, B., Doggart, N., and Lovett, J. C. (2006). Transaction costs of community-based forest management: Empirical evidence from Tanzania. *African Journal of Ecology* 44: 468–477.

Millennium Ecosystem Assessment (2005). *Ecosystems and human wellbeing: General synthesis*. Washington, D.C.: Island Press.

Miller, R. E., Williamson, R. L., and Silen, R. R. (1974). Regeneration and growth of coastal Douglas-fir. In O. P. Cramer (Ed.), *Environmental effects of forest residues management in the Pacific Northwest, a* state-of-knowledge compendium, J1–J41 (USDA Forest Service General Technical Report, No. PNW-24). Portland, Ore.: U.S. Department of Agriculture.

Murphy, M. L., and Hall, J. D. (1981). Varied effects of clear-cut logging on predators and their habitat in small streams of the Cascade Mountains, Oregon. *Canadian Journal of Fisheries and Aquatic Sciences* 38: 137–145.

Naidoo, R., Balmford, A., Ferraro, P. J., Polasky, S., Ricketts, T. H., and Rouget, M. (2006). Integrating economic costs into conservation planning. *Trends in Ecology and Evolution* 21 (12): 681–687.

Pagiola, S. (2002). Paying for water services in Central America: Learning from Costa Rica. In S. Pagiola, J. Bishop, and N. Landell-Mills (Eds.), *Selling forest environmental services: Market-based mechanisms for conservation and development*, 37–62. London: Earthscan.

Reid, W. (2006, March). Millennium Ecosystem Assessment Survey of Initial Impacts. Retrieved July 17, 2007, from www.millenniumassessment.org/documents/Document.798.aspx.pdf.

Stigson, B. (2007, February 28). Why ecosystems matter to business. *Environmental Finance*. Retrieved July 17, 2007, from www.wbcsd.org/plugins/DocSearch/details.asp?type=DocDet&ObjectId=MjMxNTU.

Tallis, H., and Kareiva, P. (2006). Shaping global environmental decisions using socio-ecological models. *Trends in Ecology and Evolution* 21 (10): 562–568.

Troy, A., and Wilson, M. A. (2006). Mapping ecosystem services: Practical challenges and opportunities in linking GIS and value transfer. *Ecological Economics* 60: 435–449.

United Nations Food and Agriculture Organization (2004). *Payment schemes for environmental services in watersheds*. Rome: United Nations Food and Agriculture Organization.

Weyerhaeuser, H., and Kahrl, F. (2005, October). An enduring match? Livelihoods, conservation, and payments for environmental services in the uplands of China's Southwest. Paper presented at the Tropentag 2005 Conference on the Global Food and Product Chain: Dynamics, Innovations, Conflicts, Strategies, Stuttgart-Hohenheim, Germany.

World Wildlife Fund and Rossing Foundation (2004, October). Living in a finite environment (LIFE) Project (Draft End of Project Report for Phase II: August 12, 1999–September 30, 2004). Washington, D.C.: U.S. Agency for International Development.

Wu, J., Sketom-Groth, K., Boggess, W. G., and Adams, R. M. (2003). Pacific salmon restoration: Trade-offs between economic efficiency and political acceptance. *Contemporary Economic Policy* 21: 78–89.

Xu, J., Yin, R., Li, Z., and Liu, C. (2006). China's ecological rehabilitation: Unprecedented efforts, dramatic impacts, and requisite policies. *Ecological Economics* 57: 595–607.

Yuan-Farrell, C., and Kareiva, P. (2006a). *Ecosystem services: Status and summaries.* Arlington, Va.: The Nature Conservancy.

Yuan-Farrell, C., and Kareiva, P. (2006b). *Payment for ecosystem services: Status and summaries.* Washington, D.C.: World Wildlife Fund.

Zavaleta, E. (2000). The economic value of controlling an invasive shrub. *Ambio* 29: 462–467.

Zhang, P., Shao, G., Zhao, G., Lemaster, D. C., Parker, G. R., Dunning, J. B., et al. (2000). China's forest policy for the 21st century. *Science* 288: 2135–2136.

PART IV LEGAL EMPOWERMENT OF THE POOR

13 | The Commission on Legal Empowerment of the Poor

NARESH SINGH

The law is not something you invent in a university—the law is something that you discover. Poor people already have agreements among themselves, social contracts, and what you have to do is professionally standardize these contracts to create one legal system that everyone recognizes and obeys.

—Hernando de Soto

This is a wholly different approach to the poverty debate . . . our Commission will focus on a unique and overlooked aspect of the problem: The inextricable link between pervasive poverty and the absence of legal protections for the poor.

—Madeleine K. Albright

Power concedes nothing without demand. It never did and it never will.

—Fredrick Douglass

The majority of the world's three billion poor spend their working lives living outside the rule of law, where their ability to create wealth

Naresh Singh is the Executive Director of the Commission on Legal Empowerment of the Poor.

from their work and their capacity to leverage their assets are unrecognized by the formal legal systems that govern their countries, and where their dignity is continually violated. Whether these poor workers are farmers or taxi drivers, cart vendors or day laborers, garment makers or shop merchants, the vast majority possess assets of some kind, but the legal system excludes them from the protections and the processes essential for growing a businesses, securing property, and retaining stability during difficult times (Atuahene, 2006). Basic foundational legal recognitions frequently taken for granted throughout the Western world, such as property title, recognition of contracts, birth certificates, and business permits are kept out of the reach of the poor. An often confusing, time consuming, and expensive patchwork of bureaucratic regulations effectively makes access to the formal legal system an impossibility (de Soto, 2000). These formalities ultimately force the poor to remain in the informal sector, where their labor, businesses, and homes are insecure and unable to appreciate in value via the market mechanisms of the formal economy.

The Millennium Development Goals (MDGs) have set an objective to eradicate extreme poverty, initially by reducing by half the number of people who live on less than one dollar per day by 2015 (UN Millennium Development Goals, 2007). In order to accomplish this goal, new thinking surrounding poverty reduction, development, and foreign investment is necessary, not only to achieve short term objectives but also for long-term and lasting development strategies. Building on the work of the Peruvian economist Hernando de Soto, the government of Norway, along with other Nordic countries, played a key role in defining the need for legal empowerment. With support from Canada, Egypt, Guatemala, Tanzania, and the United Kingdom, a proposal for a legal empowerment Commission was presented to the United Nations (UN) Secretary-General, who welcomed it as an important contribution to the fight against poverty (Legal Empowerment [LE], n.d.c). The Commission on Legal Empowerment of the Poor was launched in 2005, as an independent entity hosted by the United Nations Development Programme (UNDP).

The LE commission aims to make the legal protections and economic opportunities so integral to poverty reduction the right of all, not the privilege of the few. The Commission is the first global initiative to focus specifically on the link between informality, poverty, and the law. Legal empowerment focuses on a broad range of legal rights essential to facilitating individuals' participation in the economic activity of their country. Property rights, in particular, are an important, though not exclusive, focus for the Commission. Accessible property rights are the first fundamental legal protection necessary to incorporate poor and disenfranchised populations into an economic system governed by the rule of law. Contract rights, mortgages and credit systems, and labor rights all stem from formal recognition of the assets the poor have to leverage. To accomplish these tasks, the Commission employs both top-down and bottom-up strategies. High-level international, regional, and national political leadership is called on to implement the legislative reforms necessary to incorporate the poor into the formal legal system. On the ground, the Commission builds off the significant work already completed in this field, and takes these findings onto a global level agenda by coordinating efforts, developing best practice research, and engaging with stakeholder groups including: indigenous peoples, women, displaced populations, informal sector associations and federations, grassroots organizations, labor unions, cooperatives, and poor workers themselves. Poor people and grassroots organizations are driving the LE agenda, not only because they are the intended beneficiaries, but because their experiences, knowledge, and recommendations must inform the work of the Commission for its work to have lasting impact.

The success of legal empowerment depends on commitment across the entire spectrum. Policymakers must make the legislative reforms necessary to create a legal environment wherein the poor, many of whom are illiterate, can weave their way through complicated bureaucratic systems. Just as important, those policymakers must not only reduce the bureaucratic complexities, but also recognize and incorporate the existing informal procedures people are already using to govern their everyday lives. Merely creating a system that enables the poor, with training, to access the legal system will be less successful than giving formal legal significance to the existing informal procedures already being utilized by the poor. Those living in the informal system, who may have dual roles in the formal system as well, and grassroots organizations must be involved to educate policymakers on the informal structures poor communities have developed to solve problems ordinarily resolved via the formal legal system. International and multilateral institutions must lend their support to the work of the Commission and recognize the importance of formalization of the economic system. Finally, bilateral donors must put forward the funds necessary to enable these efforts to take root.

The Commission aims to have many diverse outcomes, including these:

1. Documentation of global experiences that brings out key lessons and variations from across the world.

2. A succinct Commission Report summarizing global experience and laying out a broad action agenda for governments.

3. A more detailed reform "toolkit" for policymakers.

4. A proposed Charter on Legal Empowerment of the Poor that can be tied to relevant human-rights conventions.

5. A new body or institution with a concrete mandate to monitor progress and accountability in this area, or a mandate incorporated into an existing body or institution.

6. A specific funding mechanism.

These outcomes, while not having an immediate direct impact, would set the stage for the policymakers and institutions that have the power to make the necessary changes to move quickly and efficiently toward legal empowerment.

Structure and Organization

The role of the Commission in accomplishing the ultimate goal of legal empowerment is twofold. First, the Commission draws on global knowledge from international institutions, governments, and civil society organizations to formulate new policy recommendations based on empirical examples and experiences in the developing world. Second, the Commission leverages its access to high-level officials to generate political will among global policymakers for reforms that will reduce the scope of the informal economy and expand legal empowerment.

The Commission is co-chaired by Madeleine Albright, former Secretary of State for the United States under the Clinton administration, and Hernando de Soto, the Peruvian economist and author whose work has been integral in identifying the specific problems of exclusion of the poor from the formal legal system. The Commission's membership includes former heads of state (President Zedillo of Mexico, President Mkapa of Tanzania, President of Ireland and former High Commissioner of Human Rights Robinson, President Cardoso of Brazil, and former Prime Minister of New Zealand Mike Moore), current high court justices (Justice Kennedy of the Supreme Court of the United States and Clotilde Medegan Nougbode of the High Court of Benin), as well as academics, current and former government ministers, and a Nobel Laureate (LE, n.d.a).[1]

The Commission members are supported in their work by a board of advisors consisting of representatives from the major international development organizations and banks, UN specialized agencies, and international nongovernmental organizations (NGOs) working on development (LE, n.d.a). The Commission has partnered with such organizations as the Indonesian Legal Aid Foundation, the International Institute for Environment and Development, the Rural Development Institute, the World Resources Institute, and Women in Informal Employment Globalizing and Organizing (LE, n.d.e).

The bulk of the Commission's work is accomplished via two different mechanisms: five working groups organized around the five thematic areas that the Commission has identified for its agenda; and national and regional consultations processes, which are informed,

---

1. A complete list and biographies of all Commission members are available on the Commission's website: http://legalempowerment.undp.org/who/members.html.

facilitated, and attended by the partner organizations mentioned earlier.

Working Groups

Each of the Commission's working groups spearheads one of its five thematic areas, which are: access to justice and rule of law; property rights; labor rights; legal mechanisms to empower informal businesses; and road maps for implementation of reforms. In addition, the Commission recognizes three issues that intersect closely with its objectives and will, therefore, be essential to their promotion and achievement: gender equality, the differing circumstances of indigenous populations' relationships with national governments, and environmental sustainability.

Women face particular exclusions from the legal system through unequal treatment before the law. Such exclusion persists both in formal systems and in customary procedures, which frequently stand contrary to international human rights standards. Women are commonly less educated, in general, and less likely to be informed of their legal rights than men are. Reducing and eliminating these barriers is a necessary component of ensuring access to justice.

Indigenous people face many similar barriers. Many communities exist outside the reach of an effective state legal system, making access to the legal structures prohibitively expensive. Discrimination by judges and lawyers, language barriers, illiteracy, and the high cost of litigation all fuel further exclusion from the legal system. These obstacles must be addressed to ensure that meaningful access to justice and the rule of law are secured for indigenous populations.

The Commission is also cognizant of sustainable development efforts in formulating its recommendations. Fitting sustainable development into the Legal Empowerment agenda requires a careful balancing of the law-as-obstacle narrative. Environmental regulations that present significant obstacles to formal sector entry, by virtue of the costs of compliance or the lack of clarity in the regulation, can push people to remain in the informal sector. Importantly, this does not mean that unsustainable practices do not occur; rather, it simply leaves them unchecked outside the rule of law. By promoting and facilitating entry into the formal sector, the Commission's recommendations will therefore also promote adherence to national environmental regulations and international norms. Environmental justice is part of the legal empowerment narrative as well, particularly in relation to indigenous populations who have seen their property taken or destroyed for development projects. Ensuring affordable access to justice enables people in poor communities to seek judicial oversight for the environmental impacts of new development projects. All three cross-cutting issues are carefully treated and considered in the working groups to achieve harmony between the legal empowerment agenda, women and indigenous people's equality, and sustainable development.

Access to Justice and Rule of Law

The goal of the Commission on Legal Empowerment of the Poor, as its name suggests, is to enable the poor to use the law and the legal system to realize their full human potential. Many development theorists and practitioners, as well as the poor themselves, recognize the important role that the legal and judicial system plays in either facilitating or hindering development efforts.[2] Frequently, the poor experience the law as an obstacle rather than a support structure. The law, and the costs associated with using the legal system, can make it difficult or impossible to run a legitimate business, to secure redress for exploitation by the powerful, or even to participate as a full member of the community. Many poor communities experience law and law enforcement only as instruments of repression; instead, the law should be a source of opportunity—for expanding access to economic benefits, for ensuring government accountability, and even for effecting broader social change and promoting sustainable development.

---

2. See, for example, Carothers (2006); Dam (2006); Golub (2003); Sen (1999); Shah and Mandava (2005); and Trubek and Santos (2006).

The challenge for the Access to Justice and Rule of Law Working Group is to affect a transformation from law-as-obstacle to law-as-opportunity for the poor and disempowered. Four key subthemes within Access to Justice and Rule of Law have been identified: lack of legal identity, ignorance of legal rights, unavailability of legal services, and unjust and unaccountable legal systems.

Lack of Legal Identity

Some 50 million children born each year do not have birth certificates (UNICEF, 2006). In the world's least developed countries, up to 70 percent of children are not registered (UNICEF, 2006). A birth certificate is the first link between a person and government. A birth certificate proves that a person exists, and all protections of the legal system and participation opportunities flow from that fundamental recognition. From the perspective of the legal system, people lacking such basic documentation simply do not exist. They cannot vote, legally own property, or file a lawsuit. Their lack of legal identity may make it impossible to receive basic services, like water, sewage, mail delivery, and garbage collection. For many living in wealthy countries who take their registered legal identities for granted, this may seem shocking, but the problem of wholesale legal exclusion may be more widespread than is appreciated or publicly acknowledged.

In Peru, approximately one million indigenous Peruvians living in the highlands have no legal identity and no legal rights (Axworthy and Bielsa, 2006). In theory, they can register with the government, but in practice registration is difficult, expensive, and discouraged by an array of formal and informal practices. Similar problems are present in other Latin American countries and in other regions of the world. The Roma in Eastern Europe are often excluded from the recognition or full protection of the formal legal system (Cameron, 2003). Many indigenous tribes in rural India are unregistered, making them ineligible for important government services, including emergency relief following natural disasters. Denial of a legal identity is also an important problem for noncitizens or internally displaced persons (Refugees International, 2005). Although particularly widespread in certain parts of the developing world, it may also afflict indigenous populations in wealthy countries.

The denial of legal identity to large numbers of people is a serious and neglected problem, and no attempt at legal empowerment of the poor can succeed without tackling it. Some reforms have attempted to address this issue by facilitating programs for registration and issuing birth certificates; however, many have not been fully implemented or fully effective (Axworthy and Bielsa, 2006). Registration can be politically sensitive because granting legal recognition to large numbers of previously excluded individuals in democratic countries may alter the balance of political power, but it is crucial to the Commission's agenda.

Ignorance of Legal Rights

Acquisition of formal legal identity may be necessary for legal empowerment, but it is not a sufficient condition. Failure to inform poor people of their legal rights and obligations can be as substantial a barrier to access to justice. Perhaps even more distressing, many of the people tasked with making legal decisions or providing representation also do not understand what the law allows or requires. Thus, any effective legal empowerment strategy has to address the lack of adequate knowledge of the law.

The problem of legal ignorance may result from inadequate dissemination of information or deliberate obfuscation. A lack of legal knowledge can arise due to the inscrutable nature of legal jargon, particularly for poor and undereducated populations. Furthermore, in countries or regions characterized by linguistic diversity, certain groups may be disadvantaged because their language is not the official or de facto language of government and law. In South Africa, despite a constitutional provision establishing eleven official languages, individuals who speak neither English nor Afrikaans are often at a huge disadvantage when trying to understand their rights and navigate the legal system (Seligson, 2001).

Certain strategies appear to have had success in improving legal knowledge among poor communities. Some of these strategies involve training legal advocates and service providers (Cantrell, 2004). Likewise, training laypeople in basic legal skills, such as writing a valid will, allows the poor to gain access to legal enforcement mechanisms while avoiding the high cost of formal legal representation (Rhode, 1996). Strategies incorporating broad-based popular education, including the use of popular entertainment (radio, television, street theater, comic books, and so on) can be utilized to convey important messages about legal processes, rights, and obligations to large groups of people (Special Court, 2003). In countries or regions where school enrollment and literacy rates are sufficiently high, basic legal education can also be integrated into the school curriculum. Legal curriculum building has proven an effective way to communicate legal understanding not only to school-age children, but to their families as well.

The obstacles-to-opportunities narrative for pursuing legal empowerment of the poor embraces any strategies that increase legal knowledge among the disenfranchised. In an information-poor environment, the law is more likely to seem distant, arcane, and hostile, which leads to a situation in which the poor are unlikely to rely on legal mechanisms to enforce their rights and are more vulnerable to exploitation. A poor population more knowledgeable of legal rights and obligations may begin to see law as presenting opportunities for defending against exploitation and furthering economic interests. In addition, insofar as the poor recognize that the law-in-action falls short of the legal system's professed ideals, they may create more pressure for improving the system.

Unavailability of Legal Services

One of the most serious barriers preventing access to justice for the poor is the unavailability, or expense, of obtaining legal representation or other forms of legal assistance. Even when individuals are not formally excluded from the legal system, and are generally aware of their legal rights, they may remain outside the formal legal system because they lack access to legal services.

Unavailability has several dimensions. First, institutions critical to the provision of legal services to the poor may be nonexistent. Second, necessary legal services may be prohibitively expensive for poor people. Finally, legal institutions and services may be geographically distributed in such a way that, for practical purposes, they are inaccessible to poor and rural populations. The physical absence of formal legal institutions and services often allows non-state actors to assume a dominant role in providing dispute resolution mechanisms and other services typically associated with the justice system. That the only contact many poor individuals have with the legal apparatus of the state is, therefore, through the (often repressive) police, only further undermines confidence in the legal system.

Suggested strategies for how to provide legal services vary from allowing greater participation in the legal system by non-lawyers, such as paralegals, law students, and other professionals, to utilizing local bar associations to encourage or even require pro bono commitments, to eliminating the barriers to market representation such as laws forbidding lawyers to work on a contingency basis or forbidding attorneys from advertising. Again, the topic of improving access to legal services can be framed as part of a larger obstacles-to-opportunities narrative. Without adequate legal assistance and representation, the poor cannot hope to use the legal system to advance their welfare. The law instead becomes a barrier, rife with snares for the unwary. With access to legal expertise, however, the poor can navigate the law and use the legal system for their protection and benefit. The obstacles-to-opportunities narrative is also relevant to prospective providers of legal services to the poor. Frequently, offering legal services to the poor involves great risks and difficulties, with little possibility of substantial benefit. If this situation can be changed to make offering legal services to the poor easier and more rewarding (financially or otherwise), the prospects for legal empowerment may expand dramatically.

Unjust and Unaccountable Legal Systems

Improving access to the legal system does not improve access to justice unless the legal system actually provides justice. This final stage of access to justice, if omitted, would leave unresolved problems that plague many countries' bureaucratic and judicial systems.

Throughout the world, clogged court dockets can prevent the poor from gaining access. The graft and corruption often required to go through the motions of registration must be put to an end. Expanding and incorporating the customary and the informal dispute resolution systems that are the exclusive judicial mechanism available to the poor in rural areas, and thereby making them accountable to the government, may be necessary to provide meaningful access to justice.

Property Rights

The core of property is not the simple possession of an object, but the abstract rights bestowed upon the possessor by the legal system. The conceptualization of property as a bundle of legally defined rights is the linchpin for transforming tangible assets into abstract relations of credit, security, and exchange. If an effective market is the key to the creation of prosperity, then the greatest injustices to the poor are the constraints that prevent them from transforming their assets into property regimes and deny them access to the market. The legal empowerment narrative is focused on releasing the potential that resides within poor countries and among poor people. The key to unlocking dead capital—"assets [that] cannot readily be turned into capital, cannot be traded outside narrow local circles where people know and trust each other, cannot be used as collateral for a loan, and cannot be used as a share against an investment"—is a process of transformation of assets into fungible capital (de Soto, 2000). This transformation is key to moving poverty reduction from mere slogan to concrete plan of action and commitment.

The poor control major assets in the urban and periurban areas. Unrecognized, asset utility and marketability are severely undermined. For example, the security of assets held informally is tenuous at best, leading to property that cannot be relied upon, invested in securely, or borrowed against. Secure title and ownership rights enable people to extract greater short-term as well as medium- to long-term benefits from their assets. Property rights that convey secure tenure may enhance access to credit by the poor. Access to credit accelerates a family's or community's ability to make an expensive investment, such as purchasing materials for a home or farm equipment. Without credit, such a purchase would likely require decades of savings or be wholly unattainable. The proceeds of a loan can be used to build or improve a structure, start a business in a home or elsewhere, or be invested in education. Where property is used as collateral, families can benefit from the value of their property without selling it and, thereby, capture its economic benefit at an earlier stage in life. Other ways in which property title enables the legal empowerment of the poor include these:

- It enhances of the poor's ability to exercise agency

- It provides higher value returns to the owners of those assets

- It enables the leveraging of property to gain access to capital and to create wealth

- It creates responsibilities (obligations), leading to greater accountability

There is some legitimate concern that granting land title to poor people accustomed to a community ownership model may lead them to give up their property to cover immediate needs and forgo the long-term benefits of property ownership. To counter this concern, there is need for creative legal and social thinking to ensure adequate protections are in place. Property protections and principles of communal ownership are particularly important to be cognizant of when considering the effects of property title in the context of indigenous populations. Women's rights to property are even more severely unrecognized by the dual failure to recognize women's access to property title in the formal system and by cultural barriers preventing women from owning property. Such barriers require a careful analysis of how to integrate women's access to property into the cultural constructions existing throughout the world. The key to granting property rights is not a simple matter of conducting land surveys and providing paper title unfamiliar to people who have lived and worked their entire lives via informal procedures. It requires a careful understanding of the practices the poor themselves have created and then building on those practices to implement a property rights regime in a manner that allows the poor to leverage and protect their assets within a collaborative framework on which all people agree.

Last, securing property rights has tangible benefits for sustainable development as well. The MDG for Ensuring Environmental Sustainability includes two key targets that are reinforced by providing the poor with legal title to their property: reducing by half the proportion of people without sustainable access to safe drinking water, and achieving significant improvement in the lives of at least 100 million slum dwellers by 2020 (UN Millennium Development Goals, n.d.). Property title, as with legal recognition of individuals, makes the provision of services to the poor, particularly urban slums, significantly easier. Improved sanitation services such as clean drinking water, sewage, and garbage collection can aid sustainable development efforts by utilizing more environmentally sound practices. Likewise, providing urban slum dwellers with access to property title enables them to improve their own lives by becoming part of the formal market economy. As conditions in slums improve, their environmental impact is lessened. These goals are further reinforced by the incentives to protect the sustainability of personal property and the surrounding area that property ownership creates. In rural areas, farmers with secure title to their farmland may choose to use more sustainable farming practices to protect the soil.

Enforcement of property rights is key to the protection of poor people's assets and to their ability to move or leverage them in the marketplace. Where the legal system does not provide adequate enforcement of the property rights of the poor, particularly in rural areas where people have land to farm, but little money to buy seed and equipment and few ways to protect their harvest, the legal system needs to be developed to aid the poor in enforcing their rights.

Labor Rights

One-half to three-quarters of nonagricultural employment in the developing world is secured through the informal sector (International Labor Organization [ILO], 2002). That percentage is considerably higher for women than for men (ILO, 2002). While the reasons that workers and businesses remain in the informal sector are complex, formal labor regulations that are inappropriate for the economic realities of the country, are burdensome, or impose excessive costs, play a significant role. The challenge is to create a legal infrastructure that enables all workers to obtain legal recognition and protection, as well as access to the services and resources necessary to obtain work, while ensuring that labor laws are not so onerous as to cause businesses to resist and remain outside of the formal legal system.

Workers in the informal sector face a different working environment than those in the formal sector. Outside of the formal legal system, the regulations meant to protect workers from exploitative work hours, low pay, and unsafe working conditions and to protect children from exploitative labor are unenforceable. Instead, people's job security and working conditions are left subject to the tenuous customary protections of the informal employment system. While the informal sector is able to operate "cheaply," in part due to avoiding the costs of these regulations, the cheapness of business operations does not necessarily connote efficiency. Informality undermines the productivity and quality of employment by failing to uphold the standards of labor and care enforced in the formal sector.

Because labor is not a commodity, labor markets must be viewed and treated differently from markets for goods and services. Labor rights include roles for both state and non-state actors, inasmuch as there is a need to encourage the state to create proper legislative regimes that account for the processes through which informal businesses function and the different market realities they face. There must be recognition that a "one size fits all" labor regulation approach does not adequately take into account the diversity of situations and circumstances of the business world. In particular, reforms must also be evaluated for their impacts on disadvantaged populations. Relaxing labor laws, as is currently done in many countries to lure foreign investment and to compete in the global marketplace, disproportionately effects women and the poor. Likewise, discrimination against both indigenous people and women in the labor market undermines both equality and economic growth. Labor organizations, grassroots organizations, and NGOs must be included in the reform process to ensure that the rights of all workers are respected and protected.

Legal Mechanisms to Empower Informal Businesses

The fourth working group examines ways to provide broad access to legal tools, instruments and organizational forms suited to informal businesses. Doing so enhances their opportunities for growth, by enabling poor people to combine labor, technology and investment in order to raise their own productivity, limit risk exposure, and access credit and capital. This mandate requires significant research into how the entrepreneurial spirit, innovation, and creativity found among informal business owners can be channeled to reduce the barriers they face to sustained involvement in the formal economy.

Without sufficient recognition and reforms in the legal system, the formal sector will continue to prove too costly for informal businesses to make the transition. Legal tools, such as business permits that provide protections and privileges for business owners, must be tailored to make them easier and cheaper to apply for and receive. Furthermore, these permits must be structured to serve the types of businesses that are prevalent in the informal economy. Properly opening or closing down a business can be complicated by a frequently overwhelming number of bureaucratic steps that prove unfeasible and inefficient for the informal sector (de Soto, 2000). Investigating and investing in the local financial institutions that are capable of providing access to credit, equity, and capitalization on reasonable terms, and that are based on existing informal businesses and the assets they possess, is another important step. Access to credit enables the scope and operation of businesses to expand beyond the narrow confines of personal savings and risk. The ability to incorporate a business, so that shares can be sold and the ultimate risk of investment spread to a larger number of individuals, is essential for growth.

It is equally important to ensure that the legal system is able and willing to enforce the contracts entered into by informal businesses, to allow them to operate and deal with a larger customer base, beyond those who can pay cash or are in the narrow group of individuals a business owner may know and trust personally. Conflict resolution mechanisms must be reviewed to ensure maximum simplicity and comprehension of the contracting mechanisms used by those operating in the informal sector. Creation of business protections, such as microinsurance policies, that are appropriate for the size of businesses in the informal sector also encourages business growth. All of these reforms facilitate the transition from the informal to the formal sector, and enable business owners to expand their businesses beyond the practical limitations of informality.

Ensuring that women and indigenous people's businesses are recognized by the formal system is a distinct issue requiring particular attention. Domestic work is still largely considered women's work and there is continued pressure on girls to drop out of school earlier than their male peers to help take over household responsibilities. Providing women, in particular, with the means to invest in their own businesses can effect positive changes in the perception and reality of women's participation in the economic system. The innovative nature of indigenous people's practices is as often overlooked. While globalization has generally fueled demand for new approaches, many barriers still limit indigenous entrepreneurship. Microcredit and microfinance programs should be expanded, therefore, since they are frequently the only source of credit for indigenous people and women alike (Yunus, 1999). All reform interventions need to be informed by a deep understanding of cultural and social structures, in particular the traditional roles and restriction of women, and indigenous conceptions of economic growth and participation. In addition, recognition of the disruptive potential of formalization must be addressed before beginning the process in any community.

Last, just as small businesses provide substantial gains for the economic system, their collective impact on the environment is no less substantial. While small businesses provide good jobs, localized products, and targeted, responsive innovation, they often do not follow environmentally sound practices. Bringing these businesses into the fold of the formal economy, where government regulations can be enforced, is vital to long-term sustainable development. Properly structuring regulations so that they do not present insurmountable barriers for small businesses will encourage formalization and extend the reach of government regulations for environmental protection.

Access to legal tools for businesses and an understanding of market regulation have proven essential throughout the development of Western markets, but they have not appeared in many developing countries. Reforming their business statutes and structures is essential to transforming informal enterprises into participants in the formal economy.

Road Maps for Implementation of Reforms

The final working group synthesizes the key practical outputs of the four other working groups and focuses on the implementation of reforms. Critical in this regard is the analysis of both the gaps between the working groups' thematic areas and the synergies and potential conflicts between their recommendations. The goal is to identify possible road maps for reform, policy options, toolkits, and indexes for use by policymakers around the world in implementing governmental and institutional reforms.

The group is also considering the creation of an inventory of possible "do's" and "don'ts" and a checklist for policymakers, and developing indicators that will facilitate monitoring of such reform processes. Its work draws on existing guidelines, frameworks, manuals, indexes, indicators, and other related aides, and it will assist the Commission's key audiences to implement its policy recommendations. The work of the Commission, particularly in the context of these toolkits, is particularly cognizant of potential impacts on the environment and adherence to countries' existing environmental regulations. This toolkit will support policymakers in proposing reforms and, once implemented, in measuring their results.

National and Regional Consultation Processes

There are similarities and differences in the ways countries around the world, and over time, have approached the challenge of legal empowerment. These experiences are only partly understood, in terms of both their key attributes and their outcomes. Furthermore, there has been no systematic effort to compare or synthesize lessons learned across global experience in this area. The national and regional consultation process is an integral part of one of the essential tasks of the Commission: examination of a representative selection of such experiences in order to identify promising reform paths and appropriate policy instruments. Creating a centralized location where these experiences can be referenced allows the Commission and the policymakers it aims to assist to draw upon these experiences to achieve real outcomes on the ground.

At the same time, the Commission's presence and partnership building with grassroots organizations inspires those at the local level to exert pressure and start creating demand for change. These partnerships are another central component of the Commission's mission to build bridges to enable dialogues between those in high-level government positions and the poor.

The consultation process began in 2006 with consultations taking place on the regional and national level in Central Europe and Asia, Brazil, East Africa (Ethiopia, Kenya, Tanzania, and Uganda), and Indonesia. These consultations were facilitated by commission members from the region and by collaboration with grassroots organizations. In 2007, national consultation processes were held in Bangladesh, India, Pakistan, Sri Lanka, Mexico, Guatemala, Thailand, and the Philippines, West Africa (Benin, Burkina-Faso, Mauritania, and Mali), Southern Africa (Mozambique, Namibia, South Africa, and Zambia), the Middle East and Northern African countries (Egypt, Jordan, Morocco, and Yemen), as well as a combined consultation for island states that focused on the specific challenges they face (LE, n.d.b).

Making a Difference

The Commission's approach is novel in that, rather than identifying needs that can be alleviated through outside support, it looks to and builds on existing assets to solve many of the economic problems faced. This is not a rejection of the international, regional, national, and subnational institutions whose work has functioned on alternate theories, but rather a complement to those strategies that focus on encouraging markets as they exist to work better for and enrich the poor. It is important to recognize that the free-market system need not be the zero-sum game its critics so often portray. Empowering the poor to begin using their existing assets, growing their businesses, and participating in formal economic structures does not require the rich to concede any power or wealth. The goal of the Commission is not redistribution, but growth. It seeks to expand the economic pie in ways that reinforce sustainable development efforts and enable a greater number of people to participate in the benefits of formal markets.

The Commission has the support of more than twenty world leaders, high-level ministers, and international, regional, and national development organizations, all of which are able to generate a greater level of political will than has been harnessed by previous legal empowerment efforts. The world leaders and supporters of the Commission are building momentum for local and global change and, in turn, fueling other initiatives within the grander development agenda. Nonetheless, there are risks that come with any effort at development. Even well-intentioned, well-reasoned, and well-articulated reforms can fall short of their intended outcomes and, by doing so, further exacerbate poverty. The poor who gain title to their property for the first time may quickly sell it off to cover short-term needs and ultimately be left in worse condition than before. Women who take out loans may find themselves unable to make repayments and lose what few assets they have as a result. Indigenous people and other vulnerable groups face objectively different situations that, if not adequately studied and understood, could lead to ineffective, or even harmful, reforms and outcomes. The Commission will take careful note of the risks associated with reform, and with its recommendations in particular, to ensure proper monitoring and evaluation frameworks, and special protections for women and indigenous peoples, and environmental sustainability.

Conclusion

It has been more than sixty years since the Bretton Woods institutions were first created to facilitate international trade, invest in development, and integrate the global marketplace. Since that time, development programs have expended millions of man-hours and billions of dollars annually on the goal of development. Nonetheless, more than three billion people still lack access to the formal legal and economic structures that have been proven to work throughout the West. Likewise, it has been more than thirty years since the Stockholm Conference put protection of the environment on the development agenda, yet environmental degradation is continuing at a rapid pace. The Commission on Legal Empowerment of the Poor aims to improve this situation by placing legal empowerment at the top of the global development agenda and building bridges for dialogue and communication between high-level government institutions and the poor.

The Millennium Development Goals have set an ambitious agenda for eradicating poverty throughout the world and promoting sustainable development, but their results, if not focused on the fundamental question of access to justice and rule of law, property rights, a proper understanding of the role of labor rights, the informal sector of the economy in which the vast majority of the world's poor live and work, and the link between legal empowerment and sustainable development, will ultimately be short lived and fail to deliver real economic development that benefits the poor and protects vital environmental resources. It is time for a new approach to development that specifically focuses on empowering the poor to participate in the global market. It is time for legal empowerment.

References

Atuahene, B. (2006). Land titling: A mode of privatization with the potential to deepen democracy. *St. Louis University Law Journal* 50: 761–782.

Axworthy, L., and Bielsa, R. (2006). Presentation to the permanent council of the OAS of the chief of the electoral observation mission to the second round of the presidential elections in Peru. Retrieved January 30, 2007, from www.oas.org/speeches/speech.asp?sCodigo=06-0148.

Cameron, L. (2003). Right to an identity. Retrieved January 30, 2007, from www.errc.org/cikk.php?cikk=1066.

Cantrell, D. (2004). The obligation of Legal Aid lawyers to champion practice by non-lawyers. *Fordham Law Review* 73: 883–899.

Carothers, T. (Ed.) (2006). *Promoting the rule of law abroad: In search of knowledge.* Washington, D.C.: Carnegie Endowment for International Peace.

Commission on Legal Empowerment of the Poor (n.d.a). Board of advisors. Retrieved January 30, 2007, from http://ment.undp.org/who/board.html.

Commission on Legal Empowerment of the Poor (n.d.b). Consultation processes. Retrieved January 30, 2007, from http://legalempowerment.undp.org/what/consultation.html.

Commission on Legal Empowerment of the Poor (n.d.c). Funding. Retrieved January 30, 2007, from http://legalempowerment.undp.org/who/funding.html.

Commission on Legal Empowerment of the Poor (n.d.d). *Legal empowerment: Unlocking human potential.* New York: Legal Empowerment.

Retrieved January 30, 2007, from http://legalempowerment.undp.org/pdf/Legal_Empowerment_Brochure.pdf.

Commission on Legal Empowerment of the Poor (n.d.e). Partnerships. Retrieved January 30, 2007, from http://legalempowerment.undp.org/challenge.

Dam, K. (2006). *The law-growth nexus: The rule of law and economic development*. Washington, D.C.: Brookings Institution Press.

de Soto, H. (2000). *The mystery of capital: Why capitalism triumphs in the West and fails everywhere else*. New York: Basic Books.

Douglass, F. (1849). Letter to an abolitionist associate. In K. Bobo, J. Kendall, and S. Max (Eds., 1991), *Organizing for social change: A mandate for activity in the 1990s*. Washington, D.C.: Seven Locks Press.

Golub, S. (2003). *Beyond rule of law orthodoxy: The legal empowerment alternative* (Carnegie Paper No. 41). Washington, D.C.: Carnegie Endowment for International Peace.

International Labor Organization (2002). *Women and men in the informal economy: A statistical picture*. Geneva: International Labor Organization. Retrieved January 30, 2007, from www.ilo.org/public/libdoc/ilo/2002/102B09_139_engl.pdf.

Refugees International (2005). *Lives on hold: The human cost of statelessness*. Washington, D.C.: Refugees International. Retrieved January 30, 2007, from www.refugeesinternational.org/content/publication/detail/5051.

Rhode, D. (1996). Professionalism in perspective: Alternative approaches to nonlawyer practice. *New York University Law Review* 22: 701–711.

Seligson, M. (2001). Lay participation in South Africa from apartheid to majority rule. *International Review of Penal Law* 72: 273–283.

Sen, A. (1999). *Development as freedom*. New York: Knopf.

Shah, P. J., and Mandava, N. (Eds.) (2005). *Law, liberty and livelihood: Making a living on the street*. New Delhi: Academic Foundation.

Special Court of Sierra Leone (2003). *Special court made simple*. Retrieved January 30, 2007, from www.sc-sl.org/specialcourtmadesimple.pdf.

Trubek, D. M., and Santos, A. (Eds.) (2006). *The new law and economic development: A critical appraisal*. New York: Cambridge University Press.

UNICEF (2006). Birth registration fact sheet. Retrieved January 30, 2007, from www.unicef.org/protection/files/Birth_Registration.pdf.

United Nations Millennium Development Goals (n.d.). Retrieved Feb. 8, 2007, from www.un.org/millenniumgoals.

Yunus, M. (1999). *Banker to the poor: Microlending and the battle against world poverty.* New York: PublicAffairs.

14 Opportunities in Environmental Stewardship

Climate Change and Legal Empowerment of the Rural Poor

OLAV KJØRVEN AND ESTELLE FACH

The advent of the modern global community brought about a world in which any notion of existing in splendid isolation has yielded to one of unparalleled interdependence. From this connectedness emerged three enduring and entwined issues. At the center of development efforts is the challenge of improving the lives of three billion poor people, including more than one billion living on less than a dollar a day. Environmental challenges such as climate change, the depletion of biodiversity, and unsustainable patterns of extraction and production have also become multidimensional issues calling for global solutions. Reinforcing and feeding on these are political challenges such as conflict, war, security, and governance. At the core of these three issues is a lack of a fair and inclusive rule of law.

The predominant reliance on top-down policy approaches has often created conditions in which large numbers of people are compelled to exit formal structures of law. Most of the world's poor live without even the most basic legal protection for their homes, assets, or sources of livelihood (Commission on Legal Empowerment of the Poor [LE], n.d.). As part of the informal economy, they occupy land they do not own, work in small, informal businesses, and rely on

Olav Kjørven is the Assistant Secretary-General and Director of the Bureau for Development Policy at the United Nations Development Programme (UNDP). Estelle Fach is an international-law specialist and a consultant with UNDP.

friends and informal money lenders for loans (LE, 2006). Informality limits the working poor's access to broader economic opportunities and makes them vulnerable to the uncertainty, corruption, and violence prevalent outside the rule of law. Be it in business, labor, justice, or land ownership, informality carries with it high social, economic, and environmental costs. It is intrinsically linked to poverty and undermines the full political empowerment of the poor. It does not solely affect individuals but also entire societies. The large numbers of people that are excluded from legal systems lack the ability to organize, incentives toward environmental stewardship, and opportunities for social and economic development.

It would be simplistic to equate informality with poverty; informality is a complex phenomenon in which causalities are often difficult to prove, but it limits the poor's horizon of opportunity. By contrast, a "legal" empowerment of the poor provides tools, supported by mechanisms under international and national law, for poor people to move out of poverty and obtain the benefits and protections that real citizenship under an inclusive legal order can provide. With a focus on the rural poor in developing countries, this chapter details how innovative international and national legal regimes in carbon finance can simultaneously promote environmental stewardship and provide a way out of poverty.

Legal Empowerment of the Rural Poor Using International Regimes of Carbon Finance

There is fortunately little remaining debate on the reality of climate change. Overwhelming amounts of scientific evidence support the conclusion that observed changes in the global climate are, in large part, due to human activities. The Stern Review Report on the Economics of Climate Change (Cabinet Office, 2006) and the recent reports from the Intergovernmental Panel on Climate Change (IPCC) (2007) have done much to convince the international community and public opinion worldwide that delaying climate change mitigation is dangerous and more costly than inaction. For the poorer developing countries, the potential impacts of climate change will undermine progress toward the Millennium Development Goals and exacerbate many of the poverty and environmental issues they already face (African Development Bank [ADB] et al., 2003). Among the anticipated adverse impacts are the rise of sea levels, harm to ecosystems, loss of biodiversity and extinction of certain vulnerable species, a decrease of yields in tropical and subtropical regions, increased frequency and severity of storms, floods and droughts, variability in precipitation, and shortages of water (IPCC, 2001b). All of these will lead to food scarcity and threats to human health and livelihoods, particularly for the working poor living in tropical regions (IPCC, 2001b). Climate change will have a disproportionately adverse affect on the poor, as other catastrophic events such as conflicts, epidemics, or natural disasters already do (Oxfam, n.d.). Recent widespread flooding in Haiti, Kenya, and Ethiopia has already led to tragic deaths and losses of livelihoods. Unless immediate action is undertaken, the world's poor will suffer the brunt of climate change's social and economic ramifications.

Climate Change and the Rural Poor

While the increasing share of international attention devoted to slum dwellers (United Nations Secretary-General, 2001) demonstrates a concern about the growing rural-to-urban drift, it should not distract from the fact that three quarters of the world's poorest currently live and work in rural areas, where they face specific challenges. The unique rural dimension of poverty is characterized by a lack of income and productive assets, a lack of access to essential economic and social services, but also a lack of power, participation and respect (United Nations Economic and Social Commission for Asia and the Pacific [UNESCAP], 2007). The rural poor are often adversely affected by policies taken by decision makers in urban areas that fail to reflect their realities and priorities (UNESCAP, 2007). The informal nature or remoteness of rural settlements also means that they receive inadequate warnings of impending disasters and little information on adaptation strategies (Oxfam, n.d.).

The adverse effects of global climate change will likely affect the rural poor disproportionately. A heavy dependence on environmental resources and undermined capacity to adapt renders people in poor rural areas exceptionally vulnerable and perhaps the least resilient to global climate changes. For instance, while farmers in the developed world can make up for short rainy seasons by using artificial water sources, African farmers who rely on rainwater and do not have access to basic systems of irrigation lack that option and will face much more severe consequences for their crops. The drylands, where the poorest and most marginalized people live, face increased desertification which will have dramatic consequences for livelihoods (United Nations Development Programme [UNDP], n.d.). Compared to the contribution of greenhouse-gas emissions from the industries and activities of wealthy countries, the least developed countries and their rural poor have barely contributed to the climate change crisis (IPCC, 2001a). While there is no time for finger-pointing, one must recognize that the social and economic effects of climate change could disproportionately harm the parts of the world that have least contributed to the problem. Therefore, the whole international community should undertake the challenge of assisting the least developed nations to adapt to the effects of climate change.

As a parallel, simultaneous and complementary effort, the mitigation of climate change is a global responsibility. The international community's success or failure to address climate change in the next few years will have a significant impact on the decades to come. Governments, nongovernmental organizations, and intergovernmental institutions are increasingly recognizing the necessity of combined top-down and bottom-up approaches. What, if anything, can vulnerable rural populations do to mitigate the effects of global climate change? How can legal empowerment support their efforts and what benefits can they reap?

Forestry and Rural Land: Valid Carbon-Sequestering Areas

Mitigating global warming necessitates efforts from both developed and developing countries. The former are expected to achieve drastic reductions in greenhouse gas emissions and to strive to undertake strategies that seek to avoid reliance on a carbon economy. The 1997 Kyoto Protocol to the 1994 United Nations Framework Convention on Climate Change (UNFCCC) assigns mandatory targets to Annex 1 (industrialized) countries for the reduction of greenhouse gas emissions, and allows them to subtract appropriate sinks—any process or activity that removes a greenhouse gas or its precursor—in their calculations of total emissions. Introduced under the Kyoto Protocol, the Clean Development Mechanism (CDM) is one of the three market-based mechanisms of carbon trading and the only one that involves non–Annex 1 (that is, developing) countries (UNFCCC, n.d.). It allows Annex 1 countries to earn carbon credits—certificates equivalent to a reduction of one ton of carbon dioxide—by investing in project activities that reduce greenhouse gas emissions and contribute to sustainable development in non–Annex 1 countries. From a pragmatic point of view, industrialized countries buy emission credits at a lower price than costly changes in industrial activities at home, while developing countries benefit by receiving new financing for sustainable development. The net total emissions of greenhouse gases are not expected to change as a direct result of the CDM. Nonetheless, in addition to augmenting financial flows to developing countries, the mechanism has the potential to transform markets by assisting with technology diffusion, changing purchasing habits, and promoting environmental awareness and education. Existing CDM projects include energy efficiency, renewable energies, methane gas or fuel substitution, and carbon capture by certain biological processes described herein (UNDP, 2006b).

Yet, so far, carbon trading under the CDM has benefited middle-income industrializing countries and has largely overlooked poor developing countries. India and Brazil alone account for half of all confirmed projects, and by the end of 2006 nine countries accounted for 82 percent of all projects that had reached the registration stage (UNDP, 2006b). Only four out of the forty-eight least developed countries had confirmed CDM projects at that time, representing less than 2 percent of total project numbers (UNDP, 2006b). The complexity and cost of legal and technical skills necessary to put forward project proposals to the CDM board has thus far prevented many developing countries from participating. Yet poor countries, and in particular sub-Saharan countries, can have their fair share of carbon finance. To facilitate investment, the UNDP Millennium Development Goals Carbon Facility (MDGCF) was recently created (UNDP, 2007). As part of the Nairobi Framework, a United Nations partnership between six development institutions (UNFCCC, 2006), the MDGCF seeks to provide assistance to least developed countries so that they may join the trading market.

Strategies to broaden the participation of poor countries in the CDM could rely on increasing the share of natural carbon sinks activities worthy of carbon credits. Developing countries often have a vast wealth of the biological resources (plants or trees) that capture or sequester greenhouse gases. It is clear that biological sequestration of carbon benefits the atmosphere: the IPCC has estimated that biological sinks have the potential to capture up to 20 percent of the excess carbon released into the atmosphere over the first fifty years of the twenty-first century (2001). Another study by Japan's National Institute for Public Health and the Environment suggests that up to 40 percent of the mitigation achieved worldwide could come from biological sinks (2003). Yet under the current CDM guidelines only a handful of projects qualify in the area of land use, land use change and forestry (commonly abbreviated as LULUCF).[1] These include the biological sequestration of carbon by *afforestation* activities—converting open formerly forest-land into a forest—and *reforestation* activities—restocking of trees on forest land that has been depleted. *Conservation* of existing forests does not at this point qualify to earn carbon credits under the CDM. The mechanisms are accordingly different: benefits from reforestation are based on absorption of atmospheric carbon over time, while preventing deforestation refers to avoided emissions by protecting the carbon that is already stored in mature forests.

---

1. In addition, the CDM caps LULUCF to 1 percent of total CERs (Certified Emissions Reductions). At the time of writing, however, this number had been far from attained.

Avoiding deforestation, however, carries an enormous environmental potential: it is estimated that 25 percent of global greenhouse gas emission is due to the burning and decay of cleared forests and woodlands, in particular from the conversion of tropical forests (Fearnside, 2000). Overall deforestation rates have declined in the last few years, in part due to the increase of plantations, yet conversion of natural forests continues to take place at an annual rate of seven million hectares a year (Food and Agriculture Organization [FAO], 2005). But from a sustainable development perspective, the most exciting reason that forest conservation should be included as a carbon-credit worthy activity under the scope of the CDM may be that it represents an opportunity for poor rural populations to engage in mitigating climate change. Deforestation often occurs to make way for agriculture and urban development or to generate revenues from the sale of timber. If forest conservation qualifies to receive carbon credits, and if these credits become a tangible alternative source of revenue[2] for the rural poor, incentives toward preserving carbon sinks will naturally grow. These new mechanisms could provide benefits that compete with the attractiveness of unsustainable slash-and-burn practices.[3]

Other biosequestration activities such as soil and water conservation, management of grazing land or sustainable agriculture are currently excluded from the CDM, although they fall within the scope of the Kyoto Protocol.[4] Agricultural land is, for example, an important part of the carbon cycle, and management practices determine whether lands are carbon sources or carbon sinks (IPCC, 2000). For example, burning crop residues, human-induced soil erosion and desertification, grassland degradation, wetland reclamation for agriculture, low water-use efficiencies, organic matter and fertility loss, excessive tillage, and the like are practices that promote net carbon

2. This would likely necessitate higher prices for CERs than are currently occurring.
3. The social and environmental sustainability aspects of compensation for avoided deforestation are clear; the economic sustainability, political will and subsidies aside, remains to be established given current pricing of CERs.
4. The Kyoto Protocol, Article 10.b (i), lists agriculture as one of the regional programs containing measures to mitigate climate change.

emissions. Similarly, poor manure management and overuse of fertilizers release nitrous oxide and methane, two gases with global warming potential of twenty and more than three hundred times that of carbon dioxide, respectively. Alternatively, land management under good conservation and nutrient management, conservation or zero tillage, crop rotations, good management of fallow lands, or preservation of wetlands favors the enhancement of carbon sinks. In addition, sustainable agricultural practices create more resilient and productive systems. Indeed, increasing the organic matter in the soil through carbon sequestration stimulates soil biology and biodiversity, improving, in turn, soil fertility and production. Organic agriculture illustrates an economically and ecologically sustainable practice: the growing demand worldwide represents an added opportunity for developing countries exports and promotes healthier ecological systems. A study presented at the United Nations Conference on Trade and Development (UNCTD) recently highlighted that under certain conditions organic agriculture creates crops and soils that are more resistant to droughts and floods, thus further contributing to preventing or mitigating the catastrophes that could result from climate change (UNCTD, 2004). Adding to the mix the potential to mitigate climate change by biological sequestration, this case becomes a win-win-win situation that carries the triple benefits of poverty alleviation, climate change adaptation, and climate change mitigation (Niamir-Fuller, 2007).

While the environmental and social benefits of the carbon sequestering practices above are clear, the specific circumstances in which carbon finance improves the internal rate of return to a level that makes it economically justifiable for farmers and local communities to engage in such projects remains to be determined on a per-country basis. At this point the few successful examples have been buttressed by subsidies and political will. Providing support to governments who encourage such measures is one option; mechanisms in carbon finance need to be further negotiated to delineate others.

State parties to the Kyoto Protocol have entered negotiations toward defining a post-2012 commitment period. They will soon decide whether the current system should remain as is, be simplified or improved. On this latter path, integrating a valorization of the role of rural lands as carbon sinks would create a fantastic opportunity to exploit synergistic environmental and social measures. If and when rural lands and forests are widely valued and legally accepted as carbon capture activities, the rural poor will be recognized as the guarantors of an enormous wealth of carbon credits. The flow of resources under the CDM will also help investments in resource conservation and land management activities, which are prerequisites for the development of sustainable agriculture in developing countries.

The potential benefits for the livelihoods of the rural poor and the environment are extraordinary. Enabling the poor to tap that new potential income, however, implies providing the legal tools and protection to reap the benefits of their good stewardship. Thus far neither the Kyoto Protocol nor the CDM have specified criteria for equitable sharing of benefits deriving from greenhouse gas-reducing activities. Legal empowerment needs to happen simultaneously at the international level, by engaging parties towards clarifying the question of the ownership of carbon credits, and at the national level, by addressing the key and connected issue of secure land rights.

Key Challenges and Opportunities

The majority of research on carbon sequestration has thus far focused on technical elements such as carbon accounting tools, enforcement, and monitoring, effectively eluding a detailed examination of social issues. For compensation for carbon sequestration to work for the rural poor, a few precautions will need to be in place.

Who Owns the Carbon Credits?

CDM credits can be exchanged through bilateral or unilateral agreements. The first kind of agreements occurs between countries; the second allows project proponents to be governments, nongovernmental organizations, private entities, or individuals who register the project activity with the CDM board. A project proponent for a rural activity would conduct measurements of carbon soil and would be handed an equivalent number of titles to carbon credits, or certified emission reductions (CERs), each corresponding to the equivalent of a reduction of one ton of carbon dioxide. CERs can then be traded or held. Under this construction, ownership of the certificates is specified upon registration on an ad-hoc basis, thus avoiding solid and systematic legal measures to ensure an equitable process.

Ownership of the CERs is at times granted to the registering enterprise. While this model can be valid for investments in energy plant or fuel switching, it would fail to foster participation from the rural poor: lacking organizational, financial, and technical capacities, they will likely be deprived of the opportunity to reap any financial benefits from their stewardship. Alternatively, some have proposed that the owners of the land should be the owners of the carbon-sequestrating process, and thus the carbon credits (Kägi and Schmidtke, 2005). Yet this proposal simply leaves aside the large majority of the rural poor who lack formal land rights. Indeed, in many developing countries much of the land is held under customary tenure, which may or may not be recognized by formal laws. In West Africa, for example, it is estimated that less than two percent of the land has any formal paper documentation, with most rights being claimed through unwritten forms of tenure (Toulmin, 2006).

The lack of consistency on the issue of legal ownership of CERs affects the chances for project approval and leads to confusion on how to interpret the local CDM rules (ICF International, 2006). Maybe more importantly, it fails to create safeguards for rural populations. In Brazil, for example, several NGOs have noted that the Plantar project, a combined fuel switching and eucalyptus plantation supported by the World Bank, may have neglected social sustainability and local poor communities (May et al., 2005).[5] In the Ecuadorian Andes, carbon forestry projects have been accused of abrogating the rights of peasant and indigenous communities, leaving them uninformed, manipulated, indebted, and altogether worse off (Granda, 2005). In addition, several countries are already treating carbon as a mining

---

5. See also "Broken Promises: The World Bank and Forests," available at www.fern.org/media/documents/document_3625_3626.pdf (accessed July 30, 2007).

resource, and therefore the property of the state.[6] This creates clear disincentives for farmers or local communities to engage in the carbon market. By contrast, a pro-poor national agenda should ensure that legal mechanisms are in place to vest carbon ownership in rural populations, hence formally recognizing their role as environmental stewards. Insuring that the poor receive monetary rewards for good management practices would strengthen their livelihoods and increase their incentives to continue providing these services (Boyce and Pastor, 2001). Monetary compensation for environmental services has been shown, when incorporating the social component, to simultaneously advance the promotion of environmental sustainability and contribute to poverty reduction (Rosa et al., 2004). At the international level, discussions on equitable benefits under the Kyoto Protocol should be accelerated. Such mechanisms are, for example, already in place and continuously improved under the framework of the Convention on Biological Diversity and other international agreements (Secretariat for the Convention on Biological Diversity, 2002).

Securing Land Rights

In addition to the economic rewards, receiving ownership of carbon credits could give the rural poor mechanisms and incentives to organize and petition their governments to obtain a formal acknowledgement of rights to their lands. Land is the primary means by which many in the world earn a living and invest, accumulate, and transfer wealth between generations, yet informality in land tenure continues to prevail in developing countries (LE, 2006). A life in fear of eviction, lack of incentives to invest in one's land, and the particular predicament of women and other marginalized groups are common concerns to those lacking formal recognition of their land rights. In this light, secure tenure arrangements constitute a legal necessity to protect the rural poor should they start receiving benefits under the CDM. As land value rises rapidly with the foreseen amplification of carbon finance, unwritten claims to land place rural populations at an increasing risk.

---

6. Australia is such an example.

Farmers with few official claims to their land could indeed be easily evicted by governments seeking to reap the dividends from natural carbon sinks. Outside groups such as large-scale forest-carbon enterprises may also have an interest in gaining new property rights to manage these biosequestration services, the result of which could be to plunge the rural poor into deeper poverty and insecurity.

By contrast, and in a reinforcing loop, secure property or tenure allows rural communities to organize and gain strength in their negotiations during contractual agreements with carbon credit traders. Additionally, land rights can improve the issues of nonpermanence and liability detracting investors from biosequestration projects. Indeed, so far afforestation and reforestation have generated little investment interest and represent less than one percent of the projects under the CDM (UNDP, 2006b). Long timelines, the temporary nature and vulnerability to a variety of natural or anthropogenic risks, and uncertainties commonly referred to as "non-permanence issues" (Ellis, 2001), have discouraged investors.[7] In addition, afforestation/reforestation projects encompass transaction costs in negotiating land rights: most enterprises have reached for the low-hanging fruits (e.g., denuded state land) rather than attempting to negotiate with communities. While several NGOs are attempting the latter, their projects have not yet been registered with the CDM. Secure tenure or property rights, by contrast, offers stronger and longer-term guarantees: the formal land owner and manager can commit to carbon sequestration activities and strive to protect his or her carbon-capturing assets from accidents, such as fires, for the duration of the agreement. Since the very populations that live and depend upon forestry resources not only have a direct interest in their conservation but also a physical presence to ensure it, their legal empowerment via secure land rights represents a way to mitigate risks and, as a consequence, increase the number of biosequestration projects. Toward this goal, and hand in hand with economic rewards, secure land rights are essential in guaranteeing

---

7. Some of those uncertainties concern a post-2012 commitment period, which limits carbon contract projects.

that rural populations conserve resources to ensure long-term economic returns and environmental services.

The realization that securing land rights provides both socioeconomic and environmental benefits is increasingly widespread. In Niger, the *New York Times* reports, empowerment of rural populations has resulted in spectacular environmental results, with bottom-up initiatives by farmers allowing for more than seven million acres newly covered by trees. This was made possible by a gradual realization, later backed by the government, that farmers could own the trees, which were previously regarded as a property of the state—a legacy of colonial times (Polgreen, 2007). Incentives to protect and plant trees have grown, resulting in better top soil and less erosion. Poverty reduction has simultaneously taken place, since increased revenues from the sales of branches, pods, fruit, and bark have replaced illegal logging.

Yet the answer is by no means as simple as privatizing all land and handing over titles to the poor. Strengthening the property rights of the poor is a vast endeavor, complicated by the diversity of social and ecological systems in rural lands. While many systems may be valued as natural carbon sinks, they require different and context-adapted property-rights arrangements.

Land-registration processes vary worldwide in their scope and scale, ranging from short-term certificates of occupancy to titling procedures (Toulmin, 2006). The ways in which tenural arrangements can be set up to guarantee rights under the CDM are not limited to leasing or privatizing lands. Indeed, as economic incentives grow, clarifying property and user rights to common resources becomes even more essential. Common property areas such as forests, grazing lands, and peatlands (partially decomposed organic matter) are essential components of rural landscapes and display a great potential for achieving large volumes of carbon sequestration. Preserving common property agreements is also vital to certain populations, such as pastoral herders (Niamir-Fuller, 2000). Representing 10 percent of Africa's rural population and key contributors to meat and dairy production, as well as to the protection of biodiverse systems, pastoralists need secured access to grazing land and water. Examples of successful resource management strategies include the West African Sahel, Ethiopia, and Mozambique, where locally agreed rights of passage along agreed pathways were established (Toulmin, 2006). More generally, pluralistic approaches to land tenure, under a coherent national legal framework, will be necessary to foster adapted solutions, such as the policies in eastern and southern Africa (Adams and Turner, n.d.). Here, new legal developments provide for customary land tenure while simultaneously offering clear and secure paths to modern formats as they are identified. Conceived of as evolutionary processes, these formats adapt to the evolving needs identified by individuals or communities, demonstrating in turn that flexibility does not contradict formality.

The key to success in tenural arrangements, in short, lies first in locally adapted and flexible arrangements obtained through participative processes (Adams and Turner, n.d.). These, in turn, must be backed by strengthened governance systems that can combat corruption. There is increasing evidence of local chiefs, who receive payments for environmental services (PES) under watershed management schemes, selling off land that they are supposed to steward on behalf of the community.

Land Tenure and Women's Rights

Any attempt to develop the potential for carbon finance through secure tenural arrangements must take gender equity issues into careful consideration. Women are often the main custodians of land, yet customary land rights often disadvantage them. Existing rules in many developing countries prevent women from owning property or receiving land as inheritance. In many nations the high incidence of AIDS-related deaths increasingly leaves destitute widows and orphans cut off from land owned by their dead husband's or father's kin (Human Rights Watch, 2003). Only recently has the particular vulnerability of rural areas to AIDS been studied (FAO and UNAIDS, 2001), highlighting the fact that the epidemic's impact on agriculture and rural development could further compel gender and tenure inequalities. Certain regions, however, have instituted new and progressive laws to counter this phenomenon. For instance, the regional government of Amhara, Ethiopia, supported by legislative federal measures in 2005, undertook new land registration and certifications that led to the creation of laws requiring that the certificate contain the names of both the head of the household and the spouse (in practice a picture of husband and wife is attached) so that they own the land jointly (Deininger, Zevenbergen, and Ali, 2006). This illustrates how legal systems can be locally adapted to empower marginalized segments of the population, represented here by poor rural women, and thus improve security of tenure and livelihoods.

Comprehensive and Comprehensible Environmental Approaches

Existing compensation schemes that reward rural communities for managing ecosystems have generally revolved around conservation of biodiversity, scenic beauty, and watershed management, and have provided insightful lessons (Landell-Mills and Porras, 2002). These include, for example, the necessity to focus not only on payments but also on enhanced social status; to build the compensation schemes with a primary objective of strengthening livelihoods; and in that respect, to avoid seeking economies of scale and concentration of powers in the hands of one sole intermediary. Approaches should, in addition, be comprehensive, so that certain ecosystem services are not favored to the detriment of others. For example, although management of plantation forestry may be cheaper and easier to monitor, it should not be undertaken at the expense of the biodiversity-rich ecosystems of natural forestry (Rainforest Foundation et al., 2005). Instead, and to maximize social and environmental benefits, watershed management, biodiversity conservation, and carbon sequestration should be synchronized efforts. Furthermore, the international community and national governments have to break away from the old and persisting paradigm that considers "agricultural landscapes" and "natural landscapes" as separate siblings. The two are not only interlinked with regard to farmer/herder production, but also connected in terms of carbon sequestration.

In order to vest the benefits of comprehensive approaches in the rural poor there must be education programs that explain the concept of "carbon," and its sequestration and release, to local players. While the economic incentives will be tangible, explaining the link to carbon itself to farmers and herders in developing countries—and the rationale for protecting soil sinks, regulating the percentage of harvested biomass, and so forth—will require significant and long-term national efforts that should be supported by the international community.

A Hidden Opportunity

Biological sequestration can therefore generate rural income, reduce large volumes of greenhouse gas emissions, and promote sustainable agriculture and forestry practices. It can also provide additional benefits such as protection and improvement of soils, water resources, habitats, and biodiversity (Amano and Sedjo, 2006). The commercial and legal boundaries in carbon finance are currently being tested. Powerful and organized forestry groups in developed countries are contributing to the debate on carbon rights and shaping new legal mechanisms (Kägi and Schmidte, 2005), but the voice, rights, and interests of the rural poor in developing countries with regard to carbon finance need to be fully embraced. Carbon markets, created by global and national institutions, will ultimately involve rethinking long-established property-rights regimes, and need to be harnessed through rules and laws that ensure that the benefits flow to the poor. Lack of legal ownership, political marginalization, and exclusion from the decisions that affect how ecosystems are managed could be obstacles to the poor's enjoyment of and access to these benefits (UNDP et al., 2005). By contrast, formalizing land rights and delineating ownerships to carbon credits will be indispensable to protecting and benefiting the rural poor. Allocating ownership to carbon in a regulated, equitable and transparent manner under the flexibility mechanisms of the Kyoto Protocol or under voluntary carbon trading schemes is, thus, a hidden opportunity to tackle the synergistic challenges of land ownership, revalorization of agricultural activity, lessening of climate change through carbon sequestration, and ultimately rural poverty reduction.

As a final remark, while this section has focused on rural areas, the urban poor can also directly benefit from carbon market mechanisms. A housing project designed under the CDM with the assistance of SouthSouthNorth, an NGO involved in climate change and social development, involved equipping more than two thousand low-income houses in Kuyasa, South Africa, with energy-efficient lighting, insulated ceilings, and solar water heaters (SouthSouthNorth, n.d.). This project became the developed world's first CDM project validated by a "Gold Standard," an independent recognition that distinguishes best practice methodologies delivering high-quality carbon credits (ibid.).

Political, Judicial, and Financial Challenges and Opportunities

Establishing the international and national frameworks that will make carbon ownership and land tenure a reality for the poor can take place only with solid legal structures that take into account political components, such as conflicts, and that are backed by formal and reliable judicial and financial systems.

Land and Conflict

Land is a key component in today's interconnected poverty and environmental challenges; the interconnectivity between land and conflict is profound, and it has the potential to affect carbon finance directly. Indeed, in a crisis context, and in particular a conflict over land and its resources, several factors may undermine biological sequestration and the mechanisms of carbon financing. For example, land and carbon sinks deteriorate because of extractive tendencies of warring parties searching for resources. A conflict also increases the risks of nonpermanence, discouraging investors for long periods of time.

Much of today's political volatility and conflicts, locally, regionally, or internationally, arise from tensions and competition in access to land and resources. The unstable peace situation in Ivory Coast, a country de facto split in two, is a prime example of the consequences of sudden changes in the property rights regime. To exploit ethnic tensions for political reasons, rural workers of foreign origin were abruptly deprived of property rights to lands they had been harvesting for generations. This precipitated the tensions and led to the current crisis. In Sudan, a history of successive conflicts erupting in rural areas in the wake of droughts and famine pitted pastoralist herders against sedentary farmers. In the absence of government reaction and systems to handle disputes, the population armed itself, giving rise to decades of civil war (Flint and de Waals, 2005).

The effects of climate change will likely cause the situation to deteriorate: decrease in farmable land, drought and other climate-related shocks have a potential to accelerate conflict and violence. Countries in the regions of West Africa and the Nile basin may become particularly vulnerable given their high water interdependence, although the 2006 Human Development Report stressed that until now cooperation on water issues has prevailed over confrontation worldwide (UNDP, 2006a). Moreover, the interplay between land and conflict goes both ways, since conflict often entails large displacements of people, which then accelerate land disputes. The continued threat of conflict in several African countries is a significant contributing factor in the proliferation of slums in urban areas. In Sudan, again, the prolonged crisis in the southern region led to the mass exodus of rural communities to the capital Khartoum, which accommodated almost half of the more than six million internally displaced persons during the late 1990s (Moreno and Warah, 2006).

To avoid conflict ignition or escalation, the issues of land and property rights advocated here must be addressed through careful and inclusive processes. Land rights must be perceived as *legitimate* before they are made *legal* by the national and local governments (Toulmin, 2006). Public, open and transparent dialogues with a variety of stakeholders are essential to achieve a locally owned land rights agenda (Toulmin, 2006). The process should clarify the relationship between new laws and existing customary tenure and take into account both technical and political considerations. It must also be adapted to the national and local contexts. Indeed, while individual land-titling initiatives—issuing a formal piece of documentation that establishes ownership—have proved successful in India (Toulmin, 2006), they have been inadequate in much of Africa and failed to generate sufficient tenure security or stimulate investment (Mwangi, 2001). Additionally, titles do not necessarily confer rights to occupy, use, or develop the land (Toulmin, 2006), and have led to large numbers of disputes, as is reportedly the case in Kenya (Adams and Turner, n.d.). By contrast, in Mozambique, Uganda, South Africa, and Lesotho, strengthening and affirmation of group rights and land administration are now concomitant with legal space to register individual titles. New mechanisms of equitable distributional processes must be supported by comprehensive dispute manage systems.

Securing Rights by Appropriate Formal Judicial Systems

To bear fruit for the poor and prevent conflicts, distributional justice in land/carbon credit holdings needs to be paired with procedural justice (Brown and Cordeba, 2003). When rights cannot be defended against challenges, they provide little incentive and no security (Bruce, 2004). As a result, people are pushed into developing their own informal systems of justice distribution. These range from simple peer pressure to organized crime and are most likely to affect the disenfranchised poor disproportionately. Sometimes formal dispute settlement mechanisms are simply inaccessible: courts are often located in urban sectors, removed from slums or rural areas where the poor live. Adequate dispute settlement mechanisms should replace the often corrupt, unprofessional, or unpredictable judicial systems of many developing countries. Access to justice and the rule of law necessitates mechanisms to strengthen the voices of the poor—such as providing them with a legal identity and informing them of their legal rights—supported by the responsiveness and accountability of judicial service providers.

Alleviating Informality in Financial Systems with Secure Tenure

Finally, achieving distributional equity of carbon credits will have to be supported by a reduction in the informality in financial systems. It is already clear that informality in land tenure issues is often aggravated by extralegal financial systems in many developing nations. First, insecure tenure and fear of eviction deter populations from investments in longer-term yields, locking them into poverty and making them more vulnerable to all kinds of unfavorable change. Second, financial systems are often too immature and inflexible to offer viable services to potential customers from the ranks of poorer segments of society. Insecure tenure often limits the ability to build up assets to escape poverty and protect against external shock, thus exacerbating the issue. Banks are indeed reluctant to offer loans to those who have no titles to land or housing that can be used as collateral. As an example, 95 percent of loans are secured by land assets in Zambia and 80 percent in Indonesia, clearly delineating an opportunity gap for those who cannot claim such rights for lack of formal documentation. The disenfranchised poor are thus compelled to find informal mechanisms for storing assets and with greater risk, or to invest in forms that can be insecure—such as cattle, jewelry, or cash—and do not guarantee return rates. Investment opportunities originate from friends or unscrupulous moneylenders. Inadequate access to credit has been identified as one of the greatest obstacles for small entrepreneurs in developing countries.

Land rights may give people not only incentives to invest in their land but also assets against which to borrow, to create, or to expand business opportunities. The connection between tenure and investments is clear in urban contexts, but less so in rural areas. Studies have shown that land registration has not led to increased investments in Kenya, Senegal, Somalia, or Uganda (Platteau, 1996). Formal registration, it has been argued, is poorly fitted to commonly held rural lands, and, perhaps more importantly, does not work in isolation. More intricate mechanisms that take into account common ownership will be needed in some instances to design a locally adapted formalization of land rights.

Conclusion: Toward a "Right to Carbon"

Climate change is now understood as part of the larger challenge of sustainable development. Its mitigation using the CDM can also contribute to poverty alleviation, as long as it is supported by land rights that are adapted to local and national contexts and are secure and enforceable. Naturally, legal and institutional changes are complex and time-consuming, and although their impacts are cumulative, they often take place on a time scale greater than election cycles. Yet global and local leaders from industrialized and developing countries are now converging around the idea that extending legal protections and property rights holds unrealized potential for poverty reduction and sustainable development. Given appropriate legal frameworks and proper oversight, new mechanisms in carbon finance have a tremendous potential to alleviate rural poverty. The opportunities for improved livelihoods, growth, development, and greater environmental good are genuine. Furthermore, the opportunity costs—that is, failing to formalize carbon and land tenure rights—are high. If carbon finance takes off as anticipated, disengaging the rural poor from its potential revenues will inevitably lead to exploitation, loss of livelihoods, further marginalization, and a plethora of other social and environmental problems.

We have examined here one possible form of legal empowerment of the rural poor, which derives from international mechanisms and necessitates supporting action at the national and local levels. This discussion does not pretend to be an alternative to other developmental approaches, but to provide a complementary strategy, if enacted in time. The need for a global, regulated, transparent, and equitable framework under the regime of the Kyoto Protocol and the Clean Development Mechanism is pressing. Legal measures to empower the rural poor will have to be undertaken through a rigorous and agreed upon system that should be ready at the onset of the second commitment period in 2012. The international community should undertake a global discussion on the matter at the next Conference of the Parties to the United Nations Framework Conference on Climate Change, and simultaneously explore voluntary, innovative, and experimental actions in carbon finance.

References

Adams, M., and Turner, S. (n.d.). Legal dualism and land policy in eastern and southern Africa. Retrieved July 11, 2007, from www.capri.cgiar.org/wp/..%5Cpdf%5Cbrief_land-03.pdf.

Africa: Climate change becoming a matter of life and death (2005). *IRIN News*, April 4, 2005. Retrieved July 11, 2007, from www.irinnews.org/report.aspx?reportid=53719.

African Development Bank, Asian Development Bank, Department for International Development: United Kingdom, Directorate-General for International Cooperation: the Netherlands, Directorate General for Development: European Commission, Federal Ministry for Economic Cooperation and Development: Germany, et al. (2003). *Poverty and climate change, reducing the vulnerability of the poor through adaptation*. Paris: Organisation for Economic Co-operation and Development.

Amano, M., and Sedjo, R. A. (2006). Forest sequestration: Performance in selected countries in the Kyoto period and the potential role of sequestration in post-Kyoto agreements (Resources for the Future Report). Retrieved July 11, 2007, from www.rff.org/rff/Documents/RFF-Rpt-ForestSequestrationKyoto.pdf.

Boyce, J., and Pastor, M. (2001). *Building natural assets: New strategies for poverty reduction and environmental protection*. Amherst, Mass.: Political Economy Research Institute. (Natural Assets Project Report). Retrieved July 11, 2007, from http://cjtc.ucsc.edu/docs/RR3.pdf.

Brown, K., and Cordeba, E. (2003). A multi-criteria assessment framework for carbon-mitigation projects: Putting 'development' in the centre of decision-making. (Tyndall Centre for Climate Change Research, Working Paper 29). Retrieved July 11, 2007, from www.tyndall.ac.uk/publications/working_papers/wp29.pdf.

Bruce, J. W. (2004). Strengthening property rights for the poor. In *Collective action and property rights for sustainable development* (2020 Focus 11, Brief 16). Washington, D.C.: International Food Policy Research Institute.

Cabinet Office, Her Majesty's Treasury (2006). *The Stern review: The economics of climate change*. Cambridge: Cambridge University Press.

Commission on Legal Empowerment of the Poor (n.d.). *Legal empowerment: Unlocking human potential*. New York: Legal Empowerment. Retrieved August 10, 2007, from www.undp.org/legalempowerment/pdf/Legal_Empowerment_Brochure.pdf.

Commission on Legal Empowerment of the Poor (2006). Overview paper. Retrieved July 11, 2007, from http://legalempowerment.undp.org/pdf/HLCLEP_Overview.pdf.

Deininger, K., Zevenbergen, J., and Ali, D. A. (2006). Assessing the certification process of Ethiopia's rural lands. Retrieved July 11, 2007, from www.mpl.ird.fr/colloque_foncier/Communications/PDF/Deininger.pdf.

Ellis, J. (2001). Forestry projects: Permanence, credit accounting and lifetime (OECD/IEA Information Paper COM/ENV/EPOC/IEA/SLT (2001)11). Retrieved July 11, 2007, from www.oecd.org/dataoecd/4/58/2467909.pdf.

Fearnside, P. M. (2000). Global warming and tropical land-use change: Greenhouse gas emissions from biomass burning, decomposition and soils in forest conversion, shifting cultivation and secondary vegetation. *Climatic Change* 46: 115–158.

Flint, J., and de Waals, A. (2005). *Darfur: A short history of a long war.* London: Zed Books.

Food and Agriculture Organization of the United Nations (2005). Extent of forest resources. In *Global forest resources assessment 2005, progress towards sustainable forest management* (FAO Forestry Paper 147) (11–46). Rome: Food and Agriculture Organization. Retrieved July 11, 2007, from ftp://ftp.fao.org/docrep/fao/008/A0400E/A0400E03.pdf.

Food and Agriculture Organization of the United Nations and Joint United Nations Programme on HIV/AIDS (UNAIDS) (2001). Sustainable agricultural/rural development and vulnerability to the AIDS epidemic. Retrieved July 11, 2007, from http://pdf.dec.org/pdf_docs/PNACK239.pdf.

Granda, P. (2005). Carbon sink plantations in the Ecuadorian Andes, impacts of the Dutch FACE-PROFAFOR monoculture tree plantations' projects on indigenous and peasant communities. *World Rainforests Movement Series on Tree Plantations* 1.

Human Rights Watch (2003). Policy paralysis: A call for action on HIV/AIDS-related human rights abuses against women and girls in Africa. Retrieved July 11, 2007, from www.hrw.org/reports/2003/africa1203/index.htm.

ICF International (2006). Nigeria: Carbon credit development for flare reduction projects. Retrieved July 11, 2007, from http://siteresources.worldbank.org/EXTGGFR/Resources/NigeriaGGFR Guidebook °IC F.pdf.

Intergovernmental Panel on Climate Change (2000). *IPCC special report on emissions scenarios.* Geneva: Intergovernmental Panel on Climate Change. Retrieved July 11, 2007, from www.grida.no/climate/ipcc/spmpdf/sres-e.pdf.

Intergovernmental Panel on Climate Change (2001a). *Climate change 2001: The scientific basis* (Contribution of Working Group I to the Third Assessment Report of the Intergovernmental Panel on Climate Change). Cambridge: Cambridge University Press.

Intergovernmental Panel on Climate Change (2001b). *Climate change 2001: Impacts, adaptation, and vulnerability* (Contribution of Working Group II to the Third Assessment Report of the Intergovernmental Panel on Climate Change). Cambridge: Cambridge University Press.

Intergovernmental Panel on Climate Change (2007). *Climate change 2007: Impacts, adaptation and vulnerability* (Contribution of Working Group II to the Fourth Assessment Report of the Intergovernmental Panel on Climate Change). Retrieved July 11, 2007, from www.ipcc-wg2.org/.

Kägi, W., and Schmidtke, H. (2005). Who gets the money? What do forest owners in developed countries expect from the Kyoto protocol? *Unasylva* 222 56 (3): 35–38.

Kyoto Protocol to the United Nations Framework Convention on Climate Change (1997). Retrieved July 11, 2007, from http://unfccc.int/resource/docs/convkp/kpeng.pdf.

Landell-Mills, N., and Porras, I. (2002). *Silver bullet or fools' gold.* London: International Institute for Environment and Development.

May, P., Boyd, E., Chang, M., and Veiga Neto, F. C. (2005). Incorporating sustainable development into carbon forest projects in Brazil and Bolivia. *Estudos Sociedades e Agricultura* 1.

Moreno, E. L., and Warah, R. (2006). The State of the World's Cities Report 2006/7, Urban and slum trends in the 21st century. United

Nations Chronicle 43 (2). Retrieved July 11, 2007, from www.un.org/Pubs/chronicle/2006/issue2/0206p24.htm.

Mwangi, E. (Ed.) (2001). *Land rights for African development: From knowledge to action* (Consultative Group on International Agricultural Research (CGIAR) System-wide Program on Collective Actions and Property Rights (CAPRi) Policy Brief). Washington, D.C.: CAPRi.

National Institute for Environmental Studies and the Institute for Global Environmental Strategies (2003). Climate regime beyond 2012: Incentives for global participation. Retrieved July 11, 2007, from www.nies.go.jp/social/post2012/pub/jrr2003–1.pdf.

Niamir-Fuller, M. (2000). *Managing mobility in African rangelands: A legitimization of transhumance.* Bourton on Dunsmore, England: Practical Action.

Niamir-Fuller, M. (2007). Land degradation. Unpublished internal document, United Nations Development Programme, Global Environment Facility.

Oxfam (n.d.). Climate change. Retrieved July 11, 2007, from www.oxfam.org.uk/what_we_do/issues/climate_change/introduction.htm.

Platteau, J.P. (1996). The evolutionary theory of land rights as applied to sub-Saharan Africa. *Development and Change* 27 (1): 29–86.

Polgreen, L. (2007, February 11). In Niger, trees and crops turn back the desert. *New York Times.*

Rainforest Foundation, CDM Watch, Global Witness, Sink Watch, Forest Peoples Programme, Environmental Defense, et al. (2005). Broken promises: How World Bank policies fail to protect forests and forest peoples' rights. Retrieved July 11, 2007, from www.fern.org/media/documents/document_3625_3626.pdf.

Rosa, H., Barry, D., Kandel, S., and Dimas, L. (2004). Compensation for environmental services and rural communities: Lessons for the Americas. *Political Economy Research Institute Working Papers Series* 96. Retrieved July 11, 2007, from www.ccmss.org.mx/documentos/payment_of_envirnmental_services.pdf.

Secretariat for the Convention on Biological Diversity, United Nations Environment Programme (2002). Bonn guidelines on access to genetic resources and fair and equitable sharing of the benefits arising out of their utilization, convention on biological diversity.

Retrieved July 11, 2007, from www.cbd.int/doc/publications/cbd-bonn-gdls-en.pdf.

SouthSouthNorth (n.d.). SSN activities. Retrieved August 17, 2007, from www.southsouthnorth.org/ssnactivities.asp.

Toulmin, C. (2006). Securing land rights for the poor in Africa—Key to growth, peace and sustainable development (Commission for the Legal Empowerment of the Poor Thematic Position Paper). Retrieved July 11, 2007, from http://legalempowerment.undp.org/pdf/Africa_land_2.pdf.

United Nations Conference on Trade and Development (2004). *Trading opportunities for organic food products from developing countries*. New York: United Nations.

United Nations Development Programme (n.d.). Drylands and climate change. Retrieved July 11, 2007, from www.undp.org/climatechange/adap05.htm.

United Nations Development Programme (2006a). *Human development report 2006, beyond scarcity: Power, poverty and the global water crisis*. New York: Palgrave Macmillan.

United Nations Development Programme (2006b). *The clean development mechanism: An assessment of progress*. New York: United Nations Development Programme. Retrieved July 11, 2007 from www.energyandenvironment.undp.org/undp/index.cfm?module=Library&page=Docum ent&DociumentID=5883.

United Nations Development Programme (2007). *MDG carbon facility, leveraging carbon finance for sustainable development*. New York: United Nations Development Programme. Retrieved July 11, 2007, from www.undp.org/mdgcarbonfacility/docs/brochure-eng-29may07.pdf.

United Nations Development Programme, United Nations Environment Programme, World Bank, and World Resources Institute (2005). *World resources 2005—The wealth of the poor: Managing ecosystems to fight poverty*. Washington, D.C.: World Resources Institute. Retrieved July 11, 2007, from www.wri.org/biodiv/pubs_description.cfm?pid=4073.

United Nations Economic and Social Commission for Asia and the Pacific (UNESCAP) (2007). *Persistent and emerging issues in rural poverty reduction.* Bangkok: UNESCAP. Retrieved July 11, 2007, from www.unescap.org/pdd/publications/RuralPoverty/ChI.pdf.

United Nations Framework Convention on Climate Change (n.d.). Clean development mechanism (CDM). Retrieved July 11, 2007, from http://cdm.unfccc.int/index.html.

United Nations Framework Convention on Climate Change (1994). Art. 1.8. Retrieved July 11, 2007, from http://unfccc.int/resource/docs/convkp/conveng.pdf.

United Nations Framework Convention on Climate Change (2006, November 15). UN Secretary-General announces "Nairobi Framework" to help developing countries participate in the Kyoto Protocol. *UNFCCC Press Release.* Retrieved July 11, 2007, from http://unfccc.int/files/press/news_room/press_releases_and_advisories/application/pdf/061115_cop12_pressrel_1.pdf.

United Nations Secretary-General (2001). Road map towards the implementation of the United Nations millennium declaration: Report of the Secretary-General (UN General Assembly Doc. No. A/56/326). Retrieved July 11, 2007, from www.un.org/documents/ga/docs/56/a56326.pdf.

CONCLUSION

PAOLO GALIZZI AND ALENA HERKLOTZ

As its title indicates, the essays in this volume have sought to advance understanding of the relationship between poverty eradication and the environment. To this end, the book has led the reader through emerging paradigms in integrating environmental conservation and economic development, analyzed the international natural disaster risk reduction regime and two of its most challenging catastrophes, surveyed recent developments in the fields of conservation information and development decision making, and explored legal empowerment as a means to preserve and protect natural resources while alleviating poverty.

Although no attempt has been made to analyze all the innumerable areas in which development and conservation intersect, this volume offers a guide to many of the most promising and significant opportunities for both sustaining and improving life on earth. As asserted in Part I, a new paradigm has emerged that views poverty eradication and sound environmental management as not only nonexclusive, but also completely dependent on one another. Investing in environmental conservation is essential for successful long-term development; conversely, environmental sustainability cannot be achieved without parallel gains in poverty eradication. In-depth analyses of the importance of environmental wealth to poor people and nations, as well as of the potential of environmental investment, ecoagriculture, sustainable production and consumption patterns, and ecosystem renewal, demonstrate the many options available for addressing the needs of the poor and of the planet.

Environment and development are, likewise, essential to the success of efforts to reduce and recover from natural disasters. In Part II, the reader learns that natural disasters are more often the product of converging environmental degradation and poverty-driven vulnerabilities than inescapable acts of nature. Furthermore, natural disasters have undermined advances made in reversing poverty and ecosystem decline time and again, often quite literally wiping them out. The lessons we can learn from the profound tragedies of the South Asian tsunami of 2004 and Hurricane Katrina center on the intersection of economic and environmental degradation. Proper management of natural resources is critical to protecting the most vulnerable and safeguarding their path out of poverty.

Part III addresses the reality that knowledge about the role of the environment in meeting poverty alleviation goals will not lead to their achievement per se. Knowledge can bring about change only when it is learned and applied. The democratization of knowledge creation and exchange is shown to be important, particularly the development of a learner-centered collaborative approach. A strong case is made for granting poor nations and people free and open access to environmental information in order to support and enhance their vital conservation and development efforts. In particular, knowledge of social impacts, financial viability, design innovation, and balanced leadership is indispensable to the entrepreneurs and investors who can drive viable, scalable, and sustainable social change. Economic valuation of ecosystem services is, likewise, essential for bringing natural capital and conservation into the cost-benefit calculations that inform public decision making and therefore affect the environment and the communities it sustains.

Part IV explores legal empowerment of the poor as yet another avenue for pursuing the joint goals of development and conservation. The Commission on Legal Empowerment of the Poor aims to empower the poor to participate in the global market by addressing the fundamental issues of access to justice and rule of law, property and labor rights, and the informal economy. By focusing on the link between sustainable development and legal empowerment as well, the Commission sets an ambitious agenda for unleashing the potential of lower-income countries and communities to grow in harmony with the ecosystems on which they depend. The reader is also introduced to the potential mechanisms for legal empowerment available under the international climate regime. Carbon and other environmentally derived opportunities provide promising policy options for creating legal structures that harness the potential of secure ownership of economic assets and ensure justice and opportunities for the poorest while promoting environmental sustainability for all.

The intellectual and implementation challenges raised in the pursuit of truly sustainable development are exciting, diverse, and among the principal issues dominating international efforts and awareness today. This volume has aimed to shed light on the emerging paradigm of the unparalleled interdependence of people and the environment. The need to conserve natural resources and safeguard the health of our planet is urgent. Demonstrating the roles the environment can and should play in poverty alleviation, the authors advance a worldview in which the environment is not only the necessary backdrop for human growth and development, but also has an integral role to play. That we need the environment to survive is long settled and beyond dispute; this volume has sought to contribute to our understanding of a newfound, but vital, dynamic that promises far more: the many ways in which nature can help us to thrive.

INDEX

abundance of resources, rate of income growth and, 35
accounting, natural capital and, 17
Acumen Fund, 253
Adam and Brittany Levinson Fellow, 145
African Americans in New Orleans, 200
Agenda 21, 7
agricultural biodiversity, 66
agricultural development, climate change and, 72
agricultural production, 65
    conversion due to food needs, 70
agricultural products, global demand for, 64
agricultural sustainability, ecosystem management and, 71
agricultural systems, diversification, 72–73
agriculture
    biodiversity and, 73
    ecoagriculture, 73
    ecological degradation and, 68
    ecosystem services, 73
    forests, 70
    industrial, climate change and, 72
    sustainability, 71
    threats to ecosystem, 67

agriculture-conservation landscapes, integrated, 65
Agriculture Street Landfill (New Orleans), 199
Albright, Madeleine, 305, 309
Annan, Kofi
    MA (Millennium Ecosystem Assessment), 11
    Millennium Project, 5
ants, 243
Apex Joint Women's Committee, 132
APPTA (Smallholder Association of Talamanca), 75–76
Asia earthquake, 145
Asociación Ak' Tenamit, 117
asset base of the poor, 31–33
    increasing, 38–39
asset control, 316
asset dissipation, family size and, 37
assets, children as, 32
asthma, 19
Atchafalaya River, Mississippi River and, 189
Ax, Floridalma, 105

Bacher, Fr. Hermann, 125
bananas. *See* Chiquita

360 | Index

Base of the Pyramid (BoP), 254
biodiversity, 65
  agriculture and, 73
  conserving, 71
  cork production, 101–3
  ecosystem services and, 73
  Equator Prize, 21
biodiversity conservation, 64
biodiversity loss, 6
biofuels, 69
Biological Conservation, 244
biological coridors, 71
biological sequestration, 344
biotechnology products, 42
Bishop, Joshua, 30
Bitter Bamboo Project case study, 223–25
BMZ (German Federal Ministry of Economic Development and Cooperation), 22
BoP (Base of the Pyramid), 254
Bourne, Joel K., Hurricane Katrina "prediction," 187
Breaux, John (Senator, Louisiana), 196
British Petroleum, 273
Budapest Open Access Initiative, 238
Budowski, Tamara, 113
bushmeat, 71
business
  ecosystems and, 274
  informal, empowering, 320–22
  profit margins, climate change and, 18–19

canals, New Orleans, 190–91, 196
cap-and-trade systems, 15
capital
  dead capital, 316
  natural capital, 17, 272
carbon credits, 15
  distributional justice, 347
  gender issues, 342–43
  land rights and, 339

carbon dioxide emissions, climate change and, 19
carbon finance, boundaries, 344
carbon markets, 344
carbon-sequestering areas, 332–37
carbon sequestration projects, 15
carbon storage and sequestration, 42
carbon trading, CDM and, 333
Caribou Coffee, 94
Carmelita forestry cooperative, 104
Carrera, José Román, 105
case studies
  CEC (Commission on Education and Communication), 225
  IUCN Bitter Bamboo Project, 223–25
  IUCN Livelihoods and Landscapes Strategy, 221–23
  social entrepreneurship
    international development enterprises in India, 256–62
    Jamii Bora, 262–69
CBD (Convention on Biological Diversity), 238–39
CCTs (continuous contour trenches), 126
CDM (Clean Development Mechanism), 15, 333
  carbon trading, 333
  credits, exchanging, 337
  poor countries' participation, 334
  soil conservation and, 335
  water conservation and, 335
CEC (Commission on Education and Communication), 219
  case study, 225
Cenicafe, 93
CER (certified emission reduction), 338
  ownership, 338
CERES, climate change and, 20
Certified Sustainable Products Alliance, 94
children
  as assets, 32
  natural disasters and, 150

China
  flood, 292–93
  forests, 292–93
Chiquita, 98–101
  corporate responsibility, 100
climate change
  agricultural, industrial, 72
  agricultural development and, 72
  carbon dioxide emissions, 19
  CERES and, 20
  climate change adaptation, 19–20
  developing countries and, 18
  economy and, 18
  ecosystems and, 18, 19
  environmental management and, 18
  heat waves, 146
  human health and, 18
  infectious diseases and, 19
  investment, 41
  land and conflict, 346
  MDGs, 331
  profit margins, 18–19
  respiratory diseases and, 19
  rural poor and, 331–32
  Stern Review Report on the Economics of Climate Change, 330
  temperature rise and crop yields, 72
  weather, extreme events, 19
Climate change futures: Health, ecological, and economic dimensions, 18
Clinton, Bill (President), Indian Ocean tsunami, 171–72
Coastal Wetlands Planning and Restoration Act of 1990, 196
CODATA (Committee on Data for Science and Technology), 238
coffee farming, 88–93
  Kraft and, 95–98
Commission on Education and Communication (CEC), 219
  case study, 225
Commission on Legal Empowerment of the Poor, 305
  board of advisors, 309
  grassroots organizations, 307
  legal system, 307
  leverage, 307
  membership, 309
  organization, 308–10
  outcomes desired, 308
  the poor and the legal system, 311–12
    legal identity, 312–13
    legal rights, 313–14
    legal services, 314–15
    unjust, 315–16
  property rights, 307
  structure, 308–10
  working groups, 309–10
    access to legal system, 311–16
    empowering informal businesses, 320–22
    labor rights, 319–20
    property rights, 316–18
    reform implementation, 322–23
    thematic areas, 310
Committee on Data for Science and Technology (CODATA), 238
communal property, 46
  success of, 46
Community Based Natural Resource Management Programme, 23
community ownership, *versus* land title and, 317
community water harvesting in Rajasthan, India, 76
conflict, land and, 345–47
conservation
  access to information, 230–31
  accessing, 235–37
  ants, 243
  Budapest Open Access Initiative, 238
  CBD, 238–39
  ConserveOnline, 240–42

362 | Index

cost of, 233–35
fragmentation, 231–33
IUCN, 239–40
maps, 242–43
publishing, 233–35
long-term, 37
wild species, 75
Conservation Biology, 244
Conservation for Poverty Reduction Initiative (CPRI), 218
Conservation Geoportal, 242
Conservation in Practice, 244
Conservation International, 22
ConserveOnline, 240–42
continuous contour trenches (CCTs), 126
Convention on Biological Diversity, 11, 22
2010 Biodiversity Target, 6
Convention on Biological Diversity (CBD), 238–39
Convention on Migratory Species, 11
Convention to Combat Desertification, 11
coral reefs, 16
benefits to local communities, 34
conservation, costs and benefits, 42
tsunami and, 173
cork production, 101–3
corporate responsibility, Chiquita, 100
cost-benefit analysis, environmental policy and, 274, 276
Costa Rica
Association for Tourism Professionals (ACO-PROT), 115
ecotourism, 114
fuel tax for forests, 294–95
National Biodiversity Institute (INBio), 115
Costa Rica and Panama transboundary comanagement, 75–76
Couton, Adrien, 253
CPRI (Conservation for Poverty Reduction Initiative), 218
Crasborn, Carlos, 104

credit, access to, 317
credit and resource-rights policies, 37
crop yields and temperature rise, 72
Crudgington, Elisabeth, 217
Cuba, 206–8
culture of safety, Cuba, 206–7

Daily, Gretchen, 272
Damaraland Camp, 23
de Soto, Hernando, 305, 306, 309
dead capital, 316
death, developing countries, natural disasters, 148
deforestation
greenhouse gases and, 335
Sumatra, 176
deltaic process, wetlands and, 191–93
Deromedi, Roger (Kraft), 95
Dervis, Kemal (Administrator of UNDP), 7
desertification, 332
developing countries
climate change and, 18
deaths, natural disasters, 148
ecosystem services and, 16
environmental assets and, 16
development resources, natural disasters, 149
Direct Relief International, 169
dirty dozen pesticides, 98
disaster preparation, Cuba, 206–7
disaster resilience of the poor, 148
disaster resilient construction methods, 148
disease, natural diasters and, 151
distributional justice, land/carbon credit holdings, 347
diversification of agricultural systems, 72–73
diversity of wealth, 32
Domtar, Inc., 110
Douglass, Frederick, 305

drinking water, investing in, 16
drip irrigation, 258–59
drylands, 332
D'Souza, Marcella, 121
dump sites, New Orleans, 199–200

early warning systems, natural disasters, 150
EarthChoice papers, 110
ecoagriculture, 64
    biodiversity and, 73
    community water harvesting in Rajasthan, India, 76
    ecosystem services and, 73
    indigenous honeybee conservation in the Hindu Kush Himalayas, 76
    landscapes, 73
        agricultural production areas, 73
        farming communities, 77–78
        institutional mechanisms, 73
        natural areas, 73
        policies, 79–80
        scale, 78–79
    resource management, 74–75
    stakeholders, 78–79
    transboundary comanagement in Costa Rica and Panama, 75–76
Ecochard, Jean-Louis, 230
Ecological Complexity, 244
ecological degradation, 68–69
ecological footprint, 20
Ecological Indicators, 244
Ecological Informatics, 244
ecological poverty, 68
ecological production functions, 277–80
Economic and Social Council (ECOSOC), 155
economic development
    environmental assets and, 31
    natural disasters and, 149
economic growth
    governance and, 45
    India, 121
    food security, 122
    natural disasters and, 149
economic poverty, ecological poverty and, 68
economic rate of return on environmental investments, 16
economic threat of natural disasters, 146
economic values of forests, 41
economy
    climate change and, 18
    the poor and, 306
ECOSOC (Economic and Social Council), 155
ecosystem management
    agricultural sustainability and, 71
    rural communities, 343
ecosystem services, 65
    agricultural biodiversity, 66
    agriculture and, 73
    biodiversity and, 73
    cultural services, 12, 66
    degradation of, 13, 66–67
        reversing, 13
    developing countries and, 16
    management, MDGs, 8
    monetary value, 14
    natural capital, 17
    nonmarket benefits, 40
    provisioning services, 12, 66
        subsistence farmers, 66
    regulating services, 12, 66
    subsistence farmers and, 65
    supporting services, 12, 66
    wild biodiversity and, 66
ecosystems, 11
    agriculture, threats, 67
    business and, 274
    changing, 13
        gains from, 13
    climate change and, 18, 19
    cork production and, 101

infectious diseases and, 19
intact, value of, 16
managing, fighting poverty and, 21
natural disasters, 151
Ecosystems and human well-being: Synthesis, 11
ecotourism, 114
education
   natural disasters and, 150
   risk awareness of the poor, 148
   rural poor, 343–44
emergency disaster relief, 149
enforcement of property rights, 318
Enríquez, Francisco, 117
environment
   as luxury good, 31
   as natural protectorate during tsunamis, 173
   the poor and
     livelihood and, 16
     vulnerability and, 16
   recovery from disaster and, 174
environment governance, devolution, 21
environmental assets, 16
   developing countries and, 16
   economic development and, 31
   investing in, 21
   natural capital, 17
   sacrificing, 31
   vulnerability of, 34–35
environmental conservation, as luxury good, 16
environmental degradation, natural disasters, 147
environmental destruction, poor as agents of, 36
environmental hazards, women and, 35
environmental health, India, 122
environmental impact, income and, 18
environmental investments
   budget, estimating global, 38–39
   climate change, 41
   coral reefs, 42
   economic rate of return, 16
   forests, 42
   prices and, 46
   rates of return, 39–44
   social worth, 40
   wetlands, 42–43
   wildlife conservation, 43
Environmental Kuznets Curve, 18
   environment as luxury good, 31
environmental management, climate change and, 18
environmental policies
   cost-benefit analysis and, 274, 276
   market-based, 46
   resource management and, 274
environmental sustainability, achieving, 6
environmental wealth, total wealth and, 34
EPA (Environmental Protection Agency)
   New Orleans dump sites, 199
   surface-water supplies and, 289–90
   water, New York City and, 290
Equator Dialogues, 22
Equator Initiative partnership, 22
Equator Knowledge, 22
Equator Prize, 21
   Namibian Torra Conservancy, 23
Equator Ventures, 22
evacuation
   New Orleans
     cars and, 200
     consideration for poor residents, 200
   the poor's ability to act, 148
expropriation, risk of, 37
extensive risk, 162

Fach, Estelle, 329
Families, size, 32, 37
farmers
   coffee farming, 88–93

ecoagriculture, 77–78
small-scale
 India, 261–62
 social entrepreneurship and, 255
 subsistence farmers, 65
Federación Nacional de Cafeteros de Colombia, 88
fertilizer, greenhouse gases and, 336
Fiji, Locally-Managed Marine Area Network, 24
financial mechanisms, 286–88
Finca Morros, 93
fisheries, 71
 benefits of reducing catch effort, 43
 Indian Ocean tsunami, 178
Fix, Lewis, 110
flood. *See also* Hurricane Katrina
 China, 292–93
flood control, Napa Valley, 290–92
flood management, New Orleans, 201–3
food
 lower-productivity lands, 70
 population growth and, 70
 temperature rise and crop yield, 72
 wild products, 71
food demand, growth, 69
food production, 70
food security, 64, 65
Fordham University, 22
 Leitner Center for International Law and Justice, 145, 187
Forero, Margarita, 113
forest management rights, 287
Forest Protection Committee (FPC), Village Watershed Committee and, 130–31
Forest Stewardship Council, 87
 cork production, 101
 Domtar, Inc., 110
 Indonesian teak plantations, 108
 Pictorial Offset Corporation, 112–13

forest use, traditional *versus* unsustainable timber harvest, 17
forestry
 rural land and, 332–37
 sustainable *versus* small-scale farming, 17
forests
 agriculture, 70
 biotechnology products, 42
 carbon storage and sequestration, 42
 China, 292–93
 cork, 101
 Costa Rica, 294–95
 economic values, 41
 genetic resources, 42
 income and, 34
 manchiche, 105
 pharmaceutical products, 42
 pucté, 105
 Sumatra, 176
formal economy, the poor and, 306
fossil fuels, 19
free grazing, Mhaswandi, India, 125
fuel
 fuel tax for forests, Costa Rica, 294–95
 gathered wood, 72
fungicides, Chiquita, 100

Garcia, Benedin, 106
gathered wood, 72
GBIF (Global Biodiversity Information Facility), 231
gender differences, 35
geographic information systems (GIS), 242–43
German Federal Ministry of Economic Development and Cooperation (BMZ), 22
Gibson Guitars, 105
GIS (geographic information systems), 242–43

366 | Index

GLCF (Global Land Cover Facility), 242–43
Global Biodiversity Information Facility (GBIF), 231
global environmental investment budget, 38–39
Global Information Commons for Science Initiative, 238
Global Land Cover Facility (GLCF), 242–43
Gómez, Raúl, 99
governance, economic growth, 45
government investments, 286
grain monocultures, 70
Grandoca-Manzanillo National Wildlife Refuge (Costa Rica), 75–76
Grant, Joy, 272
greenhouse gases, 333
   deforestation and, 335
   fertilizer and, 336
GROOTS International, 22
growth of income, resource abundance and, 35
Gulf of Honduras, tourism, 116–19

Hacienda Tijax Jungle Lodge, 118
Halverson, Elspeth, 3
Hartwick rule for sustainability, 17
Harvard Medical School, 18
Harvard University, 11
harvesting rainwater, 127
HASHI project, 23–24
health, human, climate change and, 18
heat waves, 146
Herklotz, Alena, 145
HFA (Hyogo Framework for Action), 147
Hindu Kush Himalayas' indigenous honeybee conservation, 76
honeybees, indigenous to Hindu Kush Himalayas, 76
Horizontes Nature Tours, 113
   professionalism of naturalist guides, 115
   scholarships, 116

Hortua, Daissy, 91
Hotel MarBrissa, 118
household income, natural disasters and, 149
household-level wealth, 34
human impact on environment, addressing, 7–8
Hurricane Betsy (New Orleans), 191
Hurricane Katrina, 145. *See also* New Orleans
   Bourne, Joel's "prediction," 187
   canal breaches, 196
   cost, 196
   environmental impact, wetlands, 197–98
   Lake Pontchartrain and, 196
   lessons learned, 203
   New Orleans, pumping water out of, 198
   oil industry and, 197
   wetlands and, 188, 197–98
   restoration, 198
Hurricane Rita
   oil industry and, 197
   wetlands, restoration, 198
hurricanes, 145
   Cuba and, 206
   Louisiana, 193
Hyogo Framework for Action (HFA), 147
hypothetical antidesertification policy, 41

IDEI (International Development Enterprises), 257
IDNDR (International Decade for Natural Disaster Reduction), 151
   Framework of Action, 152
IDRC (International Development Research Centre), 22
IEA (International Energy Agency), 197
IGWDP (Indo-German Watershed Development Program), 125

IIED (International Institute for Environment and Development), 11
illiteracy, India, 122
ILO (International Labor Organization), 319–20
incentive-based mechanisms, 287
income
   environmental impact and, 18
   forests, 34
   growth, resource abundance and, 35
   wealth, difference, 33
India
   agricultural land, productivity, 122
   economic growth
      food security, 122
      rate, 121
   as emerging economy, 254
   environmental health, 122
   farmers earning less than USD 1 per day, 259–61
   illiteracy, 122
   infant mortality rate, 122
   international development enterprises, 256–62
   land degradation, 122
   malaria, 122
   maternal mortality, 122
   Mhaswandi, 123–24, 124–30
      maintenance fund, 135
      PGR (Peer Group Review), 134
      QAM (Qualitative Assessment Matrix), 134
      Quality and Impact audit, 134
      sustainability, 136–38
      women, 131–32
   rainfall, 123
   resources, 122
   small-scale farmers, 261–62
   soil erosion, 123
   symbiotic relationship between environment and communities, 123
   tuberculosis, 122
   water management, 256–57
   drip irrigation, 258–59
   waterborne diseases, 122
Indian Ocean earthquake, 146
Indian Ocean tsunami, 169
   civil servants, 179
   Clinton, Bill (President), 171–72
   coordination, 177–79
   coral reefs and, 173
   environment as natural protectorate, 173
   fisheries, 178
   governments, 180
   leadership, 180
   local capacity, 179–81
   MFF (Mangroves for the Future), 181–82
   rebuilding, urgency, 174–77
   recovery, environment and, 172
   response
      financial, 169
      government and agency, 173–74
   TAFLOL, 171
   TAFREN, 171
   TAFRER, 171
   TEC, 172
   Tsunami Green Reconstruction Policy Guidelines, 174
Indo-German Watershed Development Program (IGWDP), 125
Indonesia, teak plantations, 107–10
industrial agriculture, climate change and, 72
infant mortality rate, India, 122
infectious diseases, 19
informal businesses, empowering, 320–22
informality, 330
   in financial systems, 347–48
information access
   policy decisions and, 281–82
   risk awareness of the poor, 148

368 | Index

insecurities of the poor, 306
institutional mechanisms, 286–88
insurance, 48
integrated agriculture-conservation landscapes, 65
intensive risk, 162
International Decade for Natural Disaster Reduction (IDNDR), 151
   Framework of Action, 152
International Development Enterprises (IDEI), 257
international development enterprises in India, 256–62
International Development Research Centre (IDRC), 22
International Institute for Environment and Development (IIED), 11
International Labor Organization (ILO), 319–20
International Panel on Climate Change (IPCC), 18
International Strategy for Disaster Reduction (ISDR), 155
International Union for the Conservation of Nature (IUCN), 11, 217
Internet, availability, 236–37
Investing in development: A practical plan to achieve the millennium development goals, 6
Investing in environmental wealth for poverty reduction, 15
investment
   environmental (*See* environmental investments)
   wealth and, 31
investment in environment as driver of development, 15–18
investments, government investments, 286
IPCC (International Panel on Climate Change), 18
   reports, 282

irrigation water, rivers, 67
ISDR (International Strategy for Disaster Reduction), 155–57, 161–62
IUCN (International Union for the Conservation of Nature), 11, 22, 30, 217
   Bitter Bamboo Project, 223–25
   Conservation Commons, 239–40
   Livelihoods and Landscapes Strategy, 221–23
IWRM (integrated water resource management), 203–4

jade palm leaves, 105
Jamii Bora, 262–69
Jiménez, Carlos, 116
Johannesburg Plan of Implementation (JPOI), 8
Journal of Insect Sciences, 238
JPOI (Johannesburg Plan of Implementation), 8
Juan Valdez, 89

Kaputiei Town, 265–69
Kareiva, Peter, 272
Kehutanan, Dinhas, 180
Kjørven, Olav, 329
Knowing Knowledge, 225
knowledge, networking, 225
knowledge necessary for poverty alleviation, 217–28
Koperasi Hutan Jaya Lestari (KHJL), 108–9
Kraft, 94
   sustainable coffee production and, 95–98
Kyoto Protocol
   carbon storage and, 280
   CDM (Clean Development Mechanism), 15, 333
   unintended costs, 280

La Floresta, 90
labor rights, 319–20
Laguna del Tigre National Park, 106
Lake Pontchartrain (New Orleans), 188
　Hurricane Katrina and, 196
land and conflict, 345–47
land degradation, India, 122
land-registration processes, 341
land rights, 339, 341
land title, *versus* community ownership, 317
landscape, 73
Lao People's Democratic Republic (Lao PDR), 223
legal empowerment. *See* Commission on Legal Empowerment of the Poor
　property title and, 317
legal identity, 312–13
legal rights, 313–14
legal system, the poor and, 306
Leitner Center for International Law and Justice (Fordham University School of Law), 145, 187
levees, New Orleans, building of, 188
livelihood, environment and the poor, 16
Livelihoods and Landscapes Strategy (LLS), 218
　case study, 221–23
Llach, Diego, 95
LLS (Livelihoods and Landscapes Strategy), 218
Lobo, Crispino, 121
Locally-Managed Marine Area Network (Figi), 24
location of the poor, natural disasters and, 148
Louisiana
　coastal plains' loss of land, 194
　Cuba and, 208
　hurricanes and, 193
　Louisiana Coastal Wetlands Conservation and Restoration Task Force, 196
　oil industry, 194–95
　　Hurricane Katrina and, 197
　　Hurricane Rita and, 197
　　wetlands and, 195
　racism, 199–200
　regional depressurization, 195
　resources, 193
　tourism, 193
　wetlands
　　deltaic process, 193
　　oil industry and, 195
　　restoration attempts, 196
luxury good, environment as, 31
Lyme disease, 19

MA (Millennium Ecosystem Assessment), 11
　agriculture, impacts on terrestrial land and freshwater use, 64
maintenance fund, Mhaswandi, India, 135
malaria, 19
　India, 122
manchiche, 105
mangroves, benefits to local communities, 34
maps
　conservation information, 242–43
　natural capital providers and users, 282–86
Marenco Lodge, 113
market-based environmental policies, 46
maternal mortality, India, 122
Maxwell, Annie, 169
Maya Biosphere Reserve (Guatemala), 104
　Rainforest Alliance, 104–5
McNeely, Jeffrey A., 64
McNeill, Charles, 3
MDG Carbon Facility, 15
MDG Support Initiative, 9–10

MDGs (Millennium Development Goals), 4–5
  climate change and, 331
  extreme povery eradication, 306
  MDG 1 (eradication of extreme poverty and hunger), 150
  MDG 2 (universal education), 150
  MDG 3 (empowerment of women), 150
  MDG 4 (child mortality), 150
  MDG 5 (maternal health), 151
  MDG 6 (chronic disease), 6, 151
  MDG 7 (sustainability), 6, 151
  MDG 8 (global partnership for development), 151
  men and women, 150
  natural disasters, 150
  poverty reduction, natural resources, 8
  sustainability and, 30
Mehers, Gillian Martin, 217
Méndez, Nirma, 118
Menses, Juan Carlos, 87
methane, 336
MFF (Mangroves for the Future), 181–82
MFIs (microfinance institutions), 262–69
Mhaswandi, India, 123–24, 124–30
  maintenance fund, 135
  PGR (Peer Group Review), 134
  QAM (Qualitative Assessment Matrix), 134
  Quality and Impact audit, 134
  sustainability and, 136–38
  women, 131–32
microfinance institutions (MFIs), 262–69
Millennium Ecosystem Assessment (MA), 11
Millennium Project, 5–6
Millennium Review Summit, 10–11
Millennium Summits
  2000, 7
  2005, 7
Mississippi River
  Atchafalaya River and, 189
  deltaic process and, 191–93
  levee breaches in 1927, 189
  sediment deposits, 194
Monteverde Cloud Forest Reserve, 115
Moreno, Frank, 109
Munro, Ingrid, 263

NAFRI (National Agriculture and Forestry Research Institute), 223
Nairobi Framework, 334
Nam Pheng, 223
Namibia, wildlife management, 293–94
Namibian Torra Conservancy, 23
Napa Valley, flood control, 290–92
National Agriculture and Forestry Research Institute (NAFRI), 223
National Coffee Fund, 91
National Federation of Coffee Growers of Colombia (FNC), 88
natural capital, 17, 272
  accounting practices, 17
  exhaustible, 17
  information, 281
  maps of providers and users, 282–86
  New York City and, 288
Natural Capital Project, 272
natural disasters, 145
  Asia earthquake, 145
  children and, 150
  Cuba, 207
  deaths in developing countries, 148
  disadvantaged persons and, 147–48
  disaster resilient construction methods, 148
  disease and, 151
  early warning systems, 150
  economic development and, 149
  economic threat, 146
  ecosystems and, 151
  education and, 150
  elimination of development gains, 149
  emergency disaster relief, 149

environmental degradation, 147
environmental resources and, 151
frequency, rise in, 146
global community plan of action, 147
heat waves, 146
household income and, 149
Hurricane Katrina, 145
hurricanes, 145
IDNDR, 151–52
Indian Ocean earthquake, 146
Indian Ocean tsunami, 169
   civil servants, 179
   Clinton, Bill (President), 171–72
   coordination, 177–79
   environment as natural protectorate, 173
   financial response, 169
   government and agency response, 173–74
   governments, 180
   leadership, 180
   local capacity, 179–81
   MFF (Mangroves for the Future), 181–82
   TEC (Tsunami Evaluation Coalition), 172
location, 148
low disaster resilience, 148
MDGs, 150
Pereira, Columbia earthquake, 149
poor planning, 147
population growth, 147
recovery and environment link, 174
reduction, 151
response, development resources and, 149
risk awareness, 148
severity, rise in, 146
sexual violence against women, 150
sustainability and, 151
tsunamis, 146
Tsunami Green Reconstruction Policy Guidelines, 174
UN Millennium Declaration, 150
United Nations Millennium Summit, 149
urbanization, 147
WCNDR, 153
natural resources management
   MDGs, 8
   poverty reduction, 8
   wealth from, 33
nature-based tourism, 113
Nature Conservation Act of 1996, Namibian Torra Conservancy, 23
networking knowledge, 225
New Orleans. *See also* Hurricane Katrina
   African Americans, 200
   Agriculture Street Landfill, 199
   Atchafalaya River and, 189
   canals, 190
      breaches during Hurricane Katrina, 196
      freshwater/saltwater buffer, 195
   dump sites, 199–200
   evacuation of
      cars and, 200
      consideration for poor residents, 200
   flood management, 201
   groundwater pumping, 190
   Harbor Navigational Canal, 190
   Hurricane Betsy, 191
   Hurricane Katrina
      lessons learned, 203
      pumping water out, 198
   Industrial Canal, 190
   IWRM (integrated water resource management), 203–4
   Lake Pontchartrain, 188
   levees, building of, 188
   Mississippi River Gulf Outlet (MRGO), 190

372 | Index

Mississippi River levee breaches in 1927, 189
regional depressurization, 195
settling of, 188
shoreline erosion, 190–91
sinking of, 189–90
wetlands, 188
New York City
  natural capital and, 288
  water, 289
    EPA and, 290
Ng, Helen, 253
Ngitili system of land management, 24
NIH (National Institute of Health), information availability, 238
nitrous oxide, 336
Norway, legal empowerment, 306
nutrient-driven eutrophication, 278
Nyae Nyae Conservancy (Namibia), 293–94

OARE (Online Access to Research in the Environment), 244
ODA (official development assistance), 6
OECD (Organisation for Economic Co-operation and Development), 238
oil industry, Louisiana, 194–95
  Hurricane Katrina and, 197
  Hurricane Rita, 197
  wetlands and, 195
Old River Control Structure, Mississippi and Atchafalaya rivers, 189
One Laptop Per Child (OLPC), 245
Online Access to Research in the Environment (OARE), 244
Organisation for Economic Co-operation and Development (OECD), 238

PalNet (Protected Areas Learning Network), 241
paper, Pictorial Offset Corporation, 112–13

Participatory Net Planning Method (PNP), 133
PAs (protected areas), 71
Pearce, David, 15, 30
Peer Group Review (PGR), Mhaswandi, 134
per capita wealth, 32, 35
PES (payments for environmental services), 47
pesticide-impregnated plastic bags, 98
pesticides, dirty dozen, 98
PGR (Peer Group Review), Mhaswandi, 134
pharmaceutical products, 42
Pictorial Offset Corporation, 111–13
Plan Grande Quehueche, 118
plastic bags impregnated with pesticides, 98
plastic in lieu of cork, 101
PNP (Participatory Net Planning Method), 133
poaching, ecological poverty and, 68
the poor. *See also* rural poor
  as agents of environmental destruction, 36
  asset base, increasing, 38–39
  formal economy and, 306
  information access in natural disaster, 148
  insecurities, 306
  legal system and, 306
  risk awareness in natural disaster, 148
  working lives, 305–6
population
  change, 37
    resource degradation and, 37
  growth, 69
    food and, 70
    natural disasters, 147
    technological change and, 37
    water and, 70

Index | 373

poverty
    disaster resilient construction methods, 148
    ecological poverty, 68
poverty alleviation, 65
    knowledge necessary, 217–28
    as systems problem, 219
Poverty-Environment Facility, 7
Poverty Environment Partnership, 8
    Investing in environmental wealth for poverty reduction, 15
poverty persistence, rural poor, 36
precautionary principle, 274
private property rights, 46
profit margins, climate change and, 18–19
property, as collateral, 317
property rights, 37, 307, 316–18
Protected Areas Learning Network (PalNet), 241
protected areas (PAs), 71
public disclosure, 133
Public Library of Science, 238
public policy, social capital and, 45
publishing conservation information, 233–35
pucté, 105

QAM (Qualitative Assessment Matrix), Mhaswandi, 134
Q'eqchi Maya village, 118
Quality and Impact audit, Mhaswandi, 134
Quito, water and, 292

racism in Louisiana, 199–200
RADA (Authority for Reconstruction and Development), Indian Ocean tsunami, 171
rainfall in India, 123
Rainforest Alliance, 87, 94
    Carrera, José Román, 105

certification
    benefits, 92
    Finca Morros, 93
certified coffee, 89
certified forests, Pictorial Offset Corporation, 112–13
Chiquita, 98–101
cork production, 101–3
Gulf of Honduras, 117
Juan Valdez and, 89
Kraft and, 95
La Floresta, 90–91
Maya Biosphere Reserve, 104–5
Ramirez, Edgar and, 90
Santander, 92
Sierra Nevada farmers and, 90
Silva, Gabriel, 91
Sustainable Agriculture Network and, 88
tourism and, 114
rainwater, harvesting, 127
Rajasthan, India, community water harvesting, 76–77
Ramirez, Edgar, 90
Ramsar Convention on Wetlands, 11
rebuilding, Indian Ocean tsunami, urgency, 174–77
recycling, Pictorial Offset Corporation, 112–13
reform implementation, 322–23
regional depressurization, Louisiana, 195
regulations, 286
resource management
    environmental policy and, 274
    mechanisms for, 287
resource-rights policies, 37
resources
    abundance, rate of income growth and, 35
    degradation
        family size and, 37
        population change and, 37

development, natural disasters and, 149
forest genetic resources, 42
India, 122
Louisiana deltas, 193
natural disasters and, 151
value of, usage rights and, 45
respiratory diseases, climate change and, 19
reversing ecological degradation, 69
Ricketts, Taylor, 272
Rio Earth Summit, 7
Rios Tropicales, 113
risk analysis, 274
risk awareness of the poor, 148
rivers, 67
rural communities, ecosystem management, 343
rural development, 64
rural poor
  climate change and, 331–32
  education, 343–44
  poverty persistence, 36

Sachs, Jeffrey, Millennium Project, 5
Sadangi, Amitabha, 261
Sampada Trust, 121
Samuels, Donald, 111
Samyuktya Mahila Samittee (SMS), 132
San Pondsak National Wildife Refuge (Panama), 75–76
Sangamner Cooperative Sugar Factory (SSCF), 126
sanitation systems, investing in, 16
Santander, 92
Santos, Jorge Julian, 93
SCB (Society for Conservation Biology), 244
Scherr, Sara, 64
scientific information, 237–38
  CBD, 238–39
  CODATA, 238
  costs, 243–46
  Global Information Commons for Science Initiative, 238
  Journal of Insect Sciences, 238
  OECD, 238
  Public Library of Science, 238
SEEA (system of integrated environmental and economic accounting), 17
self-insurance, 48
sexual violence, women, natural disasters and, 150
Shames, Seth, 64
shoreline erosion, New Orleans, 190–91
short-term view of life, 37
Siemens, George, 225
Silva, Gabriel, 91
Singh, Naresh, 305
sinking of New Orleans, 190–91
small-scale farmers, 261–62
SMS (Samyuktya Mahila Samittee), 132
social capital
  communal property and, 46
  Cuba, 208
  public policy and, 45
social entrepreneurship, 254
  case studies
    international development enterprises in India, 256–62
    Jamii Bora, 262–69
  small-scale farmers and, 255
social organization, Cuba, 207
social worth of environmental investments, 40
socialist government of Cuba, 206
soil conservation, CDM and, 335
soil erosion, India, 123
SCSF (Sangamner Cooperative Sugar Factory), 126
Stanford University, 272
Steiner, Achim (Director General of UNEP), 7
Stern, Sir Nicholas, 18

Stern Review Report on the Economics of Climate Change, 330
Stigson, Bjorn, 274
subsistence farmers, 65–66
Sumatra, deforestation, 176
surface-water supplies, EPA and, 289–90
sustainability
　agriculture, 71
　certification, 94–95
　Certified Sustainable Products Alliance, 94
　Chiquita bananas and, 98–101
　grain monocultures, 70
　Hartwick rule, 17
　Kraft and coffee, 95–98
　MDGs and, 30
　Mhaswandi, India, 136–38
　monitoring, 17–18
　natural disasters and, 151
　property rights and, 318
　Rainforest Alliance and, 87
　tourism certification intiatives, 94
Sustainable Agriculture Network, 88
　Rainforest Alliance and, 88
　standards, 94
Swiss Re, 11, 18
synthetic cork, 103
systems problem, poverty alleviation as, 219

TAFLOL (Presidential Task Force for Logistics, and Order), Indian Ocean tsunami, 171
TAFREN (Presidential Task Force for Rebuilding the Nation), Indian Ocean tsunami, 171
TAFRER (Presidential Task Force for Rescue and Relief), Indian Ocean tsunami, 171
Tallis, Heather, 272
Tamburaka, Abdul Harris, 108
teak plantations in Indonesia, 107–10

TEC (Tsunami Evaluation Coalition), Indian Ocean tsunami, 172
technological change, population growth and, 37
temperature rise and crop yields, 72
The Nature Conservancy (TNC), 22, 230, 272
Thompson, Sacha, 187
Tikal National Park, 105
timber, Indian Ocean tsunami rebuilding and, 175–76
TNC (The Nature Conservancy), 22
tourism
　ecotourism, 114
　Gulf of Honduras, 116–19
　Louisiana, 193
　Monteverde Cloud Forest Reserve, 115
　nature-based, 113
　Rainforest Alliance certification, 114
　sustainable tourism certification initiatives, 94
tree cutting, Mhaswandi, India, 125
triple bottom line, 273
Tropical Forest Trust, 108
　KHJL funding, 109
Tsunami Green Reconstruction Policy Guidelines, 174
tsunamis, 146
　coral reefs and, 173
　environment as natural protectorate, 173
　Indian Ocean, 169
　　coordination, 177–79
　　financing of response, 169
　　government and agency response, 173–74
　　governments, 180
　　leadership, 180
　　local capacity, 179–81
　　MFF (Mangroves for the Future), 181–82
tuberculosis, India, 122

TVE (Television Trust for the Environment), 22

Uaxactún, 105
  forest concession, 106
UN Millennium Declaration, natural disasters, 150
UNCTD (United Nations Conference on Trade and Development), 336
UNDP (United Nations Development Programme), 4, 18, 329
  Equator Initiative, 3
  Poverty Environment Partnership, 8
  UNDP Regional Centres, 7
UNEP (United Nations Environment Programme), 4
  Poverty Environment Partnership, 8
  UNEP Regional Offices, 7
UNFCCC (United Nations Framework Convention on Climate Change), 333
United Nations
  Environment Programme's Ecosystems, Protected Areas, and People, 241
  Office of the Special Envoy for Tsunami Recovery, 169
  reform, 7
United Nations Foundation, 22
United Nations Millennium Summit, natural disasters and, 149
University College (London), 30
urbanization, natural disasters, 147
USAID (U.S. Agency for International Development), 94

Village Watershed Committee (VWC), 125
vulnerability
  environment and the poor, 16
  of environmental assets, 34–35
VWC (Village Watershed Committee), 125
  Forest Protection Committee (FPC) and, 130–31

water
  India, 256–57
    drip irrigation, 258–59
  New York City, 289
    EPA and, 290
    population growth and, 70
  Quito, 292
water absorption trenches (WATs), 126
water conservation, CDM and, 335
waterborne diseases, India, 122
Watershed Organization Trust, 121, 124
WATs (water absorption trenches), 126
WCDR (World Conference on Disaster Reduction), 147
  Hyogo Declaration, 157–58
  Hyogo Framework for Action, 157–58
    priorities, 159–60
WCNDR (World Conference on Natural Disaster Reduction), 153–55
wealth
  capacity to generate future well-being, 33
  components of, 30
  diversity of, 32
  environmental wealth, 34
  household-level, 34
  income, difference, 33
  information access and, 148
  investment and, 31
  from natural resources, 33
  per capita, 32
    composition, 35
  risk awareness and, 148
  technological change and, 37
The wealth of the poor: Managing ecosystems to fight poverty, 21
weather, extreme events, climate change and, 19
West Nile virus, 19
wetlands
  benefits to local communities, 34

conservation, costs and benefits, 42–43
deltaic process, 191–93
　Louisiana and, 193
Hurricane Katrina and, 188, 197–98
　restoration, 198
Hurricane Rita, restoration, 198
intact *versus* farmed, 17
Louisiana
　oil industry and, 195
　restoration attempts, 196
New Orleans, 188
Wheeler, Keith, 217
Whelan, Tensie, 87
Where is the wealth of nations: Measuring capital for the 21st century, 17
wild biodiversity
　ecological poverty and, 68
　ecosystem services and, 66
wild products, 71
wild species, conservation, 75
wildlife conservation, 43
women
　environmental hazards and, 35
　land tenure and, 342–43
　Mhaswandi, India, 131–32
　property rights and, 317
　sexual violence, natural disasters, 150
wood, 72
Woods Institute for the Environment (Stanford University), 272
working lives of the poor, 305–6
World Bank
　Millennium Review Summit, 11
　Poverty Environment Partnership, 8
　Where is the wealth of nations: Measuring capital for the 21st century, 17
World Conference on Disaster Reduction (WCDR), 147
　Hyogo Declaration, 157–58
　Hyogo Framework for Action, 157–60
World Conference on Natural Disaster Reduction (WCNDR), 153–55
World Resources Institute, 11
World Summit on Sustainable Development (WSSD), 7
World Wildlife Fund (WWF), 11, 272
WRI (World Resources Institute), 21
WSSD (World Summit on Sustainable Development), 7
WWF (World Wildlife Fund), 11

Zaidman, Yasmina, 253